Microalgal Bioengineering

Jorge Alberto Vieira Costa
Brian Gregory Mitchell • John Benemann
Editors

Microalgal Bioengineering

Springer

Editors
Jorge Alberto Vieira Costa
Federal University of Rio Grande (FURG)
Rio Grande, Rio Grande do Sul, Brazil

John Benemann
MicroBio Engineering, Inc
San Luis Obispo, CA, USA

Brian Gregory Mitchell
Scripps Institution of Oceanography
University of California, San Diego
La Jolla, CA, USA

ISBN 978-3-031-61255-8 ISBN 978-3-031-61253-4 (eBook)
https://doi.org/10.1007/978-3-031-61253-4

This Springer imprint is published by the registered company Springer Nature Switzerland AG
The registered company address is: Gewerbestrasse 11, 6330 Cham, Switzerland

Preface

It is with great pleasure and enthusiasm that we present this comprehensive work, titled *Microalgal Bioengineering*, to be published by Springer Nature. Edited and organized by esteemed experts in the field, Jorge Alberto Vieira Costa from the Federal University of Rio Grande (FURG) in Brazil, Greg Mitchell from the University of California, San Diego (UCSD) in the USA, and John Benemann representing MicroBio Engineering, Inc. in the USA, this book represents a significant contribution to the field of microalgae research and application.

Microalgal Bioengineering encapsulates a wealth of knowledge spanning 11 meticulously crafted chapters. These chapters delve into various aspects of microalgae, covering genetics, modeling, harvesting techniques, large-scale cultivation, and the industrial applications of these fascinating microorganisms. Additionally, the book explores the intricate interaction between microalgae and environmental, social, and economic dynamics, shedding light on their potential contributions to biotechnology and sustainable development. Its conception and planning took place while Professor Jorge Alberto Vieira Costa was a Visiting Researcher with Dr. Greg Mitchell's team at UCSD and in collaboration with Dr. John Benemann.

One of the defining features of this book is the diversity of topics covered. From discussions on high-value bioproducts like biofuels, bioremediation, bioplastics, and biofertilizers to insights into pharmaceuticals, antioxidants, fatty acids, polysaccharides, proteins, and superfoods derived from microalgae, each chapter presents a nuanced exploration of the subject. The authors, coming from renowned academic and industrial backgrounds worldwide, bring a wealth of experience and perspectives, making this book a truly global effort.

This volume is not merely a compilation of academic works; it is a testament to the collaborative efforts of professors, researchers, and professionals worldwide, united in their quest to unlock the full potential of microalgae for the betterment of society. We believe that *Microalgal Bioengineering* will serve as a valuable reference for students, researchers, and professionals, providing insights into the latest advancements, challenges, and opportunities in the field of microalgal biotechnology.

We are also confident that libraries, professionals, amateur scientists, and enthusiasts in the fields of Biotechnology, Biology, Oceanography, Engineering, Biochemistry, Food Sciences, Chemistry, Pharmaceutical Sciences, Aquaculture, Energy, and Nutrition, among others, constituting the wonderful world surrounding microalgae, will find in this book a reliable source of information and inspiration.

As editors, we extend our gratitude to all contributors who have enriched this book with their valuable perspectives and expertise. We hope that readers will find this volume both informative and inspiring, stimulating new ideas and fostering continued innovation in the exciting field of microalgal bioengineering.

Rio Grande, Rio Grande do Sul, Brazil — Jorge Alberto Vieira Costa
La Jolla, CA, USA — Brian Gregory Mitchell
San Luis Obispo, CA, USA — John Benemann

Contents

About the Authors

Jorge Alberto Vieira Costa, is a Full Professor of Biochemical Engineering and Food Engineering at the Federal University of Rio Grande, Brazil. He is a Researcher Level 1A at the Brazilian National Council for Scientific and Technological Development (CNPq) and serves as the Coordinator of Graduate Programs in Food Science at CAPES (Brazilian Federal Agency for Support and Evaluation). Professor Jorge specializes in industrial microalgae projects, focusing on CO_2 biofixation and bioproducts derived from microalgae, including biojet fuel, bioethanol, biodiesel, biofertilizer, and food products. He has published over 370 papers in scientific journals, contributed chapters to 52 books, and has applied for 35 patents.

Brian Gregory Mitchell, is a Research Biologist and Senior Lecturer (Emeritus) at the Scripps Institution of Oceanography, University of California San Diego and Associate Director for the California Center for Algae Biotechnology (CalCAB). He was a founding member of the Board of the Algae Biomass Organization and the Algae Foundation. He currently serves as Vice Chairman for the Board of the Algae Foundation. Dr. Mitchell is an expert in ocean photosynthesis, satellite remote sensing, numerical modeling of algae growth in the co-limitation matrix of nutrients, temperature and light, and algal biotechnology. He is currently active in coordinating research and commercial demonstration of both microalgae and macroalgae mass culture for biofuel, animal feed, and abatement of waste CO_2 and nutrients including modeling of production yields and life cycle analysis. Dr. Mitchell has published more than 130 articles in peer-reviewed scientific journals.

John Benemann, Ph.D., CEO, graduated from the University of California Berkeley with a B.S. in Chemistry and a Ph.D. in Biochemistry, researcher at U.C. Berkeley in the departments of Civil Engineering and Microbial and Plant Biology, Associate Professor at Georgia Institute of Technology and Adjunct Professor at the University of Hawaii. He has authored over a hundred peer-reviewed publications, technical reports and reviews. His business experience includes several technology startups, strategic planning and execution of major R&D initiatives by large companies and government agencies and the production and marketing of algae biomass for aquaculture feeds and nutritional products. He is a founding Director of the Algae Biomass Organization, Dr. Benemann has carried out R&D in waste treatment, biofuels and algal biomass production, including anaerobic digestion, controlled bioreactor landfills, algal mass cultivation, and municipal wastewater treatment. During the 1970s he led an independent research group at U.C. Berkeley where, with the late Prof. William Oswald, he studied bioflocculation for low-cost algal harvesting and introduced paddle wheel mixing for raceway algae ponds. He participated in the 1980-1996 US DOE Aquatic Species Program (ASP) and a main author of its final report.

Chapter 1
Microalgae as the Central Variable in Imagining a Systemic Society with the Purpose of Abundance

Lucio Renato de Fraga Brusch

Abstract This chapter explores the transformative potential of microalgae in shaping a sustainable future. It navigates the intricate connections between natural processes, scientific advancements, and societal paradigms, offering innovative solutions to global challenges. The abstract advocates for a paradigm shift toward "amplifying reform" in supply systems, using micro-photosynthetic organisms and systemic mapping to enhance production efficiency and economic value through photobiorefineries. It also proposes the development of Integrated Biosystems, fostering closer ties between communities and the environment for harmonious living and sustainable practices.

By redefining language and understanding in production systems, the chapter highlights the potential of biological solutions to reduce environmental impacts and create economic opportunities. It challenges traditional production models, advocating for circular production inspired by nature's elegance and efficiency, particularly through photobiorefineries. This vision imagines a future where human ingenuity aligns with nature's wisdom, fostering a systemic society thriving in abundance while respecting planetary boundaries. It calls for a reimagining of our relationship with the natural world, recognizing microalgae as catalysts for innovative, sustainable solutions in food, study and begin to understand energy, and materials production.

Keywords Circular economy · Innovative production · Sustainable solutions · Systemic society · Transformative potential

L. R. Brusch (✉)
ZERI Foundation—Zero Emissions Research and Initiatives,
Porto Alegre, Rio Grande do Sul, Brazil

J. A. V. Costa et al. (eds.), *Microalgal Bioengineering*,
https://doi.org/10.1007/978-3-031-61253-4_1

1.1 Our Place and Our Time

We're imprisoned, we're in a trap of the past, and at the same time, we're ready to fly like we've never been before. We live in a time of false dilemmas, our search for liberty means freeing ourselves from the solutions and technologies that taught us how we could fly, and this is not easy and will require us even more so that we can fly. We entered the transition from the solutions of strength and imposition of pragmatism on reality, going to the subtle technologies and cooperation of intelligence with the reality and complexity of the systems. Reality is no longer limiting and has become a driving force. The past has given us a built space and solutions to address our food and energy needs that, due to their perceived performance and results, in the traditional view, are everywhere and are part of a legacy of work and dedication of millions of people in the name of progress and economic development. We had times of peace and war that allowed us to learn, implement, and enjoy our planet as if resources were infinite. Now, the challenge is to evolve, integrate ourselves into the very context of reality and learn by being part, go beyond the domain and control that contract and limit reality, to learning in an expanded reality where the immersion of intelligence takes place in the microscopic tissue, the biological structures, patterns, and the dynamic changes of information. We will be the author and result of ourselves, in a space of understanding never observed before in the history of our subspecies, expanding and going beyond *Homo sapiens sapiens*.

Despite our ability to know, and know that we know, we have little understanding of the complexity of interconnected and interacting systems and, because we do not know, or choose not to want to know, pollute the soil, waters, and air we breathe. And worse, in an egotic position, we destroyed ecosystems, extinguished species, as if it had no value or recognized value. In the promise of access to consumption and with the supremacy of individualism as a model of success, in a society where scarcity is considered value, we began to limit and make many other human beings suffer, leading them to thirst, hunger, and the struggle to survive every day, without hope. In politics, the support and permanence of elites that remain for a long time in control of natural resources, production goods, and wealth generated by all has brought us to a world of injustices, marginalization, and where many people live a life without any dignity. At the same time, and in the same context, science and technology have brought us advances that can subvert this anachronistic and incompatible way of thinking with the level of knowledge we have reached, and which is unfeasible and without perspective, because within the system that has failed in its objectives, there is also imagination and ways of transgressing without ruptures or setbacks, which forwards a transition to a better world, where it is possible to bring equal opportunities, reduce environmental impacts and where all are not only consumers of a pasteurized life but also production agents and involved with the offer of food and energy production solutions in a large cooperative coexistence platform. The global village, today it is possible to intuit and understand, will be built from circular processes of production and consumption, with increased collective efficiency, bringing to the center of solutions, to provide everyone with advanced knowledge about biology. The collection developed by Nature since the beginning of life on Earth, which is largely still out there, need to be read, understood, and applied, and that, as a fundamental concept, has always expanded the amount and diversity of life. We will have to forward ways of facing the possibility of the end of the times, where fear and selfishness define the strategies of living, because we follow in a world in decline and that goes to the bankruptcy of institutions and the rule of law. It may seem like an exaggeration, but it is the knowledge and understanding of the intimacy of Nature, through biology, that we have the clues and tips that it will be possible to establish a new order and the path of a sustainable society, is for me a point of conviction and that seems plausible and feasible.

What we have today is an internship of science that shows the strength of biology as central and inspiring knowledge for changing our way of living. After more than two centuries of the predominance of physics and chemistry as hegemonic knowledge to build on economic pragmatism, we come at a time when available knowledge and able to understand and optimize interconnected and interagent systems comes from the understanding of living organisms and ecosystems. As if there was a collective intelligence plan, and which most likely exists, it was/is from the knowledge and understanding of inorganic systems and reductionist and particular solutions, that we were able to develop techniques for visualization and management of microstructures that allow us today, in 2024, to study and begin to understand the living structures and their more sophisticated and complex processes, as we never imagined before. Basic science and engineering developed the platform and theoretical and experimental matrix that built the way to understand the appreciation of biological processes as a basis, first of all, for our collective survival and to forward the evolution of our species, and collective intelligence itself, in a recurring and circular process. For this transition, we need, especially for our survival as a species, to remember and include art and perception of the spiritual and imagine that we live a time that is a source of a great creative transgression and that carries us for a better future.

Biological life seems to us a miracle, but, of course, it is a tendency of the reality, develops from an invisible language that is present in everything, which combines and creates harmony between information, energy, and matter, and that whenever conditions exist, expresses itself and creates more biological life. To follow in this text, I accept that we can intuit and define that there is a structural land organizational systemic source that precedes everything, something still abstracts to our understanding, which in this text I will designate with "Life" using the freedom of thinking as a form of referral, which is an instrument of flow and synthesis of reality and the Cosmos to create biological life. Light, as a basic source of energy, beginning of everything, was/is stored and is offered in consecutive processes, complex and dynamic, and hence emerges as a matter that replicate, evolves, and transforms. In this succession of wonders of concrete reality, it is then that we come across microalgaes, photosynthetic bacterias, microscopic beings, pickups, and energy distributors and that function as the basic writing letters of history that describes biological life on our planet.

Present here for more than three billion years, microalgae are a physical and chemical basis structured by the genius of Life, which organizes energy and distributes it in a process that expands and explodes in all other forms of biological life, in the splendor of Nature. Resuming attention to microalgae—which are living fossils with unique characteristics—such as innovative sources filled with novelties to explore, can be the greatest and best result of all the science we develop, and understand them and associate with them as pillars of a new time—can be a strange recipe, but the more we know them, the greater the recognition of their importance to inspire a sustainable society. This is the point, still with much to clarify, but this is the point of this text.

1.2 The Astonishment and the Dazzle

Every day we have woken up, in recent decades, we have come across more and more knowledge, more questions and more doubts, all intertwined and with new concerns in the same spectrum, in a growing spiral, in an exponential growth. From our perception of what is simple and straightforward, to our understanding, from our simple gaze and defined by our specific biology, our surroundings, and our culture, we are required, and we need to understand and broaden our understanding. This includes complex issues and invisible architectures that organize and present themselves in the natural world, using and enhancing imaging techniques that reveal to us the intimacy of Nature. Similarly, we have to adapt mentally to the apparatus and technological devices that we use on a daily basis and that are built and work in ways that we cannot even imagine, individually, because they are achievements of the collective, of the common effort. It is in the microscopic world, that you see the information that models this expanded conception that requires our ideas and our interpretation of biology, chemistry, and electronic circuits, mixed with intellectual tools such as algorithms, software and artificial intelligence, reflections and descriptions that allow generating meaning and, in some way, take over this knowledge. But it is still a mere adaptation and contraction of what it really is, but it is certainly a process that will continue, expand, and offer new clues to understanding what reality is. This world, which completely escapes our direct perception, does not fit in our personal experience, emotional and cognitive, so we are obliged "to inflate" our consciousness to try to achieve the breadth of the collective, cosmic, and invisible, which merge. Science does not expect us, is not interested in our difficulties in thoughtful and understanding and, swiftly, intensifies and demands our perception of how much there is to learn, knowing our limitations. Today's society, due to the annihilation of fragile certainties that offer us the direct perception of events, further expands our obsolescence, we live in perplexity, both intellectual, and emotional, and with regard to our ability to accomplish, we are in precariousness. We live in a time of disappearances of what is traditional, such as professional occupations, the understanding of economic and political relations and the vision of the Science, the Cosmos, and the Universe.

For all this, we need to build and develop a new "human being" in a world without borders, interconnected and interdependent. It is from this perspective that we live environmental disasters, new migrations and attempts to bring violence and intolerance of totalitarian systems to power over communities and countries so that we return to the past we already know, but that no longer fits. We live intensely the old phrase: I prefer the bad I already know to the good that I cannot understand or control. In this strangeness we live in a world in disenchantment that seeks one's own ignorance, become more ignorant, to diminish the strangeness of living in a world that seems hopeless, but which is full of possibilities. All this is paradoxical, because we live in a time of splendor of science and technology, because as a humankind so we can plan and do unimaginable things fifty years ago, soon we would be proud of our ability to evolve and our strength to continue with haste and wonder. Then why does it look like we are, for many, going straight to the misfortune of the annihilation of humanity?

This question is building around. I myself have reflected on this several times since my adolescence, but not as a clear and concise inquiry as we are today, because now, in these times, it is clear to so many of us that we live in a world where information is available to everyone, and everyone has the exact perception of the very limitation of understanding of the real world and existence itself. Individually, until now, living well seemed to be standardized by seeking and taking advantage of the opportunities of life, enjoying the small and great pleasures that one could access, and this would be good, would be what to seek from the perspective of the dispute for success, represented by money, power, and glory, to have and accumulate the most that someone can, this seemed to be the right thing to do, and still seems to be, for most people, in the absence of something better to believe.

In this transition of interests and values, in the collective plan, we are in transition, the concept of humanity, once a sophisticated idea, has been built and is increasingly present in our minds as an inherent concept to decode the future. Information and news come from everywhere, impact us, thrill us, and make us astonished. Climate change is in our backyard and there are those who despair, there are those who deny. The threat of terrorism comes around the corner. Poverty and marginalization may be coming with the next day.

What to do? How to react without regressing? How to trail? The choice, it seems, is almost obvious: we need to take advantage of this magical moment of human knowledge, especially our ability to "look" what is microscopic, which goes far beyond the form, but mainly understand standard, relationships, and dynamics and adopt the view of complexity as a great journey that opens to our understanding of the things in the world that surrounds us, we need to perfect our look and our contemplation, reason and heart, expand them, review and challenge what seemed already dominated, guaranteed, and risk-changing perspective and leave aside much of our mistaken knowledge, our practical summaries and distance from reality, from the basic premises, as we now know and did not know, or we could live without knowing, until very little time.

1.3 Innovation and Evolution

The solution of many of our problems to ensure our basic needs to survive and develop, were thought, throughout history, pragmatically, which means that we collectively build controlled solutions, predictable, appropriate and that could, at the beginning, develop alternatives, expansions and protect ourselves and our goods, and then, in the end, offer artifacts, technologies, and techniques that would be marketable, what was, and still is, essential for the development of the modern economy. Within this pragmatic view it was essential to value inventions and technologies for intellectual property and intensify resources and production goods, in a combined way, under the control of those who demonstrated greater ability to seek and maintain and transform science and technology into utilities and marketable services. Within this process, a priori, "living collectively" and "producing for all," the concentration of knowledge has become something fabulous and, today, no more than one hundred transnational companies dominate and control the way and forms of planetary production and, despite the much they offer, the human and environmental costs that are generated are also huge and poorly distributed, because in the pragmatic vision has value what is scarce and that can generate profit, a combination that has made products and services increasingly standardized and with the potential to be made available throughout the planet. In this system, it focuses money, power, and glory between few and its effects, negative externalities, are distributed to and supported all over the planet.

In this model and progression of applicable knowledge, we have a lot, but not for everyone, because the restriction on access remains essential in the economic growth model. Add to this revenue the weakening of governments and administrations in the face of the globalization process, so today we have a system that concentrates income and power in the hands of companies and particular interests of groups with great power and that dominate the financial market, in a world that poverty and marginalization of the economic process expands, with the great indicator being represented by unemployment, depression among young people, and pasteurization of customs and culture. At the same time, the crumbling of political systems by corruption and the acceptance of the model of economic development hegemonized by transnational companies and the financial system continues to generate conflicts and wars and the phenomenon of migration has become evident and has reached a stage that cannot be predicted when and how it will end.

The philosophy behind the concentration of knowledge and permanent search for competitive innovation has as its legacy the emergence of a widespread planetary confusion, on the social and economic level, but that, from the point of view of technology, has brought us to a world completely new and with unknown possibilities and risks.

The possibility of misfortunes and blessings go hand in hand. How can innovation will meet the satisfaction of human needs, with equality and justice for all and, not to a reality that leads the world to a situation of the existence of two human species, one part evolving and with access to everything and the full development of their capacities and another, the rest of the planetary population, enslaved in the daily search for the satisfaction of its basic needs?

This question can only be answered, optimistically and positively, if we can, from the visualization of the system, develop new ways of satisfying the need for food and energy and how to provide opportunities for education and full health to be universalized.

This text does not have the answers but indicates, through some reflections, examples, and proposals that the visualization of the microscopic world by a new look, with regard to the offer of services and products, may be one of the pillars that will allow us to reorganize the planetary system in the sense of equality and fraternity, which are the basis of harmony in human coexistence.

1.4 On the Trail of Solutions

The separation of the world between organic and inorganic is didactic and, in practice, very useful. The models and intellectual separates we make, creating differences that can qualify and quantify, is part of the scientific method and the construction of knowledge. Science, which since the middle ages develops in its cycles of competence, shapes our way of living and thinking the world, generating associated technologies that define our behavior, attitudes, and vision of the future, first with glass and optics in stained glass and spectacles; the production of bells and reaching metallurgy and mechanical devices, with perfect swatches and the development of powerful armaments; the industrial revolution, that brought thermal machines and extensive use of coal; oil, with all its applications, especially automobiles, which modeled cities and the way of living; modern wars, with the development of aeronautics, interchangeable and production in large quantities, and communication; electromagnetism, with engines, radio, and television; quantum mechanics and relativity, which brought us atomic bomb, the electron detector, and the GPS; chemistry, which brought antibiotics, contraceptives, psychotropic drugs, explosives, and plastics; electronics, which brought computers, a moon ride, microcircuits, satellites, sophisticated software, the internet and recently the invasion of the enzymes, cloning, genetic engineering, artificial intelligence, and collective processes governed by software robots for large systems. In the past we wanted a lever to move the world, then a catalyst to give speed to the processes, followed in the search for a processor to expand the capacity and speed to treat data and perform calculations, we are on the threshold of living with cyborgs and robots, and so on, it's all very fast. The metaphors of who we are, and we need to live evolved with technology.

Among all technologies was electron microscopy, which expanded and qualified the look of what we didn't even imagine, which allowed us to develop our ability to manipulate organic molecules, the technology that brought biology and medicine to the center of our attentions and, what seemed to be important but not essential to the development of humanity, knowledge of the processes of life in all its details, it becomes the main movement of science in our times and perhaps no longer leaves there, the "greatest importance" is life and unconditional love to life, no longer the dominion and control of one over the other, some over many, war and destruction gradually lose the economic sense. Health and the improvement of human skills become the great objectives, because there is no more sense in not providing any human being, which may be born, the possibility of being free of diseases and having their life full of opportunities to develop. In this new world, and in the transition to this new world, there are still many old technologies, which have been important to arrive at this stage and which have become obsolete and, more, that in many cases slow the future, from traditional management of large systems to common techniques, such as use fire for all.

Among the tasks of humanity, one must imagine ways to provide the needs that people need to have access, in particular, to food and energy, will be made possible in the most sustainable way possible, because the human collective must be rethought systemically, so that everyone has the same opportunity to be full and collaborative human beings, with equal treatment and respect for autonomy and identity, as we imagine before, from the beginning of the times, an ethical and just society where everyone has the right to enjoy the best of themselves.

We are still far from this design, or collective objective. However, it seems clear that the path, the soul of this movement, is inexorable and may not be long as it seems. Science has brought technologies to a point where there is no control over what is happening, the knowledge of molecular biology and nanotechnology imposes itself, and its applications in health and biological sciences expand exponentially, is a systemic emerging property of our time, from what we call Humanity, which is the expansion of collective intelligence, the expansion of the strength of the heart and minds of all of us, to our great collective destiny. In this context, and trying to visualize through a concrete case, is worth an example that clarifies the transition in the last hundred years and on the point, I am seeking to analyze, so we glimpse the case of biodegradable plastics.

Replace traditional petroleum plastics, or broadly, the raw material represented by hydrocarbons, of all kinds, seems to be impossible. Since the nineteenth century oil, or petroleum molecules, have been refined into more than 150,000 products, in a generic estimate, which are present in all aspects of our lives, and seem to be irreplaceable and eternalize their use. The overwhelming majority of people can't imagine a world without oil and everything it offers us. However, without major analysis, oil is a fossil organic substance, which still exists, in large quantities and is relatively much cheaper than all its competitors in the hundreds of thousands of applications, and in some it is the only option. From prospecting reserves to the pot in our kitchen, there have been and there are millions of technicians, in all areas, who have imagined and developed everything we see and that we owe to the existence of oil, there is no denying, it was and it is genius. So how could we replace "black gold"?

While it seems difficult, it may be simpler than it looks. What's more, it can be very good, for all of us, so we live better and in harmony, without the problems linked with the exploitation and use of fossil fuels.

To illustrate the point, let's think about the biodegradable plastics. The first effort to produce biodegradable plastics from renewable sources is to prove that they are really plastics, with all the utilities show and offer as the usual plastics do, and that it is really biodegradable. This effort is great. First, because knowledge is not popularized, scientific research is incipient, there are not many experts, technicians, and equipment, and there is little money for investment, compared to everything that is traditional in the oil industry. And, one question has to be answered: from what raw material can biodegradable plastics be produced? Using molecules obtained from food is at least debatable, because with a world that has 800 million people who starve, produce, for example, fuels and plastics taking advantage of food chains, it seems ethically indefensible.

Too many questions, perhaps is better to forget, but if there is the hand of the market, it is because there is the hand of destiny in multiple interconnected collective processes. Is it? Looks like it does. The interesting question is that the conditions necessary for someone to decide to invest in this alternative can come from a localized crisis, some plan that needs to be reconverted and a social crisis, with unemployment and loss of tax collection. And enjoy pioneering, imagining, and performing in a different and new way. For now, I will not name or locate where this happened, I am interested in the arguments and reasoning, but it is a case report occurred, viewed without the details.

It would take a group of entrepreneurial scientists to convince someone, to be supported in search of solutions to a crisis, and show, or signal, the technical and economic viability of this technological adventure. It would also be necessary to convince the government to provide support, funding and, mainly, patience and complicity to allow the maturing of the endeavor. Following, considering costs barriers, convince some large traditional player to deliver some oil refining facility so that it could be adapted to the new raw material. And finally, produce from a natural and renewable raw material, adequate and available in sufficient quantity to make the production flow. A herculean task, however all this was built and emerged as a viable business, with significant results, even if it is not yet comparable to the oil derivatives market, but ultimately it has showed its technical and commercial viability, because patents have been developed, from processes, machines, and everything else and its adoption by the market, and today represents an important trend and has commercial vigor.

Let's now get to the point: how to prove that biodegradable plastics are biodegradable even, and that they are useful, that is, can they replace conventional plastics, and be competitive in price and quality?

The first step is the set of national standards that will be applied in the production and characterization of products. If the demands are too harsh, impossible to start, because the curve of experience, for example, for oil, is already 130 years. What we have is thirty years of experience and using only a few sources as raw materials, essentially food. Therefore, it was necessary to develop adapted and progressive standards in time, and so it was done. Anyone who reads will ask: is it? Did someone do it? Yes, there are those who imagine, believe, and prosper.

The first generation of the company's products had to prove its usefulness and that it was guaranteed in the face of technical evaluations, and that the raw material was renewable, they used cardoon, a Mediterranean perennial plant and that it was not a food, sometimes described as weed. This was achieved, gradually, and it functioned. The most difficult movement, but after there was an explosion of solutions.

The company then went to develop new products, able to compete where oil plastic could not get good results, attacking in the fragility of competition. There are many materials that are really non-recyclable. Among them the packaging long life, as they mix metals, paper, and plastic; good to use, terrible to get rid of and impossible to put it in a circular process. Or instant coffee cups, as they mix inorganic and organic products that are actually garbage after use, because they cannot be separated. That would be good if all components were biodegradable and separable, it would be possible to recycle easily. After many years of research and testing, the company produced a biodegradable plastic capable of supporting the heat of coffee infusion, how about? It's really a success story.

The next step was to improve biodegradability, better understand how plastics can biodegrade, where they can and should degrade—in soil, water, and air—and use their own biodegradability in favor of their products. Design things like: a protection for fruits that decomposes in time so that it is not there, protection, while it is harvest time, things like that. They developed several of these ideas, with economic and commercial viability.

The fourth generation was to look for new sources of plastics, natural sources. It was in this common interest with the company that I had the opportunity to meet her and subsequently visit. From the work with microalgae, and peering into the microscopic world, I participate in scientific discussions in the research group in Microalgae of the Federal University of Rio Grande, FURG, in the south of Brazil, where I knew and could understand that some rapidly growing microalgae synthesized plastics for their cell membrane, especially the poly-hydroxybutyrate, and so it would be possible to imagine that one could grow plastics, something fantastic, but that, obviously, was very auspicious, and Novamont company, in the region of Milan, Italy, our star, also wanted to know how to do. And it did.

Since micro-organisms can create metabolic routes capable of producing plastics, one can imagine making changes to microorganisms that are suitable for the company and its interests, in enough quantities for commercial production. In this way, combined with the use of microalgae as a source of food, energy, the production of plastics would compose the set of activities that would enable the economic equation.

Novamont is a new paradigm that shows, as it moves toward scientific and technological solutions committed to sustainability and ethics, new spaces of knowledge and understanding of reality emerge, and a new economy is possible, with less environmental impact and that emulates natural ecosystems.

What does this example of success mean? That it is necessary to rethink the paradigms of production and that, by insisting on a production of biodegradable plastics from locally produced renewable organic molecules, or better adapted to the site, harmonizing variables, and revealing emerging properties, one can evolve and innovate.

1.5 Freedom for Imagination

Is reality a materiality or is it a concept? The objectivity of the search for objectivity is a premise of science. That is, it is objectivity squared, it is necessary to be objective and there is an objectivity in itself. Therefore, the "Being," the observer, who influences and is influenced by his subjectivity and the observation process itself, should be excluded from the process of knowing, because it is a source of uncertainty and masking of truth. "Objective knowledge" is the basis of science and belief in an "absolute truth" that can be known, is the basis of science and religion. Very similar.

How much is lost by losing objectivity? The absolute, as conceived, is everywhere and remains all the time, and comes from all the time, and emerges as the definition of everything and, thus, the goal of the human being will then be to know and describe it? But, if we are here, around, to help build it either? If the absolute is already what he is and our saga is to integrate with him, by knowledge or surrender, what is the true way? What if, however, if the idea of something absolute that sustains and show signals for the need to understand and understand is not finished, if it is in development? If we accept that we are the curious and observant, hungry for knowledge and, we are also believers and gullible who accept and surrender, hence there is no dilemma, because we are facing reality to perceive, admire, and idolize what really "is," but we can imagine that we can interfere, improve, and expand one's own reality. Why not?

We are and express the two forms of understanding, including all the elements that result from our exchanges inside and outside, in the balance between certainty and doubt, in a process of interactions and learning in time and space, thus we evolved.

Now, and this is very important, even if it is a little bothering, especially for our way of thinking and reflecting, the inherent uncertainty of these placements does not prevent at all, nor do they hinder thinking, only manifest the particular and simplistic character of traditional thinking. And at the root of the choice of the way of thinking, where our emotional perception is immersed, surely uncertainty generates discomfort and is not always welcome.

Thus, uncertainty does not prevent us from knowing and understanding about the process of being alive and influencing and being influenced. Still, it allows us to understand more and better about the relationships that define us and support us as beings of intellect, emotion, and desire, and includes a very strong consideration regarding the consciousness that we are not separated, we are together, and our construction is collective, not only between us, but includes reality. We are *Homo sapiens sapiens*, not only knowers about but are also builders of reality.

Directing us to concrete, microalgae are out there to agitate our souls, because it answers us by being able to imprison a photon, the energy that walks through the universe, traffics through the vacuous, and then makes available to build more life, so we have to ask ourselves: how did this happen for the first time? And after it happened, what can we not even imagine, what drove to transformation of reality? A new look at what is put requires us to, at least, trust and establish ways to follow this mystery, almost miracle, and explore the possibilities and learn how microorganisms, which we can see in action today, multiply life.

Returning for another moment to the particular and simplistic aspects of traditional thinking, we know that language, which emerges as property of the Life plan, is configured as a cultural construction and results from permanent negotiations of adequacy and adaptation to a particular view of the reality of a group, being the scenario of construction mediated by the environment, history, and due to the consensus created by personal experiences limitation of the members of this group. The perception of what the reality would be by the group is constructed by recursive processes aimed at removing the diversity of thinking, reducing the fear and risks of evaluations, and reducing the number of variables, including the reduction of evolutionary dynamics and interactions. In this context, in the traditional view, a better world to live in is a world simplified by a particular and static vision. The group's leadership is to consolidate this world and justify it, by force or by argumentation or by seduction, and in general by a mixture of these forms of limitation of expectations and control of behavior. We want security, predictability.

With the increase in interactions and the increasing exchange of information between private worlds, the current moment of civilization and humanity shows an increase in anxiety and anguish with the loss of efficiency and effectiveness of very particular visions. We are in the age of perplexity, where we emphasize the recognition of patterns of behavior resulting from simplistic, particular, and static visions, and people have nowhere else to hide, or how to hide from what is much greater than them.

The diagnosis, is itself a process derived from a particular pattern, indicates that it is necessary to defend the achievements of traditional thinking. This represents the widespread adoption of a perverse form of social conduct that is the generation of ideological fundamentalisms, something new is an enemy, that sustain and intensify violent processes to prove that one particular view is superior to another, even if they are content identically constructing knowledge and thinking, results from a same process of thinking. Mild observation with less commitment obtained from so-called neutral approaches show an intensification of processes that generate immediate anxiety satisfaction. However, as the process has feedback and reinforcement mechanisms, it develops more and more resistance to stimuli and with the achievement of smaller and smaller results, requiring more and more. In the model, said consumerist people want more and more, faster, without thinking about the consequences.

In my view, the human activity sector that has better perception and is more informed about the new time we are living is the financial sector. The concepts of network, processing, instantaneity, among others, in the planetary dimension, were internalized in this sector. The new situation is that the predominance of the financial sector is due to the fact that it works primarily with information, is the result of the cyber world, and as an evolved system of decision-making at risk, generates and determines financial flows that represent flows of materials and energy in the concrete world. And as an information system can, at times, by opportunism and greed, after all it is a human activity, operate based on the idealization of availability and infinite capacity to offer resources and recovery of natural assets around the planet. A moment of folly with decisive long-term reflexes.

At this stage of the development of the globalization of companies and consolidation of globalization, we live the culmination of irresponsibility by some agents of the financial sector in relation to natural assets and the generation of increasing risks to the sustainability of maintaining life on the planet. As the financial world learns fast and today eleven years after the global crisis in 2008, we see the effort to create conditions for a world order derived from an exceptional concentration of income among few agents, people, and organizations, which can forward globally processes of insertion and marginalization, which can almost everything, because as everything is interconnected, it seems easy to move the pieces and guarantee interests. At the same time, one cannot forget the power of hitting the wings of butterflies, there are those who like to imagine and soon it is necessary to find them, to impeach them, domesticate them, within cosmopolitan agendas, creating conditions and emerging false dilemmas within science itself, like that existing among the defenders of the planet denouncing climate change and negators, with great scientists on both sides. It is a worldwide controversy that allows, in the name of the "salvation of the planet," to create a collective feeling, of humanity, of saving the future, with threats of cataclysms, large epidemics and loss of the future itself, which can even be true. The world does not know about its future, and the fear of catastrophe lies in the discourse of young people. And, of course, the financial system acts all over the planet by placing resources in the "construction of the new order," because the conflict will have to be predictable and manageable. As we do not know, yet, the speed and extent of stimuli, with them work in large systems, megadata robots test social networks and the virtual world to understand and build guidelines from the observation of how influences generate setbacks and counterattacks, to understand how the world will be in fifty years, how to make it predictable. After the threats of terrorism, epidemics, the great topic that is shown transversal, hegemonic, is climate change, whether true, or the whole truth, because it becomes a global anguish.

The question that remains for the great world agents of our times is what to do with collective anguish, how to use it, and how to benefit from it?

At some point in my life, I had the opportunity to mingle and discuss strategies with two large investment funds. Their great concern was how to conceive guidelines and forms of action capable of generating recognition of their effort to preserve the natural environment and the proper management of planetary resources. And, of course, making a lot of money. The question is what and how to do it? The main interest is to foster and support innovative projects in the management and use of natural resources that are recognized by society, that is, to have clarity of purposes and execution, including remarkable results, and with relevance to build new references and models of development and wealth generation. It can be said that they want to create a market without lies and false statements to justify private interests. Where does this idea come from? The idea of appearing to be good, that companies are good, and that the agents of structuring the "new era" are good, are in the way of good. And that deserve the money they gain.

On the other hand, it is important to understand that despite the power of funds and the financial sector, traditional knowledge is in companies, so companies are the great depositary of this knowledge. They constitute a collection that will not be modified or replaced in the short term, and which is currently contaminated by the prevailing rules of overexploitation of planetary assets and false expectations of growth and financial gain, pressured by funds and the financial sector. Still, business schools, although more advanced in their proposals to offer cutting-edge knowledge, are dependent on traditional executives to enable their continuing education programs and continue to preserve a minimum of reliability with the old world, that is, they are braked or skating, otherwise they will not have students. And universities form, in their undergraduate courses, for a much older world even older. Therefore, none of this follows the script, the scene, of those who want the common good, the good of all.

So, it is very likely that the changes will begin to occur on a trading platform that will involve at least investment funds, large individual investors, banks, innovation intelligence dispersed in different organizations, virtual companies, insurers, and representatives of traditional extractive sectors, with the attempt to mediate international structures such as the UN and its bureaucracy, and the militancy it advocates and proposes environmental and social solutions. It will be a platform. Despite its importance, the agribusiness sector of developed countries may be more responsive to change and should participate in discussions sideways, and developing countries should go further, for reactivity and oppose, preferring to discuss improvements such as obstacles and environmental barriers to their own development.

Who can bite and blow in this configuration? It will be the financial sector, which is hegemonic, but which, even for this reason, knows the weaknesses of the system and wants advances without losses of assets and future gains.

What's necessary, then?

Firstly, create a great binder concept, which, within the offers of understanding, is within the platform that intents and builds sustainability. Policies, plans, programs, and sets of sustainability indicators have been created, and the stock exchanges and the strict financial system are created, the main stimulators and developers of these proposals.

Secondly, create a planetary urgency, including businesses, governments, the UN, and civil society, which means a punishment for those places that do not privilege sustainability, a concept that brings together in its meaning the well to do, the responsibility for the future and excellence in the management of planetary resources, being the main indicator of climate change and, as they represent two-thirds of the planet, the increase in the level of the oceans, which for all can be perceived.

This seems to be a good global strategy, because it undoubtedly creates a global feeling, an economy of climate-based attention, and that allows organizing the planet, with less losses, certainly much better than reorganizing the system through conflicts, as it would be a third world war, as others believe it should be common fear to be propagated.

The society, as a whole, still has little perception, clarity, and conviction about initiatives, but organized sectors, especially large NGOs, consultancies, and institutions focused on the training and updating of executives perceive the great opportunities for intervention, image generation, and financial gains associated with the concept of sustainability. The environmental movement, more specifically, still digested within its particular views and does not have a common agenda, but the agenda arising from climate change concerns, gradually, replaces concern about biodiversity, pollution, indigenous peoples and even this common agenda tries to incorporate all topics under the domain of climate change idea.

Seeking to integrate the already written in this text, it can be observed that with the dynamic and highly interagent behavior of international trade and the supremacy of the financial sector, which generate integrative process, generator, and evaluator of the value of things, it has been established that uncertainty becomes more evident. Various private exemptions become fragile to address the diversity of interests and are not competent to address the risks of interests and decisions. Complexity requires new broader and dynamic collective approaches, and large negotiators are those capable of modeling and simulating temporary stability, and evaluation instruments resume ethics as a guiding element. Diplomacy, the art of possible agreements, becomes the permanent source of solutions to complexity, dynamics, and evolution of this complexity and construction of collective considerations, in a context of uncertainty. The civilizing framework and clear perception of humanity, as concrete elements of planning, are imposing.

Therefore, or something like that, it is prudent for companies to realize that their longevity and success as an aggregator and builder of opportunities and evolution requires openness and availability far beyond what they imagined, or imagine their managers today.

Isolated companies, seen as a castle protected and able to remove from the middle all the materials and energy needed to maintain it, even hoping to survive as an idea and metaphor, will have difficulty resisting and growing in a much more populated, interconnected world with scattered interests, and with planetary agendas beginning to impose themselves.

1.6 Is It Possible to Understand Companies as Organic Ones on a Living Planet?

Understanding or explaining life represents a full, singular challenge, that is, it is not possible to understand or explain life. But this does not prevent us from having some ideas about what it may be, or about its characteristics, nor does it prevent us from admiring or marveling.

If the Earth is a living being, and nothing could negate it from being, and it is even quite reasonable that it is as some of Gaya's thesis supporters argue, it has an interface with the world said not alive. The materials it uses and produces in their autopoietic processes are on Earth itself, as an inheritance of its own genesis and history, and the external element that crosses the interface, the outer space, is light. Thus, the fundamental source of energy is light, and photon the primitive agent of life, and everything begins there. Light is life, at least it starts and/or started like this.

Therefore, living systems exchange matter, energy, and information with the environment. Once the process began, life begins to express itself and take advantage of the opportunities of diversity and divergence. And there are many possibilities. Fungi, for example, are absorbers and processors and do not require light in their biological processes and metabolism. Thus, for these organisms, light ceases to be a fundamental element for their survival, but without light, they themselves disappear, even if they are the last to disappear.

What limit does life have to express itself? We don't know that. The diverse number of living systems depends on our observation, and our observation of our references, the ability to recognize and the opportunities we observe. Therefore, there may be more, or much more life out there than we can perceive and there is no restriction associated with the fact that we can observe it so that it exists. Only its manifestation, which implies our ability to perceive it, will not validate its existence, only validates the existence and partiality of our perception. By recognizing the processes of life and understanding them, we are able to offer life new forms of organization. This idea is very sophisticated and frightening, consonant on our times.

If *Life* is the intelligence that taught nature how to imprison the photon and turn it into organic matter, into biological life, then this is the concrete expression of the organizing soul of *Life* that arises in nature, in the Cosmos. And biological life is that then continued to teach and propose solutions that worked and whose net result led to us thinking, reflective, and *imaginative* beings, certainly *Life* is a pure intelligence that learns, develops, and produces concrete results in reality. *Life* is the booster, it is not the fate, destiny is under construction.

Where should we look and try to understand? Where do you get clues to think about companies and ways to do business? How to think about complexity and learning in economic thinking? I believe that in the study and understanding of how microalgae emerged and how they have provided the development of ecosystems, we will have great sources of inspiration, many tips. It is in microalgae, in photosynthetic organisms, that there is a possible map of our future, we lacked the ways we access and read this map, which today we have tools that are available, more and more and better. And quickly! We have to reform concepts, build theories, and develop new ways of understanding and doing. The question that arises is: do we have the seeds of intelligence capable of fruiting for this understanding? Or will we have to build, through artificial intelligence, installed in cyborgs and robots, a new human species transmuted from biological life to electronic life, to serve the goal of *Life*, so that is the way we can understand and somehow dominate the Cosmos?

The microscopic world is much more diverse and interesting than the macroscopic world, but it is not part of our perceptual framework. If we could imagine mitochondria performing its processes continuously, quickly, and with multiple interactions and exchanges of energy, materials, and information, we would know that we have nothing similar happening in the particular world that is determined by our perception of macroscopic. And one thing certain is that in the last fifty years we have begun, even if there are still few of us, to realize the immensity of the alternatives of biological structures, the world of DNA and viruses, and recently much more supported by nanotechnology. The new world of perception has arrived and will be in biology and the study of the characteristics of biological life, is the new arena of broad and depth perception. It is clear, we are still resistant to all this and prefer our vision and our particular vision, our interior little world.

But the world of large biological sizes, which is ours, arose as the complexity of microsystems has increased and new systems, larger and more structured, independent and cooperative, have emerged, when and where exchanges of information have become preponderant, as in the case of language, culture, and human history, an advanced stage of life that can, in the near future, give speed and expand the forms of structured life in matter, said biological life, in forms and expressions that we still cannot *imagine* today. The organisms, the groups of organisms, the net of organisms, and everything else we know were formed and that in the end forms the web of life and now it seems that the web will expand.

As this text proposes to *imagine*, and going without fear, let us analyze the following: let's say that *Life*, organizing intelligence, in its many attempts and errors, after all we are facing an entity that is an apprentice, as Prigogine said, has previously had excellent ideas of structures and dynamics, biological entities, but that did not prosper due to lack of ideal conditions at any given time, but who remained in stages with the potential to go far beyond. What other bridges and lines can systems use and resume their evolutionary opportunity status? Is it possible to give a second chance for old and incomplete *Life* projects?

And we must remember: 3.5 billion years ago, *Life* managed to arrest its first photon, and everything else we've noticed, the concrete world is there, we're here, but despite that little is known. The conditions, at that time, here on Earth, were very different, probable today *Life* would not have the same conditions and timing to produce the primordial event, although today, there is life everywhere, we have a large stock of biological life, yet, to learn and understand more. And maybe it's an almost miracle that can be understood, whether it's us, or the electronic life will follow us, what we turn into. Well, if we don't understand, the novelty is to understand how *Life* did and do, and imitate everything that worked out and help develop what hasn't worked out yet. The idea of unique and isolated things is a human conception based on our perception and our desire to be unique and definitive. And our temporality. Nature tries, errs, redoes, and evolves all the time, driven by *Life*. Nothing says that the presence of the human being determines this gift and "desire" of *Life*. Or that it is possible for humans to be able to remove it, just an idea of human vanity. The animating, the driving intention, is permanent and can be supported, stimulated, by paths that we do not even imagine and that science provides us, in a defining and, for many, alarming way, when we achieve, more and more, enter the intimacy of synthesis processes on a microscopic and nanoscopic scale. I am sure we are near, in the border time and the understanding of microalgae will be the scenario that will happen in this meeting.

Society, and therefore companies, evolves and is increasingly immersed in complexity, which is increasing, and it ceases to possess characteristics of the world of our perception and our traditional management tradition. The stage of human civilization generates a progressive increase in the number of interactions, connections, and multiple interferences in a finite and limited space. And everything occurs in a process and a perception in which time passes faster and faster and where there are many changes, within a new and not yet perceived dynamics of exchanges and influences, and can only be compared with biological structures, the only ones compatible with the degree of complexity of the world we live in.

Our emotion wants simplifications, stability, and objectivity. And it is right that they are great achievements of the human mind, our programming to live and pursue. Accepting again the anguish of uncertainty, permanent instability, and the feeling that we will never know, or it will take many human generations to know, deeply disturbs us. We are no longer special individuals, we become an instrument, now definitely in the light of clarity, the evolution of our species, difficult to sleep on the night of this transition of understanding and thinking. And our mind reacts and induces our emotions to stress, suffering, and depression. And fear takes over, widely.

It is likely that our ancestors, realizing the disappearance of their food sources in the forests of Africa, about 500 generations ago, have experienced the same situation and have been obliged, since the monkeys and humanoids, to become what we are today. Our current moment is a very similar moment. If there was comparable uncertainty in the human saga occurred at that time and occurs again in the current. Facing the unknown, without human internal psychological tools to do so, yet, is what only a crisis and an imperative need could impose. The surrender and determination of those who decided to meet the unknown are unmatched.

Thus, whom to support and intensify the effort toward sustainability? There remain the individuals who *imagine* the future of organizations, the visionaries, who make broad considerations and thus include the assessment of the environment, the social context and respect for cultural and ethnic differences, capable to build a systemic map and evaluation of reality, and possibilities of reality. And who are also capable to understand the science, the biology of photosynthetic organisms, to be inspired to formulate new guiding ideas, new architectures, new structures to create value for everyone on Earth.

Resuming, circularly, considering the sectors that lead to the structured movement of increasing the competitiveness of the business sector using as a vector the expanded concept of sustainability we have the global financial sector, especially large investment funds and stock exchanges; some of the transnational companies, especially those in the mining and energy sectors; and large environmental mobilization structures such as Greenpeace and entities focused on preservation and conservation of the planet's environmental assets. These are the main articulators of the introduction and specific indicators and are those who are strongly positioned in the considerations about what sustainability is, including in the strategic planning of large organizations. In this collective arena, the debates expand and improve concepts and put them into check and thus we advance.

After the Quality movement, which selected companies that offer a higher class of products and services, it is critical to transform what has become large and with global operations into regional and community assets. It is an effort to build image and ethical reference, generating the perception by local social groups that the company's performance is a source of generating wealth and quality of local life, mainly. The social shock absorbers represented by the campaigns of advertising and marketing can support but can no longer be the main generating factor of public image, and so what is required are policies, plans, programs, and actions by sustainability, not words and exhortations. Governments will have to respond to pressures for improving quality of life and ensuring food safety and access to water and energy and will have to make choices about the best types of investment for their regions.

The advance in the evaluation of the contradictory, permanent consideration in the understanding of complex systems, it is assumed that the increase in the supply and flow of information on the effects of the production of goods and services on the environment and its reflections on climate and water sources for human consumption should be considered not only for the next generations but for the current generation that is there and is distressed.

Finally, whom to support and intensify the effort towards sustainability?

Inside the established companies and business sectors will be the individuals who *imagine* the future of organizations, the visionaries, who make broad considerations and thus include the assessment of the environment, the social context and respect for cultural and ethnic differences, capable to build a systemic map and evaluation of reality and possibilities of reality. Additionally, and specifically, who are also capable to understand the science, the biology of photosynthetic organisms, to be inspired to formulate new guiding ideas, new architectures, new structures to create value for everyone on Earth. It would request time and new competences, but it will occur, and it will function as the source for new knowledge and inspiration for action. The new manager, in this new world, will need to think and act based in a knowledge mix of pedagogy, engineering, philosophy, and biology.

1.7 A Systemic "Pedagogy"?

The mental model we have about how people can and will learn is not always known or is clear. This is important, in order to make people assimilated and can understand about something, mainly unusual thinking, a good idea is to try to understand in which way people learn. It is not the intention here, but some headlines are necessary.

For Piaget, the apprentice is a boy, a Robinson Crusoe lost alone on the island of knowledge. As much as one wish to impose, learning operates from an individual decision and process of discovery and meaning creation, which internally builds the network structure capable of decoding the world, which can be communicated and generate interactions and understanding of himself and his surroundings.

For Vygotsky the human being is his relationships, his sociocultural context, and society, because he knows much more than the individual, organizes, and refers to the more sophisticated functions of the human being, especially language. It is soon up to society to bring the human being to its greatest potential, to bring it to the realization of its full potential, because the human being alone cannot recreate all the complexity that precedes it and that comes into it and recreates every moment.

In Papert, we have an extension of the pedagogy, which advances in the conceptions of Piaget and Vygotsky, without opposing them, but expands the use of reality by the use of existing processing tools in computing that greatly supplant the human being's ability to visualize and understand relationships. The perception of the world of relations of the present we have that society has entered a new phase where the acquisition of new skills and collective skills go far beyond what we knew existed. It is the society of networks, where a small contribution can be amplified and generate much more important collective advances compared with the accumulation of content and knowledge that can occur in an isolated human being. As there is always uncertainty and movement, what occurs is a trend where the collective is the domain and sharing of a tool, the computer, the basis of the new global learning process.

Furthermore, the trend in this process is the construction of meaning increasingly in network, and in networks of networks. Recently, beyond all we expected that could occur so soon, in Web 3.0 we have the intervention of artificial intelligence organizing and impelling more relations and recurrent interferences.

A question arises soon: Who is prepared for all this news in today's world, in addition to young people and the very young?

It is likely that the greatest of all the challenges of the current phase of our society is how to bring the almost obsolete world of stability and isolated interests to the world of networks and evolutionary balances, the so-called systemic world. And those who must learn, most of the time do not desire, do not understand, and are unaware of what is happening. Those who keep knowledge are no longer the monks in their manuscripts. The book has already been invented and already exists. What will the computer convey? Who keeps the keys to the future? What will surmount artificial intelligence?

Using a simple description: the new school is a school where to get free knowledge from other student, that is "Necklace in evidence is permitted" are those who know best, copy, share and improve the knowledge is the goal, collectively. And they all win. You're not good for what you keep, you're good for what you can spread. It's the knowledge society. The Google and Credit Cards template, chaordic systems, are the models.

The question is how to reconcile this new model of learning and sharing with a centralizing economic model, based on secrecy and accumulation of personal and company knowledge? And with people belonging to a professional tradition who succeeds and are very well established in this old model. In today's world, the financial reward is given by what you know of yourself, by the way you can express your disjunction and differentiation of what is common by making yourself a way to express restraint and not *abundance*.

It is difficult to understand what financial rewards will look like in a world where the knowledge society prevails. We're going to have to learn and develop. Or perhaps try again the systems where everyone is protected according to their unique qualities in the concert of social needs, a basic income to be inside and contribute. One possibility, among others, will be to return to the attempt to create the phalansteries, the structuring of a utopian society of wise, artists, and workers living without usury and exploitation of the human being.

Or, on the other hand, we may be living in the final phase of development of capitalism where Adam Smith's ideal is realized where all information is available to everyone, all the time, and will make better of those who organize themselves and are more talented to think, to accomplish, and to keep in networks.

Well, better to gradually learn to find the way to a cooperative and even unstable society that works in peace and is progressive. There's a lot to learn.

In this objective of learning and understanding, one can examine the use of Systemic Thinking as an appropriate way to bring the human being closer to the potential of learning collectively and using the advantages obtained by using the computer for processing information, and thus structure a method capable of advancing in the search for better learning performance.

The first stage of the systemic method of learning generation is to recognize the influence of relationships between variables of the daily lives of people and organizations, and to recognize the fact that in many cases these interactions supplant the very factors that generate them. When we evaluate systems where the parts are distant from each other, and there is little interaction, understanding can occur only from the evaluation of simple and direct relationships. However, with the increase in the number of variables, and the intensification of relations between them, combined with the dynamic behavior of these variables that change in time and are dependent on the environment, we have a new situation much more complex and much more demanding for its description, consequently, so that we can learn about and understand the standard of functioning of the system as a whole, and in its local and temporal manifestations, represented by the events.

The main problems arise from the reason we try to bring to the environment of complexity the tools and techniques, in general simplistic and void, used for the solution of systems of few variables and where simplifications can be adopted that facilitate the treatment of variables and the study of their behavior. Furthermore, static evaluations are privileged and there are few considerations about the various influences of the environment and the variables on the systems themselves. As Deming said, one can know everything about water and nothing about ice, that is, transferring the classic approach of treatment and the appropriate solutions of systems of few variables, low interaction index and with predictable evolution, without a dynamic to associated, is a great risk that can lead to major disasters. Returning to Deming: great efforts and a lot of work from well-intentioned people, but who do not know the system, how it works and evolves in time, can only make the situation worse. We may ask ourselves: and luck, can't you still help these people? And then we'll have another one of Deming's maxims: if you don't know what to do, do nothing; if you are required to make a decision on a system you do not know and understand, draw the solution. So don't believe in luck, but if she comes, be grateful. But luck is part of another context.

And if the situation is really complex, let's give up for acting, is that it?

Of course, no, but sometimes doing nothing could be a good idea. What is necessary, before acting on a system, is to advance in understanding about it. And this means building the maps of the relationships between the variables of this system, that is, to visualize these relationships and find and recognize the basic forms that allow us to perceive and describe the variables and thus, when we act on the system, we will have an improvement in its functioning as a whole, improving its overall performance, and assisting in accomplishing its *purpose*. Every system has finality, a purpose, which justifies its existence. This is the reason for your expectation.

Then one can ask: What if the system is too complex? To what can be answered: all systems are complex. What has occurred so far is that we have managed, in several cases, to simplify the visualization of complex systems to adapt them to the simplified treatment models that we develop and master. And this is even the fact that tremendously complicates our understanding and acceptance of needs to learn and build viable systemic production solutions to meeting needs, without leaving a trail, a footprint of destruction in the physical and historical environment, including the economic, social, natural, and cultural system, as we can clearly perceive today.

Accepting and expanding the description and mapping of systems, one might ask: what do we have resources to face and understand complex systems?

Fortunately, when we can structure a form of thinking held collectively. When our collective mind is encouraged to visualize relationships in systems, and share with others our perception and mental models about reality, one can enjoy a free space of the usual restrictions of thinking, and thus builds a learning environment that arises from the richness of human relationships and the shared knowledge present in the objects of evaluation. Bringing together, for example, several professionals from a company to think and glimpse the possibilities of the future, the evaluation of possible scenarios, rather than learning about the future, they learn a lot about what it is and how it operates their company in the present. And they do not fail to develop strategies capable of supporting your company to face different scenarios and possibilities.

And what else do we have? There enter the computers that can, from the variables and the designs, simulate and learn about various scenarios. As the actors say about a staging, simulation allows us to live several lives in various free worlds to make mistakes and, better, learning from them without paying feathers. Better yet, profiting from them because they are sins in virtual space, no penances are paid, they do not require us to buy indulgences. Failure is the mother of imagination, they hit an obstacle, in this game of learning. As a result, we learn about what it could be and how we can behave in diversity, which represents a powerful source of creation and feeling inside, going beyond the particular and unique reality that surrounds us, allowing us to broaden our perception and consciousness.

In the end learning is consolidated by the experience in this new context of perception, building more alternatives and possibilities, which means more life and human development, *the purpose is abundance of intelligence and existence.*

1.8 Control of Central Variables

The system can reorganize from central variables, which are those that are understood by all and thus achieve their maximum performance and an evolutionary dynamic without ruptures, within an evolutionary stability. In the case of the financial system's effort to place climate change as the central variable of a sustainable future, it is a valid and winning strategy. The fundamental issue, and to this, the so-called climate negativist, the finance system may have, on the one hand, created a false central variable, that is, there is, in fact, no climate change caused by mankind, it is a natural process, or the "skeptical environmentalists," who can even accept a certain change in the climate of origin in human activities, but argue that these changes can even be beneficial to humanity and also, that there is a certain change in the climate of origin in human activities, but argue that these changes can even be beneficial to humanity and also, that there is more urgent and important problems to solve, such as poverty, the difficult access to education and benefits of the modern world, injustices against women, planetary waste, among others. Finding a central variable that mobilizes the entire system is a task and a large binder of mental models.

The system can also broaden Its own understanding from the idea that every system is composed of many resilient subsystems and organizers, which reproduce the large system locally, or what exists, in fact, are the subsystems and the large system is the result of the interaction between virtuous subsystems. This option can be very relevant, is the thesis of this text. In this approach, it is ecosystems based on everything, and are the model of a large harmonic and stable system. And in ecosystems the "energy propellant spring" are photosynthetic microorganisms, microalgae, which reproduce and organize increasing their complexity and scale, that is, we are able today to understand how life expands and spreads and we can resume the "microscopic pillars" as what should be understood and used to solve how we feed and where to seek the necessary energy, and advance in the health and well-being of all. It is a hypothesis that today, in the current stage of science, we can evaluate and test its viability, whether it can be a good choice to advance and solve the "macroscopic" problems of our time. In the purpose is abundance of intelligence and existence, imagine microalgae the model and central variable for the system creation.

1.9 What to Do with Everything We Begin to Understand: The Need for Action on Reality

Considering the complexity of the time we live in, associating with the inherent complexity of natural processes and the possibilities that today's science offers us, as described so far, we can make two approaches that may be able to forward solutions to our uncertain common future.

In first place, take advantage and gather what we have best to make an "amplifying reform" of the system of supply of products and services, especially how to provide food and energy, using in its fullness what the microscopic world offers us, which are the micro-photosynthetic organisms, and organize a systemic map around them as a *central variable* and leverage of a system of high output production performance with economic value, which are photobiorefineries. Something like a new way to produce.

Secondly, organize integrated production systems capable of restarting the proximity of people and communities with the natural environment, in what are called Integrated Biosystems, and that offer multiple opportunities and expansions of our ability to live harmoniously and with high economic and environmental performance. Something like a new planetary society of Systemic Communities.

In these times of great novelty, and considering the reflection realized so far, it is urgent that we create expansions of language and the understanding of productive systems. The new words arise, and unexpectedly traditional concepts require new interpretations and expansions of meaning.

In this context, looking at what is built includes the characteristics of large industrial plants that today constitute production centers, with an extensive range of processes and products, including material separation and synthesis, intensification of knowledge and experience, with activities supported by large structures of capital-intensive production goods and sophisticated forms of networked control and management.

Thus, when describing an oil refinery, a classic definition usually be: Refinery is the usual name for oil distilleries that carry out the chemical process of cleaning and refining crude oil extracted from wells and crude oil mines, producing various petroleum derivatives, such as lubricants, coke, diesel, gasoline, LPG, kerosene, aviation kerosene, and others.

One of the extensions of meaning, already present in the literature, is what is called Biorefineries, where the replacement of oil is considered as a basic material by biomass, derived from different sources, as in the case described above of the production of biodegradable plastics, to start the production process.

While representing an important innovation by bringing the gains of oil-associated production, which is capable of generating thousands of products from a highly aggregated and unclean starting material, this innovation is related to the full use of a raw material, but at the same time has a great environmental impact, not only on its use, but also on its own production.

A mental model that is maintained in the biorefineries is the organization of a linearly structured production process, that is, products are obtained in a sequence of operations, events, where the whole is reduced obtaining smaller and smaller parts with specific uses. The greatest emphasis occurs in the use of intensive energy and pressure processes, and on the use of combustion as a key process in production.

When we analyze the forms of production used in Nature it draws the attention it produces, and solves production problems, under normal temperature and pressure conditions, as well as using synthesis and self-organization processes that build complex structures and that take advantage of residues from a process as raw material from other processes. Thus, the ecosystems that maintain life are constituted and allow the evolution of the structures and processes of life themselves, in a circular production model.

It is interesting, and important, to note that there is a mental tendency that induces us to act always as decomposers of what nature offers us as wealth and we do not think about acting as support and facilitator of synthesis processes.

Even when we think of Agriculture, which brings us closer to synthesis, it is planned and executed in the search for monocultures, the intensive use of soil and water resources and the creation of strategies based on growth and harvesting protection schemes with the use of chemicals, whether synthetic or pesticide fertilizers. Nothing like nature that generates food and health for all actors in the ecosystem, in the long run, all the time.

Another important aspect relates to what we cultivate. We go very little beyond the plants and we don't realize that the greater diversity of living creatures is microscopic. And these beings are the reason for the creation and permanent maintenance of the conditions necessary for life on our planet. Knowing them more, knowing how to reproduce and interact, can and should be the most important choice we can make for cultivation and harvesting. They grow faster, are more robust, and offer a multitude of uses and possibilities that far outweigh even our *imagination*, but it is only our *imagination* that will access how everything happens. And they can be almost reinvented, in some cases, as I wrote before.

We produce food in the same way as 10,000 years ago. And we think the microorganisms more as enemies than as the great benefactors of the process of life. Another mental model.

What would be a great innovation? Understand how to cultivate, how to shepherd (even though it is an intriguing metaphor, I first heard from Prof. Benemann) the microorganisms to obtain all foods, including the proteins, lipids, and sugars we need, allied to drugs, cosmetics, rejuvenating ones, and energy and water.

We need to know in depth the micro-photosynthetic-organisms that allow us to consume large amounts of CO_2, the villain chosen in our times, very likely a mistake, in any thesis on climate. They are the ones who offer us the biomass that will offer us bioactives, of all kinds and for all needs, and at the end, as residual biomass, what is needed to obtain energy through fermentation and biodigestion processes, among others.

What do we need? A lot of light, that is, the light itself, and the light of human creativity and ingenuity, imagination. What we did, in one of the projects with the participation of the ZERI Brazil Foundation, by using combustion gas from coal burning, from the understanding that this gas is a rich and unique source of CO_2, which is concentrated and located and is easy to use, facts that increase the possibilities of generating wealth and value, a CO_2 mine.

And you can do a lot more. Even at the end reduce the use of coal itself, because it is possible to burn that biomass from the cultivation of the production of the organisms is used. To do so, one has to think of decomposition and synthesis links, circularities, and emerging properties of systems. And that's what you can do, without limit, by circularities.

Coal, fossil fuels are necessary for transition, but they are not a disgrace.

Where else can you think so? Where will be the most advantages available? Just look at nature, watch it and learn. The behavior that teaches us Biomimetics and use creativity to even organize and facilitate processes.

Therefore, the synthesis of organic materials occurs by the absorption of energy from the Sun through photosynthesis and it is possible to completely emulate the understanding of production processes, including the synthesis and separation of materials, in a circular process where waste becomes elements of new synthesis, in a continuous use of the Sun as a generator of order by energizing the natural processes of generation and maintenance of life on the planet, and all this can be associated with the process of production of human conception. This is what we call photobiorefinery. As they are multiple and interagent actions, they constitute a system of production of multiple inputs and outputs based on photosynthesis and synthesis and separates that occur in the microscopic world, acting integrated and reducing the entropy of the system, as Prigogine explained.

The practical result is the reduction of impacts and the aggregation of value, including to reduce the impact of traditional energy-intensive production systems, such as those we currently use, by inertia or lack of creativity. By bringing natural-based solutions to inorganic production systems, we reduce environmental impact and can add value, which is a desire and a search for everyone they produce to meet human needs. And we must not forget that there is a transition from the mental models of visualization of the world. Even if the strength and richness of microorganisms are not yet clearly understood, the world of nanotechnology will show us, as has already happened in our projects, that microorganisms are natural factories of products and raw materials on the nanometric scale, and they can be observed and understood in the observation of metabolism and structures created.

The good news, we wonder it is now possible, is the answer lies in advancing the equipment and methods of visualization of the microscopic world. We extend our gaze to the great and small distances thanks to the advancement of technology and, in particular, computational systems. Therefore, taking advantage of this opportunity can represent a revolution in the way it produces, a new time with totally new solutions and practices.

Comparing the natural production processes with the processes of production of human conception, there are fundamental differences (Table 1.1). The first is that Nature does not produce garbage, because in the evolution of ecosystems all participants are beneficiaries and all waste is reused, in permanent cycles. Another important consideration is that nature does not use fire, combustion, as a hegemonic strategy to generate high temperatures and pressures necessary to make the transformations on the materials.

Nature synthesizes and separates materials and substances into temperature and pressure environments. Also, in an ecosystem, predictability and evolution are due to diversity and multiple interactions, in a dynamic and nonlinear process. When we look at human conception processes, they use few elements, are energy intensive and avoid the diversity of influences through predictability that comes from repetition and standardization, and finally it is important to realize that human and natural design processes also differ since human conception processes occur based on the model of control of internal losses to the process and the processes in the construction of connections.

Table 1.1 Comparative descriptions of the natural production processes and the processes of production of human conception

Natural processes	Processes based in human conception
Many elements, many exchanges, diversity, balance and evolution, interdependence, dynamic processes	Few elements and few exchanges, control, stability, isolation, linear processes
Synthesis and separation of materials, at environmental temperature and pression, cyclic processes, subtle and flexible in the flow of matter and energy	Reactors and processes of high pressures and temperatures, intensive and forced processes in the flow of matter and energy
Increased life, abundance, and evolution	Impact on environmental life and debt for the next generations, processes, stagnation
Without waste production, total use of waste, expansion of ecosystems	Unused waste, waste production, ecosystem reduction
Increased entropy in the Earth-Sun system occurs in the Sun	Increased local entropy—Heating and disorder on Earth

Consider the case of a Coal Thermoelectric Power Plant, which uses mineral coal as fuel, where the main objective is the transformation of stored chemical energy from a fossil fuel, in this case the mineral coal, into electricity, in an intensive process, carried out in the shortest possible time and isolated to the maximum of the environment, and having as a reservoir of emissions all the environment around it. Any and all other process that interferes, or connects, will generate losses from the point of view of economic gains. Furthermore, the process is polluting, sending sulfur, nitrogen, and solid particulate gases into the atmosphere. A big mess that must also be controlled. Add up the carbon dioxide and we will have the greenhouse effect associated with climate change. What is the quality of this energy? The quality of intensive processes of human conception. This energy will also be used in intensive systems. It is an intensive world and, therefore, where connections are not welcome. Obviously, it seems the worst place to imagine biological solutions, because the processes are not intensive, are highly connected and interdependent, we would have to make disparate worlds live together, trying to harmonize a world of cause and effect of physicochemical systems for the world of biological networks.

Traditionally, what is done is to solve the problems within environmental legislation, with gaseous emissions and particulate emissions at reasonable and adequate levels and controlling solid and liquid waste by final disposal. The biological world is a long way from the logic and routines of a manager in the Thermoelectric Power Plant. How to create connections? How to find "something" biological capable of taking an interest in the world of combustion?

Is there any way to reconcile? I use this example, because it is in this context of total "biological difficulty" that we can *imagine* unusual relationships, which they can, if we are competent to create and achieve, what will enable the business in the long term.

There are already intensive technical solutions for sulfur, nitrogen, particulate matter, and ash, which only depend on political decisions and the coping and internalization of associated costs to forward the solution. However, a problem persists. Combustion is, nonetheless, a photosynthesis instead using oxygen and producing carbon dioxide and water. So, the solution to carbon dioxide has to be photosynthesis again, that is, to take advantage of a photosynthesis of 200 million years ago and make it happen again, since everything is there to make it happen and start over.

The CO_2 is associated with the greenhouse effect and the phenomena of climate change, especially global warming. There are many scientific and technical research and studies aimed at finding solutions on how to capture it, arrest it, and save it. CO_2 is produced by all of us, in a multitude of processes, natural and industrial. Cars, factories, burned forest sand and all kind of thermal energy produces CO_2. It is a global problem of local origin, it is a macro challenge for human survival generated in the habit so human and common place to benefit from fire, combustion, to obtain and synthesize materials and products of all kinds.

The greenhouse effect, mentioned before, stems from the fact that the Sun sends a compound of electromagnetic waves, which is characteristic to it and presents several frequencies. It includes ultraviolet, of greater energy, to visible light. Hit by this electromagnetic wind, the Earth, as a whole, absorbs this energy, in various ways and heats and, in sequence, emit heat waves, infrared, which go into the atmosphere and partly return to space. On the way back into space, infrared energy could be absorbed by elements present in the atmosphere, especially by molecules that have the structural ability to absorb it. The main structures are water vapor, most importantly, and so-called greenhouse gases, with different absorption and dispersion rates of this thermal energy, heating the atmosphere. And among them, the gases, the CO_2.

The connection between increasing temperature with increased CO_2 concentration in the atmosphere shows an important correlation. The increase in this gas is mainly due to combustion processes. However, it is also necessary to understand that the greenhouse effect is natural and positive, because if there were not, our planet would have much lower temperatures than we know. And it wouldn't be possible to feed so many people. Other interpretation is that, as the ocean is heating, and as the ocean is a sink of CO_2, more heat is present in atmosphere as the ocean is not absorbing it, as the absorption is temperature-dependent, it is also a correlation. What is the good choice?

Well, it is reasonable to think CO_2 production excesses are from our development and production model, based on extensive use of combustion as a production and transformation process. Returning to the comparison between natural processes and human conception, Nature synthesizes an enormity of substances and develops numerous technologies, without setting fire to anything. And even when fire happens, it helps Nature to recycle materials. Nature uses subtle technologies, while humans use essential men, brute force, and fire.

So again, why don't we produce nature in the way of nature? If we look at the bias of biology, CO_2 represents the mobile carbon that is offered to photosynthetic biological structures, as are many bacteria, algae, and plants, allowing to produce food and maintain and develop the various forms of life on the planet. Thus, the amounts of carbon dioxide produced in a thermoelectric plant could represent an important opportunity if we can put it within the carbon cycle in a planned way.

The combustion gas of a Thermoelectric has about 12% carbon dioxide, which is 400 times larger than we have in the atmosphere. If we can see "smoke" as a CO_2

mine, then we will have the opportunity to feed any photosynthetic biological structures with great results. What becomes necessary is to develop the technical way of coupling cultivation processes to the inorganic process. The choice is to take advantage of carbon dioxide for the cultivation of microalgae, which allows to build multiple economic opportunities from new biochemical processes for food production, energy, and biocompounds, among others. This field of research proved promising and many technologies and techniques were developed from this basic idea in the research conducted at FURG, Brazil.

The introduction of microalgae cultivation using as part of the substrate for the feeding of microorganisms the combustion gas generated by burning coal "revealed" a network of opportunities represented by various products, with direct and indirect economic value and social importance, in addition to reducing environmental impacts. The result is presented below in a systemic map that aggregates the information resulting from several research projects carried out in FURG laboratories (Fig. 1.1). This field of research proved promising and many technologies and techniques were developed from this basic idea, all of them useful and applicable.

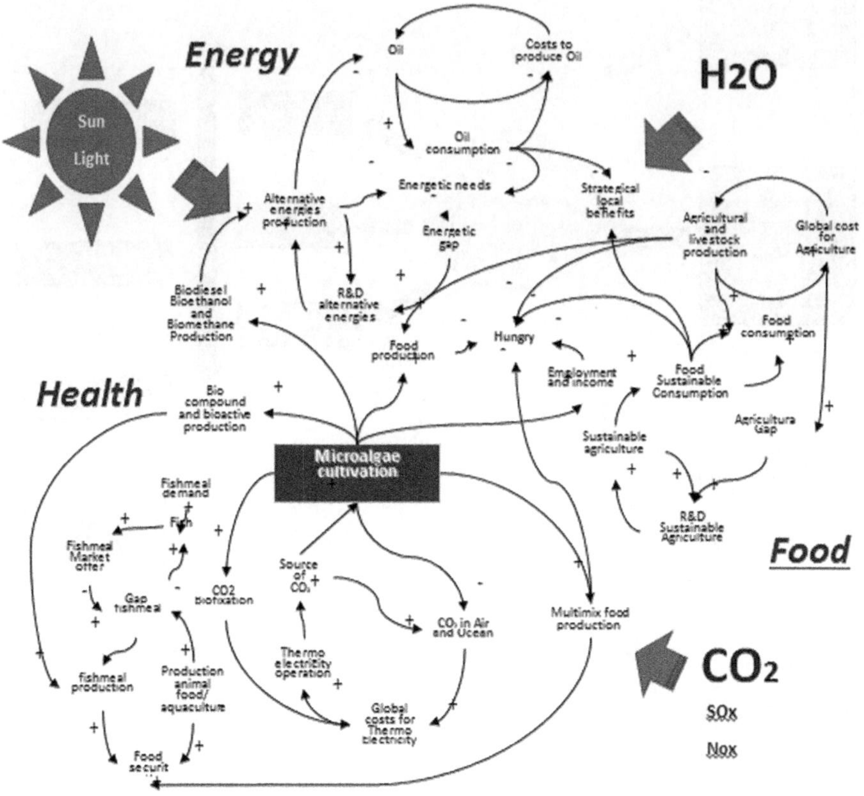

Fig. 1.1 Photobiorefinery, systemic map

It is necessary to understand that photosynthetic biological processes are a combination of light, water, CO_2, mineral salts, and heat, which give biological structures, in this case microalgae, the opportunity to reproduce and produce large amounts of biomass, which can be used in several other biological processes, in an important, and definitive, value chain.

Photobiorefineries designate the most general Integrated Biosystem, which is a cyclic photosynthetic system of multiple inputs and outputs, where the Sun is the energy base, microorganisms the transformer agents and the residues of a process the elements and substrates for new bioprocesses.

Let us remember that sources for heat generation in a thermoelectric can be completely renewable, such as using plants, such as trees and grasses or how bamboo, the possibilities are multiple (Fig. 1.2).

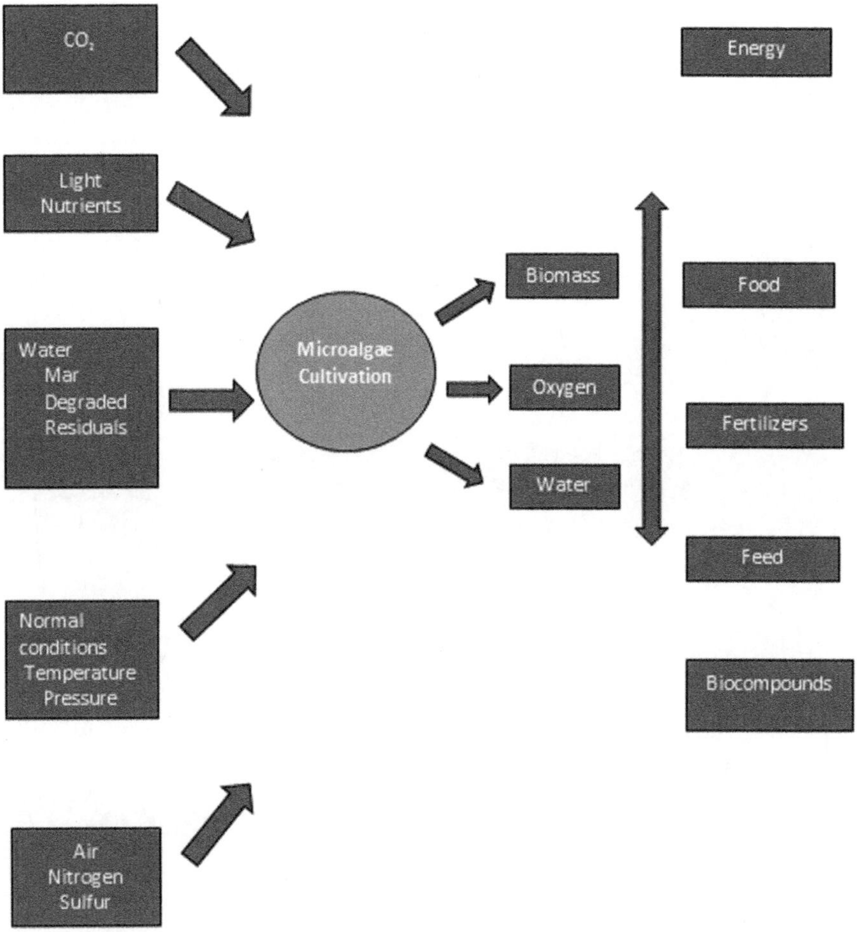

Fig. 1.2 Schematic illustration of the biorefinery process by cultivation of microalgae as the main variable of the new energy and food production systems

1.10 The Elegance of Nature

Natural processes have evolved with the planet itself since its inception. The history of the Earth, as we understand it, is a succession of stages where the conditions necessary for the emergence of life have been arising and we come to the present day. Observing nature and understanding the ways it uses to synthesize biochemically, from molecules to beings of high structural and functional complexity, is an inexhaustible source of knowledge and development of new technologies. Still, and very importantly, the processes of cooperation and all the bonds of cyclical reinforcement when there are several beings living, from microscopic to large mammals, shows that evolution occurs in a shared and cooperative way between the components of any ecosystem. This is a learning that has economic strength, especially regarding food production in agriculture, where the use of waste in new sources of raw materials for different biological processes, in chains and networks, is a great opportunity.

A major challenge for the extensive worldwide implementation of photobiorefineries relates to traditional forms of production and connections to markets. By observing traditional markets, with their large-scale production model and with high levels of standardization, the logic is to avoid any interventions and the expansion in the number of variables, new approaches that will represent additional costs, such as the acquisition of new knowledge to make the use of waste, as well as risks to predictability of financial results. The complexity of interactions, material flows, and their management; the difference in scales between processes; the creation of additional costs of non-usual accounting, especially environmental ones; and the expansion of the territory to integrate operations, cause traditional administrators to fight back and resist accepting the changes, do not even admit, in their mental models, the real possibility of gains from the use of photobiorefineries and Integrated Biosystems.

The usual fear and rules of the economy, even with a scenario of planetary hunger and inequality that are inherent to the model, will insist on rejecting a new environment and with new rules of competition. In these 23 years, I have made a commitment to myself to learn what I could learn about biomass uses, then about integrated processes, following microalgae and systemic thinking, and reaching the elegance and usefulness of the concept of photobiorefineries, I believe that I could understand more about what surrounds us and how we are at a time of rupture of knowledge, great harmonizations, and the vision of a future that will really be a disruptive evolution and full of good surprises. Microalgae were guides from the moment I realized that we were groping in a fragment of the mystery that is *Life* and its purpose.

Do we have time to learn? I believe that yes, we still have choices, the essence of the human being is to make choices and decide, Life will guide us, that's what I think and believe, in few words: *Microalgae as the central variable to imagine a systemic society with the purpose of abundance.*

To explain more, it will be necessarily more words, and time, surely. But the main idea is that Nature is the elegant demonstration about what *Life* could create, and now *Life* has us as your partner to create something new, beyond Nature.

Chapter 2
Considerations for Cultivating and Processing *Botryococcus braunii*

Daniel R. Browne

Abstract *Botryococcus braunii* is a colony-forming green microalga that is best known for its ability to produce large amounts of extracellular hydrocarbon oils. This review describes recent efforts to build a better understanding of the biological performance of *B. braunii* under various cultivation conditions, as well as methods to process the biomass and recover products with commercial potential. Challenges are identified, and strategies are discussed to overcome barriers to commercialization.

Keywords *Botryococcus* · Algae · Wastewater · Cultivation · Processing · Biotechnology

2.1 Introduction

2.1.1 Summary of Biological Properties

Botryococcus braunii (Trebouxiophyceae, Chlorophyta) is a green microalga that forms colonies composed of clonal cells embedded in an extracellular matrix (ECM) and produces large amounts of liquid hydrocarbon oils (Weiss et al. 2012). The species was first reported in 1849 and became a subject of interest when fossils of the alga were found in a wide range of coal formations (Blackburn 1936). Throughout the twentieth century, numerous hydrocarbons were extracted and identified, and researchers attempted to elucidate the underlying biosynthetic mechanisms (Metzger and Largeau 2005). There are four known clades of *B. braunii* called "races," which are distinguishable by hydrocarbon type and by rRNA phylogenetics (Kawachi et al. 2012). While the races are morphologically very similar, race B produces the

D. R. Browne (✉)
Department of Biochemistry & Biophysics, Texas A&M University, College Station, TX, USA

Artificial Intelligence & Computational Biology, LanzaTech, Inc., Skokie, IL, USA
e-mail: dan.browne@lanzatech.com

J. A. V. Costa et al. (eds.), *Microalgal Bioengineering*,
https://doi.org/10.1007/978-3-031-61253-4_2

isoprenoid botryococcene, race L produces the isoprenoid lycopadiene, race A produces fatty acid-derived alkenes, and race S produces fatty acid-derived alkanes. In all types of *B. braunii*, the hydrocarbons are sequestered outside the cells in the colonial ECM (Hirose et al. 2013). This process involves intracellular lipid bodies in a complex life cycle of formation, mobilization, and secretion (Suzuki et al. 2013).

Interest around *B. braunii* received a significant boost in 2010 when it was included in the National Alliance for Advanced Biofuels and Bioproducts (NAABB) as a core research strain under the algal biology task framework (Unkefer et al. 2017). Thanks to numerous advances over the last decade from researchers around the world, there is today much better knowledge of basic *B. braunii* biology (Jin et al. 2016b). Knowledge of hydrocarbons from *B. braunii* has also dovetailed with the field of geochemistry, which includes the study of biomarker distribution in the geological record, giving insight into the timeline for evolution of *B. braunii* hydrocarbon biosynthesis (Volkman 2014). Recent studies have continued to yield important new biological insights and create valuable resources for further study. Advanced electron microscopy techniques have enabled a better understanding of the fibrillar polysaccharide sheath that covers the surface of *B. braunii* colonies (Uno et al. 2015). Digital image analysis techniques have enabled the three-dimensional reconstruction of intracellular components (Suzuki et al. 2017). Highly sensitive mass spectrometry techniques have enabled lipidomic (Rezanka et al. 2018), metabolomic (Blifernez-Klassen et al. 2018), and glycoproteomic (Schulze et al. 2017) analyses. Next-generation sequencing has enabled the determination of the nuclear genome (Browne et al. 2017) and the chloroplast and mitochondrial genomes (Blifernez-Klassen et al. 2016).

2.1.2 Strain Selection and Development

Strain development is one of the major challenges facing the algae industry (Wijffels et al. 2013). The optimization of alga biology for a particular set of cultivation conditions and target products is a critical milestone for industrial algae production. Strains of *B. braunii* have been isolated around the world and gathered in various culture collections. Among the naturally occurring strains, there is considerable physiological variation and performance under artificial cultivation. The physiological state of an algae culture will influence the target product yield (Blifernez-Klassen et al. 2018). To obtain optimal product yields, it is critical to understand how cultivation conditions affect the physiological state of the algae.

To assess the utility of *B. braunii* for industrial cultivation, 16 strains were compared for productivity, hydrocarbon content, and carbohydrate content, using uniform cultivation conditions (Gouveia et al. 2017). Across the strains, the highest reported productivity was 0.18 g/L/day, with a maximum hydrocarbon content of 42% of carbohydrate content of 76%. However, there was a large amount of variability observed across the strains. In another study, ten strains of *B. braunii* were compared using a flask-based method (Moutel et al. 2016). The maximum reported productivity was 0.41 g/L/day and the maximum reported hydrocarbon content was

34%. Consistent with other studies, there was a large amount of variability observed across the selected strains. These screening studies are important, but they are limited in throughput when using flask-based methods for cultivation. To improve screening throughput, a microfluidic device was developed to capture single colonies of *B. braunii* and assess their growth rates and hydrocarbon contents in a massively parallel screen (Kim et al. 2014). This approach enabled the simultaneous testing of 64 different lighting conditions at 250X higher throughput compared to conventional flask-based methods.

While screening of natural strains is one strategy for industrial strain selection, it is also possible to manipulate strains. Desiccation treatment was found to be a simple but effective way of inducing temperature tolerance in *B. braunii* (Demura et al. 2013). Colonies that survived the desiccation treatment gained higher tolerance to extreme temperatures, both high (40 °C) and low (−20 °C). Genetic engineering is another possibility that is commonly practiced with other species of algae, such as *Chlamydomonas reinhardtii*. Transformation of *B. braunii* was reportedly achieved using a combination of cellulase pre-treatment and electroporation (Berrios et al. 2015). The treated cells were transformed with a plasmid containing the selective marker *AphVIII*, a gene that confers resistance to the antibiotic paromomycin. Transformants were selected on antibiotic-containing agar plates and presence of the marker was confirmed by PCR analysis. However, this result has yet to be publicly reproduced by any other scientists.

With any genetic engineering effort, it is imperative to have axenic cultures (pure cultures of a single species) of the desired strain (Vu et al. 2018). The ability to select, engineer, and maintain strains with specific traits is necessary to provide biology suitable for targeted applications. Moreover, it will enable the development of strains with superior productivity, improving the economic viability of cultivation. However, *B. braunii* cultures are often not axenic, with many associated bacterial species. Metagenomics has been utilized to assess the composition of the bacterial communities associated with *B. braunii* (Sambles et al. 2017). A better understanding of the bacterial influence on *B. braunii* productivity will enable the design of optimal bacterial consortia. Other organisms often contaminate algal cultures as well, such as rotifers, a diverse class of microscopic animals that prey upon a wide range of algal species (Abou-Shanab et al. 2016). These animals and other predators pose a challenge for the algae industry, particularly in relation to open pond operations. Strain development efforts must take into consideration the complex ecosystems that govern algal performance.

2.2 Cultivating

2.2.1 Outdoor Ponds and Photobioreactors

There has long been a debate about the superiority of open ponds or photobioreactors (PBRs) for the outdoor cultivation of microalgae. Cultivation technologies were a major focus of the NAABB program (Lammers et al. 2017). Several outdoor experiments have been conducted with *B. braunii* in ponds and PBRs. In Mysore, India, 2 strains of *B. braunii* were cultivated with 40 L outdoor raceway and circular ponds (Ranga Rao et al. 2012). With an initial density of 0.25 g/L, they reported a maximum density of 2 g/L and 30% hydrocarbon content over 18 days. In Chennai, India, 3 strains of *B. braunii* were cultivated in 2000 L outdoor raceway ponds (Ashokkumar and Rengasamy 2012). Over 15 days, they reported a maximum density of 1.8 g/L, from an initial density of 0.1 g/L. This experiment yielded a productivity of 0.1 g/L/day and a specific growth rate of 0.26 per day.

In Antofagasta, Chile, *B. braunii* was cultivated in 1000 L outdoor panel PBRs (Bazaes et al. 2012). The strain grew to a maximum density of 0.3 g/L, from an initial density of 0.1 g/L, giving a productivity of 0.02 g/L/day and a specific growth rate of 0.09 per day. In Tennodai, Japan, a strain of *B. braunii* was cultivated in 2 outdoor PBRs, a 300 L dome and a 500 L tube (Shimamura et al. 2012). Over 60 days, the dome PBR reached a maximum density of 0.6 g/L, while the tube PBR reached 3.8 g/L, with a maximum specific growth rate of 0.09 per day. These data illustrate the wide array of possible outcomes, given variable inputs and cultivation vessels.

The form of the culture container governs turbulent mixing, aeration rate, and light access, affecting algae productivity. Computational fluid dynamics were used to model flow inside a tube PBR and predict an optimal configuration (Xu et al. 2012a). This design approach was validated by growth experiments with a *B. braunii* strain in 2 L and 40 L tube PBRs, achieving a maximum density of 1.8 g/L. Feeding strategy and initial density also have a substantial impact on cultivation yield. Using a strategy of 7.5% (v/v) daily harvest and replenishment with stoichiometrically balanced media, cultures of *B. braunii* were able to reach and sustain a density of 20 g/L and productivity of 1.5 g/L/day for 3 months (Khatri et al. 2013). This groundbreaking work shattered previous reports of productivity in *B. braunii* cultures and demonstrated the importance of culture management practices for maximizing the potential of this species.

2.2.2 *Attached Cultivation with Biofilms*

In contrast to cultivation by suspension of algae in a liquid media, some algae (including *B. braunii*) have the ability to form biofilms (i.e., a thin, dense layer attached to a surface). One of the earliest biofilm cultivation studies with *B. braunii* found that the alga can achieve a productivity of 0.7 g/m^2/day, with 45% less water required in comparison to open ponds (Ozkan et al. 2012). Furthermore, harvesting was achieved by simple scraping, reducing the energy required for dewatering by 99.7% in comparison to open ponds. The surface for biofilm attachment consisted of a concrete slab. Another study compared the growth of a *B. braunii* strain using a plate PBR and a biofilm PBR under identical conditions (Pengfei et al. 2017). The results of this experiment showed very similar performance to the two systems in terms of biomass productivity, each near 4.5 g/m^2/day. The surface for biofilm attachment consisted of a glass plate covered in filter paper.

On a smaller scale, four strains of *B. braunii* were grown in biofilms attached to cellulose nitrate filter paper, with a reported productivity of 38.0 g/m^2/day (Wijihastuti et al. 2016). The highest biofilm productivity achieved 49.1 g/m^2/day using a stacked biofilm PBR configuration (Cheng et al. 2013). The surface for biofilm attachment consisted of a glass plate covered in filter paper. These studies have shown that biofilm cultivation is a strategy for *B. braunii* cultivation to achieve high productivity levels. Moreover, this method of cultivation is attractive under conditions where water availability is limited. However, significant challenges remain, as none of these studies were conducted in outdoor environments. Furthermore, these studies were not conducted for a long time, and it is unknown how this cultivation impacts *B. braunii* morphology and physiology. While this approach appears somewhat promising, it is unclear whether biofilms are amenable to large-scale cultivation of *B. braunii* under naturally occurring conditions. Besides, more work is needed to understand the biological impact.

2.2.3 Nutrient Requirements and Delivery

The primary modes of algae cultivation are: (1) autotrophic, where energy is derived from light and the carbon source is inorganic; (2) heterotrophic, where energy and carbon are both derived from an organic source; and (3) mixotrophic, both organic and inorganic carbon sources are available, as well as light energy. Autotrophic, heterotrophic, and mixotrophic conditions were compared using *B. braunii* and three other species of green algae (Gim et al. 2014). Mixotrophic conditions with 1% glucose were found to be the most effective for *B. braunii*, reaching a maximum density of 2.5 g/L and 35% hydrocarbon content over 20 days. Mixotrophic growth of *B. braunii* in 0.25 L flasks was tested with several different sources of organic carbon (Zhang et al. 2011). Utilizing glucose, an experiment was conducted in a 10 L PBR and the cultures achieved a maximum density of 4.5 g/L and 29.7% hydrocarbon content over 19 days. An axenic strain of *B. braunii* was utilized to test heterotrophic and mixotrophic growth with different organic carbon sources (Tanoi et al. 2011). The addition of glucose or mannose yielded specific growth rates approaching 0.18 per day and a maximum density of 0.37 g/L after 19 days. The effects of CO_2 concentration on *B. braunii* were tested using a 3 L PBR system (Ge et al. 2011). The range of CO_2 concentration varied between 2–20% and *B. braunii* grew under all conditions, with a positive correlation between CO_2 concentration, hydrocarbon content, and colony size. The maximum density was 2.3 g/L with 20% CO_2 after 25 days.

Of course, carbon is just one of many nutrients required for algal growth, and media formulation impact on performance. Optimization of nutrient conditions resulted in a ten-fold improvement in biomass density and hydrocarbon content for 2 strains of *B. braunii* compared to standard media formulations (Al-Hothaly et al. 2015b). Two culture media formulations were utilized to compare *B. braunii* growth under different media conditions (Díaz Bayona and Garcés 2014). Significant differences in exopolysaccharide production were observed between the two media formulations, possibly due to nitrogen availability and salinity. One of the other major elements required for cultivation is nitrogen. Adjusting the carbon-nitrogen ratio will affect the productivity of *B. braunii*, pointing to the need for tuning this parameter to achieve maximal performance (Barajas-Solano et al. 2016). *B. braunii* is capable of scouring both nitrate (commonly used in cultivation media) and ammonium from its environment (Nakamura et al. 2017). However, if the culture is not buffered, acidic pH induced by addition of ammonium can cause toxicity. Nitrate and phosphate were fed to cultures of *B. braunii* in a 1.8 L PBR to determine the optimal timing for nutrient addition and nitrate-phosphate ratio (Xu et al. 2012b). Using 2 feeding in 2-day intervals yielded a maximum density of 2.9 g/L after 19 days. The rates of nutrient uptake are related to pH of the culture media, and one study found pH of 6.5 to be optimal for *B. braunii* cultivation (Jin et al. 2016a).

2.2.4 Recovering Nutrients from Wastewater

For industrial-scale cultivation of *B. braunii*, purchasing all required nutrients is not likely economical. Thus, recent research has focused on recovering nutrients from waste materials for *B. braunii* cultivation. Livestock wastewater was used to grow *B. braunii* in a 0.8 L PBR (Lee et al. 2018). The test assessed growth rate and nutrient recovery, achieving a productivity of 0.2 g/L/day while recovering 96% nitrogen and 85% phosphorus from the wastewater. The effluent was recovered from the PBR by an integrated membrane filter. In another similar study, livestock wastewater was utilized to grow a strain of *B. braunii* in a 0.8 L PBR, achieving a maximum density of 2.6 g/L, specific growth rate of 0.2 per day, and productivity of 0.3 g/L/d over 40 days (Yu and Kim 2017). The cultures recovered 83.2% phosphorus and 94.1% nitrogen from the wastewater. One strain of *B. braunii* was grown in 0.5 L flasks using aerated swine lagoon wastewater (Liu et al. 2013). The cultures achieved a maximum density of nearly 1.0 g/L over 26 days and recovered 40.8% nitrogen and 93.3% phosphorus. These studies demonstrate the effectiveness of *B. braunii* in wastewater treatment applications for the animal husbandry industry.

Similarly, municipal sewage systems manage the wastewater that needs treatment before discharge, and *B. braunii* could serve a role in this capacity. *B. braunii* were grown in 0.6 L flasks using municipal wastewater samples collected from a treatment plant in Brazil (Rinna et al. 2017). The maximum productivity was 0.07 g/L/day, and the cultures recovered 100% of phosphorus in all trials and up to 65% of nitrogen. Aside from municipal wastewater, carbon emissions from power plants are a serious environmental hazard and a potential nutrient source for algae. A combination of municipal sewage and flue gas from a coal-burning boiler was utilized to cultivate *B. braunii*, reaching a maximum density of 1.6 g/L over eight days (Kandimalla et al. 2016). Nutrients were effectively recovered from the sewage, including nitrates (94%), sulfates (75%), and phosphates (80%). Furthermore, *B. braunii* was able to reduce contaminants in the flue gas, including CO_2 (80%),

NO_x (50%), and SO_x (45%). *B. braunii* could be a valuable tool for mitigating municipal wastewater and power plant emissions.

However, there are many other types of wastewaters suitable for algae cultivation. Wastewater from soybean curd manufacturing was utilized as a nutrient source for *B. braunii* cultivation (Yonezawa et al. 2012). It was found that addition of 2% wastewater to the medium boosted culture density by more than double compared to the control, reaching a maximum density of 2.9 g/L over 40 days and a specific growth rate of nearly 0.3 per day. This wastewater is an excellent source of organic carbon, phosphate, sulfate, calcium potassium, manganese, and sodium, but has low nitrogen content. To compare mixotrophic and photoautotrophic growth with waste-derived nutrients, a strain of *B. braunii* was cultivated using waste molasses from a sugarcane processing facility or secondary wastewater from a seafood processing facility (Yeesang and Cheirsilp 2014). The molasses provided an organic carbon source for mixotrophic cultivation and yielded a maximum biomass density of 3.1 g/L over 20 days. In comparison, the nitrate-rich wastewater supplemented with CO_2 found a maximum biomass density of 2.3 g/L while recovering 90% of the nitrate.

2.2.5 *Effects of Metals and Micronutrients*

While a substantial amount of attention has been devoted to macronutrients, such as carbon, nitrogen, and phosphorus, relatively little attention has been given to micronutrients and metals. The only analysis of metals occurring in a natural *B. braunii* bloom found detectable levels of magnesium, potassium, calcium, zinc, lead, iron, cadmium, copper, nickel, and cobalt (Wake and Hillen 1980). It is poorly understood how the concentrations of these nutrients influence *B. braunii* culture performance. One attempt to illuminate this issue employed computational modeling to predict the optimal concentrations of iron, manganese, molybdenum, and nickel for *B. braunii* cultivation (Song et al. 2012). Optimization of micronutrients affects productivity, boosting it to 1.3 g/L/day over 21 days in a 3.2 L plate PBR. Interestingly, this study tested the effect of nickel, which is not commonly included in *B. braunii* media formulations, even though it has notably high levels in a natural bloom. In contrast, the biological role of iron is well understood, and iron concentration influence in *B. braunii* morphology (Tanoi et al. 2013). Iron-rich media resulted in larger colonies with elliptical cells, demonstrating higher productivity and hydrocarbon content. Iron deficiency leads to severe morphological defects in *B. braunii*.

Certain metals are toxic, especially at high concentrations, posing an environmental health risk when present in industrial wastewaters. The effect of arsenic toxicity on *B. braunii* was investigated, and it was found that the alga can tolerate concentrations up to 2 g/L, although it impeded growth (Podder and Majumder 2017). Nonetheless, *B. braunii* showed potential as a tool for bioremediation of arsenic from wastewater. Bioremediation of zinc, copper, nickel, and iron was investigated using *B. braunii* (Areco et al. 2018). These metals were successfully removed from the media by adsorption and metabolic precipitation, demonstrating that *B. braunii* is capable of bioremediation. Considering the difficulty of disposing mine tailings and metal-contaminated industrial waste, these findings are very significant. The potential for *B. braunii* to serve as a bioremediation platform could add a new dimension to the value proposition of industrial algae cultivation. If algal bioremediation of metals could be monetized, this could become a new revenue stream and contribute to the commercialization of algae cultivation technologies.

2.2.6 Effects of Media Salinity Conditions

Although *B. braunii* is known to occur naturally in freshwater and brackish environments, the effects of salinity on morphology and physiology are still poorly understood, with only a handful of studies addressing the topic. Increasing salinity reduced the biosynthesis of exopolysaccharides by *B. braunii*, resulting in increased recoverability of hydrocarbons (Furuhashi et al. 2016a). The mean colony size also increased in response to higher levels of salinity. While seawater has a salinity of approximately 3.6%, *B. braunii* grew normally at a salinity of 0.3% and could tolerate a salinity of 0.7%, albeit with slower growth. Another study found that the addition of 25% seawater to cultures of *B. braunii* had a negligible impact on productivity, but enabled highly efficient (>90%) hydrocarbon recovery from wet biomass by hexane extraction (Furuhashi et al. 2013). In contrast, increasing salinity of the culture medium resulted in a corresponding decrease in the specific growth rate of *B. braunii* (Yoshimura et al. 2013). However, at salinity of less than 1%, the change in the specific growth rate was not significantly different from the control, suggesting that *B. braunii* can tolerate low salinity. These studies are valuable, as the ability to utilize some fraction of seawater for *B. braunii* cultivation could contribute to lower production costs in coastal areas.

2.2.7 Effects of Light and Photoperiod

Under autotrophic cultivation conditions, light is a major determining factor in algal productivity. The pigments present in *B. braunii* photosynthetic machinery include chlorophyll, lutein, violaxanthin, neoxanthin, and loroxanthin, with strong absorption between 400–500 nm and 640–700 nm wavelengths (van den Berg et al. 2017). The light quality impacted the growth, pigment composition, colony morphology, and photosynthetic activity of *B. braunii* (Baba et al. 2012). Blue and red lights were the most effective wavelengths for cultivation, with blue light promoting the highest rate of photosynthetic carbon fixation. Cells grown under red light underwent a substantial morphological shift from elliptical to spherical. Consistent with this, another study reported that *B. braunii* grew most effectively under light conditions, including blue (465 nm), red (660 nm), and far red (735 nm) wavelengths

(Kula et al. 2017). Interestingly, the omission of violet (405 nm) wavelengths from light conditions while cultivating *B. braunii* led to a slight increase in hydrocarbon content, reaching a maximum of 39% (Atobe et al. 2014).

High light intensities can be damaging to cells and *B. braunii* has several strategies for mitigating damage, including the reduction of chlorophyll content, non-photochemical quenching, colony size increase, and accumulation of extracellular echinenone (van den Berg et al. 2019). An investigation of light intensity and path length demonstrated that the best performance is achieved with medium light intensity and short to medium path length. Under high light intensity and short path length conditions, hydrocarbon content was severely reduced (Wang et al. 2019). Light scattering, diffusion, and self-shading were studied in a panel PBR (Singh Khichi et al. 2018). High light conditions promoted biomass productivity, with a maximum of 0.36 g/L/day, while medium light conditions promoted hydrocarbon content, with a maximum of 25%. Light wavelength and intensity are not the only factors to consider, as the photoperiod (light/dark cycle) impacts algal biology, particularly concerning the entrainment of circadian clocks. However, one study found that continuous illumination of *B. braunii* resulted in substantially higher productivity than a photoperiod of 12h:12h light/dark over ten days (Krzeminska et al. 2013). It is unclear how continuous illumination would affect *B. braunii* biology over a longer time. If the circadian clock is not entrained by the photoperiod, key regulatory functions may not operate normally.

2.2.8 Enhanced Growth with Stimulation

Aside from optimizing the cultivation vessel, light conditions, and nutrient conditions, it is also possible to promote growth with other biotic and abiotic factors. An extract prepared from ocean sediment stimulated *B. braunii* biomass productivity and hydrocarbon content (Gim and Kim 2018). Possible contributing factors include nitrate compounds, metal ions, and humic acids. In particular, humic acids could play an important role in increasing the bioavailability of metal ions through chelating actions. The vitamins thiamine, biotin, and cobalamin were added to the axenic cultures of a *B. braunii* strain (Ruangsomboon et al. 2018). All of the vitamins boosted biomass productivity, while thiamine also contributed to increasing hydrocarbon content. In contrast, another study found that supplementation of *B. braunii* cultures with cobalamin found little to no influence on biomass productivity and hydrocarbon content (Tanabe et al. 2013). The effects of phytohormone supplementation were tested by adding 2-azahypoxanthine to *B. braunii* cultures (Nakamura et al. 2016). The biomass productivity and carotenoid content were increased; however, the hydrocarbon content was decreased. The phytohormone 6-benzylaminopurine yielded a strong boost to the biomass productivity, hydrocarbon content, and carotenoid content of *B. braunii* cultures when applied in small doses (Liu et al. 2012). However, at higher concentrations, a lethal effect was observed. Treatment of *B. braunii* cultures with 40 kHz ultrasound stimulation for

5 min at a 4-day interval increased biomass productivity and hydrocarbon content (Xu et al. 2014). This statement is possibly due to increases in indole-3-acetic acid and membrane permeability resulting from the ultrasound treatment. The ultrasound treatment tested at a larger scale with outdoor 200 L bag PBR cultures of *B. braunii* (Wang et al. 2014b) resulted in a slightly higher biomass density but no significant effect on hydrocarbon content. Altogether, these studies provide several strategies for stimulating the biomass productivity and hydrocarbon content of *B. braunii*. For commercial cultivation, it will be important to identify more methods to improve performance.

2.3 Processing

2.3.1 Methods for Harvesting Biomass

Numerous technologies for harvesting and dewatering microalgae have been developed and evaluated within the last decade (Fasaei et al. 2018). The NAABB program included an extensive task framework involving technologies to harvest algae (Marrone et al. 2017). However, only a handful of studies have focused on methods to harvest *B. braunii*. *B. braunii* was harvested using a magnetic flocculant synthesized from iron oxide and cationic polyacrylamide (Wang et al. 2014a). This material was able to achieve a 95% harvesting efficiency with a relatively small dose and ten minutes of treatment. The primary mechanism of interaction with the algae biomass was based on electrostatic charges and surface adsorption. Another magnetic nanocomposite material was synthesized from polypyrrole and iron oxide, then used to harvest *B. braunii*, achieving nearly 99% efficiency (Hena et al. 2015). Furthermore, a method was developed to separate the nanocomposite from the harvested algae and recover it for reuse. More recently, a new method for the low-temperature synthesis of iron oxide nanoparticles and conjugation with polyethyleneimine for harvesting algae (Gerulová et al. 2018). However, this material was only able to achieve 68% harvesting efficiency of *B. braunii*, well short of the other materials discussed.

Non-magnetic harvesting methods have also been developed and tested for *B. braunii*. Several cationic starch materials were synthesized and used to flocculate cultures of *B. braunii* (Peng et al. 2017). These materials achieved up to 94% efficiency with relatively low doses and an incubation time of 20 min. They also have the benefit of being low cost; however, they are not recoverable and reusable. Glycine was considered as another low-cost flocculant for *B. braunii* (Shen et al. 2016). However, this strategy did not yield competitive results, with only 70% efficiency, high dose requirements, and a five-day treatment. Aside from chemically induced flocculation, biologically induced flocculation is another possibility. This was demonstrated by the utilization of *Aspergillus* to flocculate cultures of *B. braunii* (Al-Hothaly 2018). This strategy is relatively low cost and can achieve greater than 90% efficiency with 12 h of treatment. While the *Aspergillus* did not negatively impact the energy content of *B. braunii* in the harvesting process, this strategy may or may not be a suitable approach for harvesting, depending on the downstream application of the algal biomass and concomitant product requirements.

2.3.2 Pre-treatment for Product Recovery

Before extracting products from *B. braunii*, some pre-treatments may be necessary to improve yields. Several studies have investigated means to disrupt the colonies and cells, facilitating product recovery. Two mechanical methods were employed to disrupt *B. braunii* colonies before hydrocarbon extraction (Tsutsumi et al. 2018b). Before treatment, the extraction method recovered only 2.7% of hydrocarbons, whereas up to 82.8% were recovered after treatment. While these methods were effective, the energy consumption was much higher than the energy recovered. Colonies can also be disrupted by chemically induced release of individual cells from the colonial ECM. Simple treatment of the cultures with a 20% glycerol solution was sufficient to release up to 50% of cells in 15 min (Hou et al. 2013). This treatment enables the subsequent separation of cell-free colonies and individual cells for differential downstream processing. With individual cells of *B. braunii* isolated from the colony, it may be necessary to disrupt the cell walls for product extraction. There are numerous methods available for cell disruption, both mechanical and physicochemical, and they have been thoroughly reviewed (Lee et al. 2012). The challenge with these methods is that they are often energy intensive and unlikely to be economical, excepting high value products. Enzymatic hydrolysis of the cell wall is an attractive alternative to mechanical and physicochemical methods, as the energy consumption is much lower. The *B. braunii* cell wall was successfully degraded using an enzymatic extract from the white-rot fungi *Anthracophyllum discolor* (Ciudad et al. 2013). This approach will enable access to proteins and other intracellular products of interest.

Thermal pre-treatment of three strains of *B. braunii* was tested to determine the effect on hydrocarbon extraction with hexane (Magota et al. 2012). Without treatment, hexane extraction only recovered 1–7% of hydrocarbons. By treating the cultures at 80–90 °C for 2.5 min, hexane extraction recovered 75–95% of hydrocarbons. Further investigation of the effects of thermal pre-treatment found that it caused the release of water-soluble polysaccharides into the culture media (Atobe et al. 2015). These polysaccharides are amphiphilic and cause the formation of emulsions with water and organic solvent, which can be problematic during hydrocarbon extraction when there is a high concentration of the polysaccharides. The polysaccharides released in thermal pre-treatment originate from the fibrillar sheath that covers the surface of the *B. braunii* colonies (Furuhashi et al. 2016b). A simple solution to reduce the formation of this polysaccharide sheath is cultivation in a brackish medium. This strategy reduced the length and density of the fibrils, resulting in greater recovery of hydrocarbons during solvent extraction. Further investigation of mechanical pre-treatment found that the improved recovery of hydrocarbons is largely due to disruption of the polysaccharide fibrillar sheath at the colony surface (Tsutsumi et al. 2018a). However, this method still has the disadvantage of consuming large amounts of energy, and it is more efficient to use the brackish cultivation strategy to mitigate fibril formation.

2.3.3 Methods to Extract Hydrocarbons

Solvent extraction is a highly standard method for recovering oils from biomass. This method has been studied and applied extensively for recovering *B. braunii* hydrocarbons. Hydrocarbons were repeatedly extracted *in situ* from *B. braunii* cultures using the solvent heptane, which was not harmful to the algal cells (Moheimani et al. 2013). Continuous cultivation and repeated solvent extractions do not need harvesting and dewatering. However, this method only recovered approximately 35% of the hydrocarbons. Dimethyl ether was an effective solvent for extracting hydrocarbons from wet *B. braunii* biomass, negating the need to dewater the algae after harvesting (Kanda et al. 2013). This method enabled the recovery of 49% of the hydrocarbons but still required harvesting. Solvent extraction of wet *B. braunii* biomass with hexane or a chloroform-methanol mixture was tested with autoclave, microwave, ultrasound, and vortex assistance (Talukdar et al. 2014). These methods yielded between 25% and 38% recovery of hydrocarbons. To enhance *in situ* hydrocarbon recovery from *B. braunii* cultures, a hollow fiber membrane was utilized to improve solvent dispersion, resulting in 50% recovery of hydrocarbons without pre-treatment (Zhang et al. 2013). Given the more recent revelations about colony surface structure and fibrillar polysaccharides that impede solvent extraction, a new model was developed to predict hydrocarbon recovery from wet *B. braunii* biomass by solvent extraction (Tsutsumi et al. 2018c). Using this model to predict optimal conditions for solvent extraction, a yield of 88% was achieved using pre-treatment followed by hexane extraction.

Recovery of hydrocarbons without organic solvent extraction is also possible and has been the focus of some recent research. Electro-extraction of hydrocarbons from *in situ* cultures of *B. braunii* was achieved using nanosecond pulsed electric fields (Guionet et al. 2017). This method not only enabled highly efficient recovery of hydrocarbons, approaching 100% yield, but also extracted the polysaccharide fibrils. With the low energy usage of this non-destructive process and the recovery of two products simultaneously, this method appears very promising. Supercritical CO_2 was utilized to extract hydrocarbons from dry *B. braunii*. The performance of this method was compared against organic solvent extraction (Yildiz-Ozturk and Yesil-Celiktas 2017). The supercritical CO_2 recovered nearly four-fold more hydrocarbons than the solvents, but the need to dry the biomass first makes this method impractical. A subsequent study tested supercritical CO_2 extraction of live *B. braunii* cultures and found that the process was lethal (Yildiz-Ozturk et al. 2018). However, hydrostatic pressure treatment was not lethal and could provide another means of hydrocarbon recovery.

2.3.4 Converting Hydrocarbons to Biofuels

The ability to catalytically crack *B. braunii* hydrocarbons into fuel-range molecules using standard petroleum processing equipment has been known for a long time (Hillen et al. 1982). The highly similar processability and yields of *B. braunii* hydrocarbons and fossil petroleum have been a key driver of interest in the alga. Recently, more work has been devoted to understanding and improving the catalytic conversion of *B. braunii* oils. The density, kinematic viscosity, surface tension, and distillation properties of diesel fuel and *B. braunii* hydrocarbons were compared (Nagano et al. 2012). The density and surface tension of the two oils was very similar, but the *B. braunii* oil had a much higher viscosity and distillation temperature. Therefore, there is a need to catalytically crack *B. braunii* hydrocarbons to obtain oils suitable for injection into combustion engines. Thus, suitable catalysts are needed to achieve the desired yields of target molecules from the *B. braunii* hydrocarbons.

Squalene, which is structurally very similar to botryococcene, was cracked over a zeolite catalyst (Tracy et al. 2011). This method yielded 65% liquid, 33% vapor, and 2% coke, with the liquid fraction having an octane number of 98. A diverse array of useful molecules was obtained from the cracking, similar to what is obtained from a barrel of crude oil. While squalene and botryococcene are structurally similar, squalene may not be a suitable model, as the kinematic viscosity of botrycoccenes is much higher than squalene (Yamamoto et al. 2013). Another zeolite-catalyzed cracking method was developed, employing relatively mild conditions to convert botryococcene into diesel fuel. The process had a yield of 85% and successfully produced diesel-range molecules from botryococcene, although the cetane number was slightly lower than fossil diesel. Another class of catalyst was explored for upgrading squalene to liquid fuels and other products based on cobalt and molybdenum supported on saponite phyllosilicates (Garciano et al. 2014). This type of catalyst could be useful for a variety of reactions, including cyclisation, hydrogenation, hydrocracking, and esterification. Several nickel-molybdenum-supported catalysts were developed and utilized to test hydrocracking of *B. braunii* oils and squalene (Liu et al. 2018). Depending on the catalyst support structure, yields could be shifted between squalane, fuel gas (C_1–C_4), gasoline (C_5–C_9), aviation fuel (C_{10}–C_{15}), and diesel (C_{16}–C_{20}) range molecules.

2.3.5 Thermochemical Conversion of Biomass

Thermochemical conversion methods use high temperatures and pressures to transform biomass into usable organic products. Pyrolysis of a naturally dried *B. braunii* residue (coorongite) resulted in the production of a complex bio-oil composed of alkanes, alkenes, alkylbenzenes, ketones, alkanoic acids, phenols, and methylnaphthalenes (Garciano et al. 2012). It was suggested that pyrolysis of dry *B. braunii* could be an effective method to decarboxylate *B. braunii* bio-oil before further catalytic upgrading by hydrocracking and hydrogenation. Pyrolysis of *B. braunii* biomass harvested by flocculation with *Aspergillus* yielded 50% bio-oil and 29% bio-char (Al-Hothaly et al. 2015a). The presence of the fungal flocculant in the harvested *B. braunii* biomass had no impact on the pyrolysis yields. The reaction kinetics of *B. braunii* pyrolysis were modeled to resolve the pyrolytic mechanisms and further optimize reaction conditions (Ali et al. 2018). Modeling efforts are important because biomass composition varies greatly depending on strain selection and cultivation conditions, and there is a need for the ability to predict pyrolysis product formation. Rather than pyrolyzing the whole *B. braunii* biomass, it is also possible to first extract oil from the algae and then pyrolyze the oil-extracted residue (Watanabe et al. 2014). This has the benefit of recovering both the high-quality hydrocarbon oils, as well as pyrolysis bio-oil, bio-gas, and bio-char. These products can find valuable uses as energy and chemical feedstocks. Bio-oil from pyrolysis of the oil-extracted *B. braunii* biomass can be further reformed into C_1 gases and hydrogen via catalytic steam gasification at yields up to 91% (Watanabe et al. 2015). Considering the molecular complexity of the bio-oil and the difficulty of distillation and separation, catalytic gasification could provide a more feasible route for valorization.

2.3.6 Potential for High-Value Products

Green microalgae have potential for not only biofuels production, but also many other high-value products that could make the algae industry economically viable (Skjanes et al. 2013). *B. braunii* in particular is known to contain several high-value products, including α-linoleic acid, linoleic acid, butylated hydroxytoluene, lutein, β-carotene, and alkadienes. These molecules have applications as food additives, cosmetic additives, and possibly more. *B. braunii* contains high protein content, with several essential amino acids, and protein extracts showed high in vitro digestibility (Tibbetts et al. 2014). In addition to favorable protein content, *B. braunii* contains minerals and trace elements that are valuable components of food. The carotenoid content of *B. braunii* is another major advantage of this species, concerning lutein especially, as these molecules are valuable for nutraceutical applications (Ambati et al. 2018). Water extracts of *B. braunii* showed beneficial biological activities of dermatological interest, promoting expression of genes governing water balance, collagen maintenance, and anti-inflammatory activity in cell culture assays (Buono et al. 2012). This suggests that *B. braunii* cell extracts may have applications in skin care products.

Materials applications are another potential domain for *B. braunii* products to yield high-value products. The polysaccharides that *B. braunii* synthesis on the colony surface could be used as thickeners and gelling agents, or precursors in the production of polymers (García-Cubero et al. 2018). Polylactic acid is a particularly attractive polymer product, as it is derived from bacterial fermentation of sugars into lactic acid (Ahmed and Varshney 2011). Thus, polysaccharides from *B. braunii* could be applied as a sugar source for producing lactic acid by fermentation. Another type of polymer was produced from *B. braunii* protein by electrospinning into ultra-fine fibers (Verdugo et al. 2014). These protein-based nanofibers could be useful in tissue engineering, wound dressing, and other applications that would offer more value than nutritional applications. Polyhydroxybutyrate (PHB) is a valuable and biodegradable polymer that has been reportedly found in an isolate of *B. braunii* (Kavitha et al. 2016b). If true, this finding would constitute a significant addition to the portfolio of products that can be recovered from *B. braunii*. However, this study did not confirm the identity of the isolated strain by phylogenetic analysis, thus it is possible that the reported strain is not *B. braunii*. Nonetheless, a follow-up study demonstrated efficient production of PHB from the supposed strain of *B. braunii* using wastewater as the nutrient source (Kavitha et al. 2016a). Altogether, these studies demonstrate that *B. braunii* has significant potential as a source of high-value, bio-based materials.

2.3.7 Technoeconomic Analysis of Feasibility

One of the major questions regarding the whole of algae biotechnology is whether or not such systems are economically feasible. While a large number of techno-economic analyses have been conducted for other species of algae, very few have been conducted with *B. braunii*. A theoretical business case was developed for the industrial production of *B. braunii* oils (Shiho et al. 2012). Using a semi-open pond system on 19 hectares (47 acres) with 57 million liters of cultivation capacity, the capital cost was estimated at $2.4 million. This facility would be capable of producing 679,300 gallons of oil per year. With estimated annual operating expenses of $2.4 million, the breakeven selling price was $3.69 per gallon. Superstructure optimization was utilized to assess the viability of several models for the production of *B. braunii* (Chaudry et al. 2017). The process models revolved around the core concept of repetitively milking hydrocarbons from *B. braunii* with non-destructive solvent extraction and continuous cultivation. From the perspective of energetic feasibility, the optimal process with the stated assumptions consisted of open ponds, cylindrical sieve rotator filtration, and nanofiltration to achieve algae growth, biomass dewatering, and solvent recovery, respectively. In a follow-up study, this process was subjected to techno-economic analysis (Chaudry et al. 2018). With a facility of 544 hectares (1345 acres) containing 112 ponds and associated processing equipment, the capital cost was estimated at $71.7 million. This facility would be capable of producing 1,931,098 gallons of oil per year. With estimated annual operating expenses of $14.9 million, a minimum selling price of $12.11 per gallon is necessary to achieve a 10% annual rate of return. In the best-case scenario, it is possible that the minimum selling price could be reduced to $5.49 per gallon. Life cycle analysis of this production model estimated the greenhouse gas emissions to be −0.9 kg CO_2-equivalent per kg hydrocarbons from *B. braunii* (Chaudry et al. 2019). The ability to sequester carbon dioxide is an important aspect of sustainability for algae production.

2.4 Conclusion

Algae have great potential to serve a wide variety of markets with sustainable, high-performance products at competitive prices, but many barriers remain on the road to this vision (Kiesenhofer and Fluch 2018). Included in these challenges are the optimization of systems for cultivating and processing the algae, selecting and managing strains, mitigating contamination, developing quality standards, complying with regulations, and more. While these challenges apply to all species of algae, they are amplified especially for *B. braunii* given the relatively small amount of research and development that has been conducted with it, compared to other species, such as those in the genera *Chlamydomonas*, *Chlorella*, *Dunaliella*, *Haematococcus*, and others. Yet, among all of these species, the unique biological properties of *B. braunii* offer substantial advantages, particularly the ability to produce large amounts of extracellular hydrocarbon oils. While much progress has been made in recent years with respect to the study of *B. braunii*, much work remains to be done.

The studies of *B. braunii* cultivation have elucidated the range of productivity that is possible under various conditions. But the inconsistency of these studies poses a challenge to efforts to reproduce results. Numerous factors vary across the literature, including the cultivation vessel, the nutrient media, the light intensity, the photoperiod, the inoculation density, the experimental duration, and more. Many experiments were conducted over short periods with very low inoculation densities, which could be a factor in the low levels of productivity observed in many studies. Researchers would likely benefit from growing *B. braunii* for a long time, starting with higher density, and periodically feeding stoichiometrically balanced nutrients. More extensive evaluation of micronutrient requirements is warranted, especially for nickel. For investigation of industrial-scale cultivation, more experiments need to be conducted in outdoor environments. Of note is that *B. braunii* has never been cultivated in the current industry-standard horizontal tubular PBRs. Future research should focus on integrating the observations reviewed here to design new studies that systematically address the interactions of multiple factors in determining the level of biomass productivity and hydrocarbon content.

Achieving superior biological performance is necessary, but not sufficient, to achieve successful industrial deployment of *B. braunii*. Advances in processing *B. braunii* to recover products are essential to any commercial development. Recent research has demonstrated a significant potential for *B. braunii* to serve in the center of a biorefinery that treats multiple incoming waste streams and yields multiple outgoing valuable products. But more work is required to better characterize processes for product extraction and refinement. The algae industry would also benefit from better understanding of product performance and technoeconomic analysis. Information in both of these areas is currently very limited, which adds to the uncertainty and risk that precludes capital investment in projects to commercialize this technology and bring these products to market. However, with a concerted effort by the community of scientists who study *B. braunii*, it is possible to address the shortcomings in the scientific literature and pave a path forward for the entrepreneurs and investors aiming to commercialize this technology.

References

Abou-Shanab RAI, Singh M, Rivera-Cruz A, Power G, Bagby-Moon T, Das K (2016) Effect of *Brachionus rubens* on the growth characteristics of various species of microalgae. Electron J Biotechnol 22:68–74. https://doi.org/10.1016/j.ejbt.2016.06.005

Ahmed J, Varshney SK (2011) Polylactides—chemistry, properties and green packaging technology: a review. Int J Food Prop 14:37–58. https://doi.org/10.1080/10942910903125284

Al-Hothaly KA (2018) An optimized method for the bio-harvesting of microalgae, *Botryococcus braunii*, using *Aspergillus* sp. in large-scale studies. MethodsX 5:788–794. https://doi.org/10.1016/j.mex.2018.07.010

Al-Hothaly KA et al (2015a) Bio-harvesting and pyrolysis of the microalgae *Botryococcus braunii*. Bioresour Technol 191:117–123. https://doi.org/10.1016/j.biortech.2015.04.113

Al-Hothaly KA, Taha M, May BH, Stylianou S, Ball AS, Adetutu EM (2015b) The effect of nutrients and environmental conditions on biomass and oil production in *Botryococcus braunii* Race B strains. Eur J Phycol 51:1–10. https://doi.org/10.1080/09670262.2015.1071875

Ali I, Naqvi SR, Bahadar A (2018) Kinetic analysis of *Botryococcus braunii* pyrolysis using model-free and model fitting methods. Fuel 214:369–380. https://doi.org/10.1016/j.fuel.2017.11.046

Ambati RR, Gogisetty D, Aswathnarayana Gokare R, Ravi S, Bikkina PN, Su Y, Lei B (2018) Botryococcus as an alternative source of carotenoids and its possible applications – an overview. Crit Rev Biotechnol 38:541–558. https://doi.org/10.1080/07388551.2017.1378997

Areco MM, Haug E, Curutchet G (2018) Studies on bioremediation of Zn and acid waters using *Botryococcus braunii*. J Environ Chem Eng 6:3849–3859. https://doi.org/10.1016/j.jece.2018.05.041

Ashokkumar V, Rengasamy R (2012) Mass culture of *Botryococcus braunii* Kutz. under open raceway pond for biofuel production. Bioresour Technol 104:394–399. https://doi.org/10.1016/j.biortech.2011.10.093

Atobe S, Saga K, Maeyama H, Fujiwara K, Okada S, Imou K (2014) Culture of the green microalga *Botryococcus braunii* Showa with LED irradiation eliminating violet light enhances hydrocarbon production and recovery. Biosci Biotechnol Biochem 78:1765–1771. https://doi.org/10.1080/09168451.2014.932663

Atobe S et al (2015) Effect of amphiphilic polysaccharides released from *Botryococcus braunii* Showa on hydrocarbon recovery. Algal Res 10:172–176. https://doi.org/10.1016/j.algal.2015.05.007

Baba M, Kikuta F, Suzuki I, Watanabe MM, Shiraiwa Y (2012) Wavelength specificity of growth, photosynthesis, and hydrocarbon production in the oil-producing green alga *Botryococcus braunii*. Bioresour Technol 109:266–270. https://doi.org/10.1016/j.biortech.2011.05.059

Barajas-Solano AF, Guzman-Monsalve A, Kafarov V (2016) Effect of carbon-nitrogen ratio for the biomass production, hydrocarbons, and lipids on *Botryococcus braunii* UIS 003. Chem Eng Trans 49:247–252. https://doi.org/10.3303/CET1649042

Bazaes J, Sepulveda C, Acién FG, Morales J, Gonzales L, Rivas M, Riquelme C (2012) Outdoor pilot-scale production of Botryococcus braunii in panel reactors. J Appl Phycol 24:1353–1360. https://doi.org/10.1007/s10811-012-9787-3

Berrios H, Zapata M, Rivas M (2015) A method for genetic transformation of Botryococcus braunii using a cellulase pretreatment. J Appl Phycol. https://doi.org/10.1007/s10811-015-0596-3
Blackburn KB (1936) *Botryococcus* and the algal coals part I: a reinvestigation of the Alga *Botryococcus braunii* Kutzing. Trans Roy Soc Edin 58:841–854
Blifernez-Klassen O, Wibberg D, Winkler A, Blom J, Goesmann A, Kalinowski J, Kruse O (2016) Complete chloroplast and mitochondrial genome sequences of the hydrocarbon oil-producing green microalga *Botryococcus braunii* Race B (Showa). Genome Announc 4:e00524–e00516. https://doi.org/10.1128/genomeA.00524-16
Blifernez-Klassen O et al (2018) Metabolic survey of Botryococcus braunii: impact of the physiological state on product formation. Plos One 13:e0198976. https://doi.org/10.1371/journal.pone.0198976
Browne DR et al (2017) Draft nuclear genome sequence of the liquid hydrocarbon-accumulating green microalga *Botryococcus braunii* Race B (Showa). Genome Announc 5. https://doi.org/10.1128/genomeA.00215-17
Buono S et al (2012) Biological activities of dermatological interest by the water extract of the microalga *Botryococcus braunii*. Arch Dermatol Res 304:755–764. https://doi.org/10.1007/s00403-012-1250-4
Chaudry S, Bahri PA, Moheimani NR (2017) Superstructure optimization and energetic feasibility analysis of process of repetitive extraction of hydrocarbons from *Botryococcus braunii* – a species of microalgae. Comput Chem Eng 97:36–46. https://doi.org/10.1016/j.compchemeng.2016.11.018
Chaudry S, Bahri PA, Moheimani NR (2018) Techno-economic analysis of milking of Botryococcus braunii for renewable hydrocarbon production. Algal Res 31:194–203. https://doi.org/10.1016/j.algal.2018.02.011
Chaudry S, Bahri PA, Moheimani NR (2019) Life cycle analysis of milking of microalgae for renewable hydrocarbon production. Comput Chem Eng 121:510–522. https://doi.org/10.1016/j.compchemeng.2018.11.019
Cheng P, Ji B, Gao L, Zhang W, Wang J, Liu T (2013) The growth, lipid and hydrocarbon production of *Botryococcus braunii* with attached cultivation. Bioresour Technol 138C:95–100. https://doi.org/10.1016/j.biortech.2013.03.150
Ciudad G et al (2013) Performance of an enzymatic extract in *Botrycoccus braunii* cell wall disruption. J Biosci Bioeng 117:75–80. https://doi.org/10.1016/j.jbiosc.2013.06.012
Demura M, Ioki M, Kawachi M, Nakajima N, Watanabe MM (2013) Desiccation tolerance of Botryococcus braunii (Trebouxiophyceae, Chlorophyta) and extreme temperature tolerance of dehydrated cells. J Appl Phycol 26:49–53. https://doi.org/10.1007/s10811-013-0059-7
Díaz Bayona KC, Garcés LA (2014) Effect of different media on exopolysaccharide and biomass production by the green microalga *Botryococcus braunii*. J Appl Phycol 26:2087–2095. https://doi.org/10.1007/s10811-014-0242-5
Fasaei F, Bitter JH, Slegers PM, van Boxtel AJB (2018) Techno-economic evaluation of microalgae harvesting and dewatering systems. Algal Res 31:347–362. https://doi.org/10.1016/j.algal.2017.11.038
Furuhashi K, Saga K, Okada S, Imou K (2013) Seawater-cultured *Botryococcus braunii* for efficient hydrocarbon extraction. PLoS One 8. https://doi.org/10.1371/journal.pone.0066483.t001
Furuhashi K, Hasegawa F, Saga K, Kudou S, Okada S, Kaizu Y, Imou K (2016a) Effects of culture medium salinity on the hydrocarbon extractability, growth and morphology of *Botryococcus braunii*. Biomass Bioenerg 91:83–90. https://doi.org/10.1016/j.biombioe.2016.05.007
Furuhashi K, Noguchi T, Okada S, Hasegawa F, Kaizu Y, Imou K (2016b) The surface structure of *Botryococcus braunii* colony prevents the entry of extraction solvents into the colony interior. Algal Res 16:160–166. https://doi.org/10.1016/j.algal.2016.02.021
García-Cubero R, Cabanelas ITD, Sijtsma L, Kleinegris DMM, Barbosa MJ (2018) Production of exopolysaccharide by *Botryococcus braunii* CCALA 778 under laboratory simulated Mediterranean climate conditions. Algal Res 29:330–336. https://doi.org/10.1016/j.algal.2017.12.003

Garciano LO, Tran NH, Kannangara GSK, Milev AS, Wilson MA, McKirdy DM, Hall PA (2012) Pyrolysis of a naturally dried *Botryococcus braunii* residue. Energ Fuel 26:3874–3881. https://doi.org/10.1021/ef300451s

Garciano LO, Tran NH, Kannangara GSK, Milev AS, Wilson MA, Volk H (2014) Developing saponite supported cobalt–molybdenum catalysts for upgrading squalene, a hydrocarbon from the microalgae *Botryococcus braunii*. Chem Eng Sci 107:302–310. https://doi.org/10.1016/j.ces.2013.12.024

Ge Y, Liu J, Tian G (2011) Growth characteristics of B*otryococcus braunii* 765 under high CO2 concentration in photobioreactor. Bioresour Technol 102:130–134. https://doi.org/10.1016/j.biortech.2010.06.051

Gerulová K, Bartošová A, Blinová L, Bártová K, Dománková M, Garaiová Z, Palcut M (2018) Magnetic Fe_3O_4 -polyethyleneimine nanocomposites for efficient harvesting of *Chlorella zofingiensis, Chlorella vulgaris, Chlorella sorokiniana, Chlorella ellipsoidea* and *Botryococcus braunii*. Algal Res 33:165–172. https://doi.org/10.1016/j.algal.2018.05.003

Gim GH, Kim SW (2018) Growth factors in oceanic sediment significantly stimulate the biomass and lipid production of two oleaginous microalgae. J Appl Phycol. https://doi.org/10.1007/s10811-018-1509-z

Gim GH, Kim JK, Kim HS, Kathiravan MN, Yang H, Jeong SH, Kim SW (2014) Comparison of biomass production and total lipid content of freshwater green microalgae cultivated under various culture conditions. Bioprocess Biosyst Eng 37:99–106. https://doi.org/10.1007/s00449-013-0920-8

Gouveia JD et al (2017) Botryococcus braunii strains compared for biomass productivity, hydrocarbon and carbohydrate content. J Biotechnol 248:77–86. https://doi.org/10.1016/j.jbiotec.2017.03.008

Guionet A, Hosseini B, Teissie J, Akiyama H, Hosseini H (2017) A new mechanism for efficient hydrocarbon electro-extraction from *Botryococcus braunii*. Biotechnol Biofuels 10:39. https://doi.org/10.1186/s13068-017-0724-1

Hena S, Fatihah N, Tabassum S, Lalung J, Jing SY (2015) Magnetophoretic harvesting of freshwater microalgae using polypyrrole/Fe_3O_4 nanocomposite and its reusability. J Appl Phycol. https://doi.org/10.1007/s10811-015-0719-x

Hillen LW, Pollard G, Wake LV, White N (1982) Hydrocracking of the oils of *Botryococcus braunii* to transport fuels. Biotechnol Bioeng 24:193–205

Hirose M, Mukaida F, Okada S, Noguchi T (2013) Active Hydrocarbon biosynthesis and accumulation in a green alga, *Botryococcus braunii* (Race A). Eukaryot Cell 12:1132–1141. https://doi.org/10.1128/EC.00088-13

Hou L, Park H, Okada S, Ohama T (2013) Release of single cells from the colonial oil-producing alga *Botryococcus braunii* by chemical treatments. Protoplasma 251:191–199. https://doi.org/10.1007/s00709-013-0537-4

Jin J, Dupre C, Legrand J, Grizeau D (2016a) Extracellular hydrocarbon and intracellular lipid accumulation are related to nutrient-sufficient conditions in pH-controlled chemostat cultures of the microalga *Botryococcus braunii* SAG 30.81. Algal Res 17:244–252. https://doi.org/10.1016/j.algal.2016.05.007

Jin J, Dupré C, Yoneda K, Watanabe MM, Legrand J, Grizeau D (2016b) Characteristics of extracellular hydrocarbon-rich microalga *Botryococcus braunii* for biofuels production: recent advances and opportunities. Process Biochem 51:1866–1875. https://doi.org/10.1016/j.procbio.2015.11.026

Kanda H, Li P, Yoshimura T, Okada S (2013) Wet extraction of hydrocarbons from Botryococcus braunii by dimethyl ether as compared with dry extraction by hexane. Fuel 105:535–539. https://doi.org/10.1016/j.fuel.2012.08.032

Kandimalla P, Desi S, Vurimindi H (2016) Mixotrophic cultivation of microalgae using industrial flue gases for biodiesel production. Environ Sci Pollut Res Int 23:9345–9354. https://doi.org/10.1007/s11356-015-5264-2

Kavitha G, Kurinjimalar C, Sivakumar K, Kaarthik M, Aravind R, Palani P, Rengasamy R (2016a) Optimization of polyhydroxybutyrate production utilizing waste water as nutrient source by *Botryococcus braunii* Kutz using response surface methodology. Int J Biol Macromol 93:534–542. https://doi.org/10.1016/j.ijbiomac.2016.09.019

Kavitha G, Kurinjimalar C, Sivakumar K, Palani P, Rengasamy R (2016b) Biosynthesis, purification and characterization of polyhydroxybutyrate from *Botryococcus braunii* kutz. Int J Biol Macromol 89:700–706. https://doi.org/10.1016/j.ijbiomac.2016.04.086

Kawachi M, Tanoi T, Demura M, Kaya K, Watanabe MM (2012) Relationship between hydrocarbons and molecular phylogeny of *Botryococcus braunii*. Algal Res 1:114–119. https://doi.org/10.1016/j.algal.2012.05.003

Khatri W, Hendrix R, Niehaus T, Chappell J, Curtis WR (2013) Hydrocarbon production in high density *Botryococcus braunii* Race B continuous culture. Biotechnol Bioeng 111(3):493–503. https://doi.org/10.1002/bit.25126

Kiesenhofer DP, Fluch S (2018) The promises of microalgae-still a long way to go. FEMS Microbiol Lett 365. https://doi.org/10.1093/femsle/fnx257

Kim HS, Weiss TL, Thapa HR, Devarenne TP, Han A (2014) A microfluidic photobioreactor array demonstrating high-throughput screening for microalgal oil production. Lab Chip 14:1379–1482. https://doi.org/10.1039/c3lc51396c

Krzeminska I, Pawlik-Skowronska B, Trzcinska M, Tys J (2013) Influence of photoperiods on the growth rate and biomass productivity of green microalgae. Bioprocess Biosyst Eng 37:735–741. https://doi.org/10.1007/s00449-013-1044-x

Kula M, Kalaji HM, Skoczowski A (2017) Culture density influence on the photosynthetic efficiency of microalgae growing under different spectral compositions of light. J Photochem Photobiol B 167:290–298. https://doi.org/10.1016/j.jphotobiol.2017.01.013

Lammers PJ et al (2017) Review of the cultivation program within the National Alliance for Advanced Biofuels and Bioproducts. Algal Res 22:166–186. https://doi.org/10.1016/j.algal.2016.11.021

Lee AK, Lewis DM, Ashman PJ (2012) Disruption of microalgal cells for the extraction of lipids for biofuels: processes and specific energy requirements. Biomass Bioenerg 46:89–101. https://doi.org/10.1016/j.biombioe.2012.06.034

Lee J-C, Baek K, Kim H-W (2018) Semi-continuous operation and fouling characteristics of submerged membrane photobioreactor (SMPBR) for tertiary treatment of livestock wastewater. J Clean Prod 180:244–251. https://doi.org/10.1016/j.jclepro.2018.01.159

Liu J, Ge Y, Guangming T (2012) Enhancement of hydrocarbon productivity of *Botryococcus braunii* by an adenine type phytohormone. Adv Mat Res 512:397–400

Liu J, Ge Y, Cheng H, Wu L, Tian G (2013) Aerated swine lagoon wastewater: a promising alternative medium for *Botryococcus braunii* cultivation in open system. Bioresour Technol 139C:190–194. https://doi.org/10.1016/j.biortech.2013.04.036

Liu Y, Murata K, Inaba M (2018) Hydrocracking of algae oil to aviation fuel-ranged hydrocarbons over NiMo-supported catalysts. Catal Today 332:115–121. https://doi.org/10.1016/j.cattod.2018.07.047

Magota A, Saga K, Okada S, Atobe S, Imou K (2012) Effect of thermal pretreatments on hydrocarbon recovery from *Botryococcus braunii*. Bioresour Technol 123:195–198. https://doi.org/10.1016/j.biortech.2012.07.095

Marrone BL et al (2017) Review of the harvesting and extraction program within the National Alliance for Advanced Biofuels and Bioproducts. Algal Res. https://doi.org/10.1016/j.algal.2017.07.015

Metzger P, Largeau C (2005) *Botryococcus braunii*: a rich source for hydrocarbons and related ether lipids. Appl Microbiol Biotechnol 66:486–496. https://doi.org/10.1007/s00253-004-1779-z

Moheimani NR, Cord-Ruwisch R, Raes E, Borowitzka MA (2013) Non-destructive oil extraction from *Botryococcus braunii* (Chlorophyta). J Appl Phycol 25:1653–1661. https://doi.org/10.1007/s10811-013-0012-9

Moutel B et al (2016) Development of a screening procedure for the characterization of *Botryococcus braunii* strains for biofuel application. Process Biochem 51:1855–1865. https://doi.org/10.1016/j.procbio.2016.05.002

Nagano S, Yamamoto S, Nagakubo M, Atsumi K, Watanabe MM (2012) Physical properties of hydrocarbon oils produced by Botryococcus braunii: density, kinematic viscosity, surface tension, and distillation. Properties Procedia Environ Sci 15:73–79. https://doi.org/10.1016/j.proenv.2012.05.012

Nakamura H, Matsunaga S, Kawagishi H, Okada S (2016) Effects of 2-azahypoxanthine on extracellular terpene accumulations by the green microalga *Botryococcus braunii*, race B. Algal Res 20:267–275. https://doi.org/10.1016/j.algal.2016.10.023

Nakamura H, Shiozaki T, Gonda N, Furuya K, Matsunaga S, Okada S (2017) Utilization of ammonium by the hydrocarbon-producing microalga, *Botryococcus braunii* Showa. Algal Res 25:445–451. https://doi.org/10.1016/j.algal.2017.06.007

Ozkan A, Kinney K, Katz L, Berberoglu H (2012) Reduction of water and energy requirement of algae cultivation using an algae biofilm photobioreactor. Bioresour Technol 114:542–548. https://doi.org/10.1016/j.biortech.2012.03.055

Peng C, Li S, Zheng J, Huang S, Li D (2017) Harvesting microalgae with different sources of starch-based cationic flocculants. Appl Biochem Biotechnol 181:112–124. https://doi.org/10.1007/s12010-016-2202-9

Pengfei C, Yan W, Qiyong Y, Tianzhong L (2017) Comparison of growth, hydrocarbon accumulation and metabolites of *Botryococcus braunii* between attached cultivation and aqueous-suspension cultivation. Int J Agric Biol Eng 10:134–141. https://doi.org/10.3965/j.ijabe.20171001.3008

Podder MS, Majumder CB (2017) Toxicity and bioremediation of As(III) and As(V) in the green microalgae *Botryococcus braunii*: a laboratory study. Int J Phytoremediation 19:157–173. https://doi.org/10.1080/15226514.2016.1207601

Ranga Rao A, Ravishankar GA, Sarada R (2012) Cultivation of green alga Botryococcus braunii in raceway, circular ponds under outdoor conditions and its growth, hydrocarbon production. Bioresour Technol 123:528–533. https://doi.org/10.1016/j.biortech.2012.07.009

Rezanka T, Lukavsky J, Vitova M, Nedbalova L, Sigler K (2018) Lipidomic analysis of *Botryococcus* (Trebouxiophyceae, Chlorophyta) – identification of lipid classes containing very long chain fatty acids by offline two-dimensional LC-tandem MS. Phytochemistry 148:29–38. https://doi.org/10.1016/j.phytochem.2018.01.011

Rinna F, Buono S, Cabanelas ITD, Nascimento IA, Sansone G, Barone CMA (2017) Wastewater treatment by microalgae can generate high quality biodiesel feedstock. J Water Process Eng 18:144–149. https://doi.org/10.1016/j.jwpe.2017.06.006

Ruangsomboon S, Sornchai P, Prachom N (2018) Enhanced hydrocarbon production and improved biodiesel qualities of *Botryococcus braunii* KMITL 5 by vitamins thiamine, biotin and cobalamin supplementation. Algal Res 29:159–169. https://doi.org/10.1016/j.algal.2017.11.028

Sambles C et al (2017) Metagenomic analysis of the complex microbial consortium associated with cultures of the oil-rich alga *Botryococcus braunii*. Microbiologyopen 6. https://doi.org/10.1002/mbo3.482

Schulze S et al (2017) Identification of methylated GnTI-dependent N-glycans in *Botryococcus brauni*. New Phytol 215:1361–1369. https://doi.org/10.1111/nph.14713

Shen Y, Zhu W, Chen C, Nie Y (2016) Glycine induced culture-harvesting strategy for *Botryococcus braunii*. J Biosci Bioeng 121:424–430. https://doi.org/10.1016/j.jbiosc.2015.08.004

Shiho M, Kawachi M, Horioka K, Nishita Y, Ohashi K, Kaya K, Watanabe MM (2012) Business evaluation of a green microalgae *Botryococcus braunii* oil production system. Procedia Environ Sci 15:90–109. https://doi.org/10.1016/j.proenv.2012.05.014

Shimamura R, Watanabe S, Sakakura Y, Shiho M, Kaya K, Watanabe MM (2012) Development of *Botryococcus* seed culture system for future mass culture. Procedia Environ Sci 15:80–89. https://doi.org/10.1016/j.proenv.2012.05.013

Singh Khichi S, Anis A, Ghosh S (2018) Mathematical modeling of light energy flux balance in flat panel photobioreactor for *Botryococcus braunii* growth, CO_2 biofixation and lipid production under varying light regimes. Biochem Eng J 134:44–56. https://doi.org/10.1016/j.bej.2018.03.001

Skjanes K, Rebours C, Lindblad P (2013) Potential for green microalgae to produce hydrogen, pharmaceuticals and other high value products in a combined process. Crit Rev Biotechnol 33:172–215. https://doi.org/10.3109/07388551.2012.681625

Song L, Qin JG, Su S, Xu J, Clarke S, Shan Y (2012) Micronutrient requirements for growth and hydrocarbon production in the oil producing green alga *Botryococcus braunii* (Chlorophyta). PLoS One 7. https://doi.org/10.1371/journal.pone.0041459.g001

Suzuki R, Ito N, Uno Y, Nishii I, Kagiwada S, Okada S, Noguchi T (2013) Transformation of lipid bodies related to hydrocarbon accumulation in a green alga, *Botryococcus braunii* (Race B). PLoS One 8:e81626. https://doi.org/10.1371/journal.pone.0081626

Suzuki R, Nishii I, Okada S, Noguchi T (2017) 3D reconstruction of endoplasmic reticulum in a hydrocarbon-secreting green alga, *Botryococcus braunii* (Race B). Planta 247(3):663–677. https://doi.org/10.1007/s00425-017-2811-8

Talukdar J, Kalita MC, Goswami BC, Hong DD, Das HC (2014) Liquid hydrocarbon production potential of a novel strain of the microalga *Botryococcus braunii*: assessing the reliability of in situ hydrocarbon recovery by wet process solvent extraction. Energy Fuels 28:3747–3758. https://doi.org/10.1021/ef402298r

Tanabe Y, Ioki M, Watanabe MM (2013) The fast-growing strain of hydrocarbon-rich green alga Botryococcus braunii, BOT-22, is a vitamin B12 autotroph. J Appl Phycol 26:9–13. https://doi.org/10.1007/s10811-013-0045-0

Tanoi T, Kawachi M, Watanabe MM (2011) Effects of carbon source on growth and morphology of *Botryococcus braunii*. J Appl Phycol 23:25–33. https://doi.org/10.1007/s10811-010-9528-4

Tanoi T, Kawachi M, Watanabe MM (2013) Iron and glucose effects on the morphology of *Botryococcus braunii* with assumption on the colony formation variability. J Appl Phycol 26:1–8. https://doi.org/10.1007/s10811-013-0026-3

Tibbetts SM, Milley JE, Lall SP (2014) Chemical composition and nutritional properties of freshwater and marine microalgal biomass cultured in photobioreactors. J Appl Phycol 27:1109–1119. https://doi.org/10.1007/s10811-014-0428-x

Tracy NI, Crunkleton DW, Price GL (2011) Catalytic cracking of squalene to gasoline-range molecules. Biomass Bioenerg 35:1060–1065. https://doi.org/10.1016/j.biombioe.2010.11.027

Tsutsumi S, Saito Y, Matsushita Y, Aoki H (2018a) Effect of mechanical pretreatment on hydrocarbon extraction from concentrated wet hydrocarbon-rich microalga, *Botryococcus braunii*. Energy Fuels 32:1761–1770. https://doi.org/10.1021/acs.energyfuels.7b02753

Tsutsumi S, Saito Y, Matsushita Y, Aoki H (2018b) Investigation of colony disruption for hydrocarbon extraction from *Botryococcus braunii*. Fuel Process Technol 172:36–48. https://doi.org/10.1016/j.fuproc.2017.12.004

Tsutsumi S, Saito Y, Matsushita Y, Aoki H (2018c) Prediction of hydrocarbon yield from wet Botryococcus braunii : the influences of colony surface charge and diameter on the amount of extractable hydrocarbon. Fuel Process Technol 176:50–58. https://doi.org/10.1016/j.fuproc.2018.03.013

Unkefer CJ et al (2017) Review of the algal biology program within the National Alliance for Advanced Biofuels and Bioproducts. Algal Res 22:187–215. https://doi.org/10.1016/j.algal.2016.06.002

Uno Y, Nishii I, Kagiwada S, Noguchi T (2015) Colony sheath formation is accompanied by shell formation and release in the green alga *Botryococcus braunii* (race B). Algal Res 8:214–223. https://doi.org/10.1016/j.algal.2015.02.015

van den Berg TE, van Oort B, Croce R (2017) Light-harvesting complexes of Botryococcus braunii. Photosynth Res 135:191–201. https://doi.org/10.1007/s11120-017-0405-8

van den Berg TE, Chukhutsina V, van Amerongen H, Croce R, van Oort B (2019) Light acclimation of the colonial green alga *Botryococcus braunii* strain Showa. Plant Physiol 179:1132–1143. https://doi.org/10.1104/pp.18.01499

Verdugo M, Lim LT, Rubilar M (2014) Electrospun protein concentrate fibers from microalgae residual biomass. J Polym Environ 22:373–383. https://doi.org/10.1007/s10924-014-0678-3

Volkman JK (2014) Acyclic isoprenoid biomarkers and evolution of biosynthetic pathways in green microalgae of the genus *Botryococcus*. Org Geochem 75:36–47. https://doi.org/10.1016/j.orggeochem.2014.06.005

Vu CHT, Lee HG, Chang YK, Oh HM (2018) Axenic cultures for microalgal biotechnology: establishment, assessment, maintenance, and applications. Biotechnol Adv 36:380–396. https://doi.org/10.1016/j.biotechadv.2017.12.018

Wake LV, Hillen LW (1980) Study of a "Bloom" of the oil-rich alga *Botryococcus braunii* in the Darwin River Reservoir. Biotechnol Bioeng 22:1637–1657

Wang SK, Wang F, Hu YR, Stiles AR, Guo C, Liu CZ (2014a) Magnetic flocculant for high efficiency harvesting of microalgal cells. ACS Appl Mater Interfaces 6:109–115. https://doi.org/10.1021/am404764n

Wang SK, Wang F, Stiles AR, Guo C, Liu CZ (2014b) *Botryococcus braunii* cells: ultrasound-intensified outdoor cultivation integrated with in situ magnetic separation. Bioresour Technol 167:376–382. https://doi.org/10.1016/j.biortech.2014.06.028

Wang S-K, Guo C, Wu W, Sui K-Y, Liu C-Z (2019) Effects of incident light intensity and light path length on cell growth and oil accumulation in *Botryococcus braunii* (Chlorophyta). Eng. Life Sci 19:104–111. https://doi.org/10.1002/elsc.201800128

Watanabe H, Li D, Nakagawa Y, Tomishige K, Kaya K, Watanabe MM (2014) Characterization of oil-extracted residue biomass of *Botryococcus braunii* as a biofuel feedstock and its pyrolytic behavior. Appl Energy 132:475–484. https://doi.org/10.1016/j.apenergy.2014.07.037

Watanabe H, Li D, Nakagawa Y, Tomishige K, Watanabe MM (2015) Catalytic gasification of oil-extracted residue biomass of *Botryococcus braunii*. Bioresour Technol. https://doi.org/10.1016/j.biortech.2015.03.034

Weiss TL et al (2012) Colony organization in the green alga *Botryococcus braunii* (Race B) is specified by a complex extracellular matrix. Eukaryot Cell 11:1424–1440. https://doi.org/10.1128/EC.00184-12

Wijffels RH, Kruse O, Hellingwerf KJ (2013) Potential of industrial biotechnology with cyanobacteria and eukaryotic microalgae. Curr Opin Biotechnol 24:405–413. https://doi.org/10.1016/j.copbio.2013.04.004

Wijihastuti RS, Moheimani NR, Bahri PA, Cosgrove JJ, Watanabe MM (2016) Growth and photosynthetic activity of *Botryococcus braunii* biofilms. J Appl Phycol 29:1123–1134. https://doi.org/10.1007/s10811-016-1032-z

Xu L, Liu R, Wang F, Liu CZ (2012a) Development of a draft-tube airlift bioreactor for Botryococcus braunii with an optimized inner structure using computational fluid dynamics. Bioresour Technol 119:300–305. https://doi.org/10.1016/j.biortech.2012.05.123

Xu L, Wang F, Guo C, Liu C-Z (2012b) Improved algal oil production from *Botryococcus braunii* by feeding nitrate and phosphate in an airlift bioreactor. Eng Life Sci 12:171–177. https://doi.org/10.1002/elsc.201100110

Xu L, Wang S-K, Wang F, Guo C, Liu C-Z (2014) Improved biomass and hydrocarbon productivity of *Botryococcus braunii* by periodic ultrasound stimulation. BioEnergy Res 7:986–992. https://doi.org/10.1007/s12155-014-9441-9

Yamamoto S, Mandokoro Y, Nagano S, Nagakubo M, Atsumi K, Watanabe MM (2013) Catalytic conversion of *Botryococcus braunii* oil to diesel fuel under mild reaction conditions. J Appl Phycol 26:55–64. https://doi.org/10.1007/s10811-013-0065-9

Yeesang C, Cheirsilp B (2014) Low-cost production of green microalga *Botryococcus braunii* biomass with high lipid content through mixotrophic and photoautotrophic cultivation. Appl Biochem Biotechnol 174:116–129. https://doi.org/10.1007/s12010-014-1041-9

Yildiz-Ozturk E, Yesil-Celiktas O (2017) Supercritical CO_2 extraction of hydrocarbons from Botryococcus braunii as a promising bioresource. J Supercrit Fluids 130:261–266. https://doi.org/10.1016/j.supflu.2017.06.019

Yildiz-Ozturk E, Ilhan-Ayisigi E, Togtema A, Gouveia J, Yesil-Celiktas O (2018) Effects of hydrostatic pressure and supercritical carbon dioxide on the viability of *Botryococcus braunii* algae cells. Bioresour Technol 256:328–332. https://doi.org/10.1016/j.biortech.2018.02.041

Yonezawa N, Matsuura H, Shiho M, Kaya K, Watanabe MM (2012) Effects of soybean curd wastewater on the growth and hydrocarbon production of *Botryococcus braunii* strain BOT-22. Bioresour Technol 109:304–307. https://doi.org/10.1016/j.biortech.2011.07.090

Yoshimura T, Okada S, Honda M (2013) Culture of the hydrocarbon producing microalga *Botryococcus braunii* strain Showa: optimal CO_2, salinity, temperature, and irradiance conditions. Bioresour Technol 133:232–239. https://doi.org/10.1016/j.biortech.2013.01.095

Yu J-U, Kim H-W (2017) Enhanced microalgal growth and effluent quality in tertiary treatment of livestock wastewater using a sequencing batch reactor. Water Air Soil Pollut 228:357. https://doi.org/10.1007/s11270-017-3547-6

Zhang H, Wang W, Li Y, Yang W, Shen G (2011) Mixotrophic cultivation of *Botryococcus braunii*. Biomass Bioenerg 35:1710–1715. https://doi.org/10.1016/j.biombioe.2011.01.002

Zhang F, Cheng LH, Xu XH, Zhang L, Chen HL (2013) Application of memberane dispersion for enhanced lipid milking from Botryococcus braunii FACHB 357. J Biotechnol 165:22–29. https://doi.org/10.1016/j.jbiotec.2013.02.010

Chapter 3
Biofertilizers from Microalgae

Luísa Gouveia, Gabriel Acién, Elvira Navarro López, Hugo Pereira, Luísa Barreira, Mário Reis, and João Varela

Abstract Most chemical fertilizers are finite resources that require large amounts of energy to be either chemically synthesized or mined. Hence, agriculture needs to adopt the concept of circular economy, so that losses are avoided, nutrient recycling is stimulated, and the sustainability of the world food supply is enhanced. Microalgae are a possible solution to this problem, as they are able to uptake N and P present in wastewaters and produce biomass rich in phytohormones and amino acids able to promote plant growth when applied to the soil. To grow microalgal biomass for the industrial production of biofertilizers/biostimulants, raceways and thin-layer reactors are proposed in order to decrease production costs. However, proper control of the composition of the microalgal biomass is crucial, even though this might be difficult to achieve. In spite of this hurdle, preliminary results strongly suggest that microalgae-based biostimulants can significantly impact on plant growth and seed germination indexes when applied to soil with minimal pre-treatment. Moreover, several methodologies have been tested for producing microalgae-based biofertilizers, ranging from preparing water extracts to composting. These technologies to process algal biomass into biofertilizers are discussed.

Keywords Biofertilizers · Biostimulants · Microalgae · Bioinoculants · Sustainability · Circular bioeconomy

---

L. Gouveia
LNEG—Laboratório Nacional de Energia e Geologia, I.P./Bioenergy and Biorefineries Unit, Lisbon, Portugal

G. Acién · E. N. López
Department of Chemical Engineering-CIESOL, University of Almeria, Almeria, Spain

H. Pereira · L. Barreira · M. Reis · J. Varela (✉)
CCMAR—Centre of Marine Sciences, University of Algarve, Gambelas, Faro, Portugal
e-mail: jvarela@ualg.pt

J. A. V. Costa et al. (eds.), *Microalgal Bioengineering*,
https://doi.org/10.1007/978-3-031-61253-4_3

3.1 Introduction

Most chemical fertilizers are toxic and finite resources, which might damage the biosphere and cause several human diseases through foodstuff and ground-, surface-, and seawater pollution (Arioli et al. 2015; Mógor et al. 2018). In addition, the production costs of chemical fertilizers are rising, with the expected nominal price index to increase from 78 in 2016 to 88 by 2020 (World Bank 2019). It can be compounded by the fact that energy costs are on the rise. Besides, more than 85% of the NH_3 used in fertilizers is produced via the energy-intensive industrial Haber-Bosch process. While NH_3 is the second most produced chemical worldwide, its sustainability is highly questionable, particularly in countries that depend on fossil fuels rather than renewable sources to meet their energy demand (Pfromm 2017).

Although nitrogen can be obtained from atmospheric N_2, the other two chemicals that are widely used in agriculture, P and K, come mostly from mines. It has been suggested that the world reserves of K in the form of potash will be able to meet the current demand for several centuries. However, recent estimates by the International Fertilizer Development Center (IFDC) indicate that the world P supply in the form of phosphorus rock will last at most 300 years using the current mining technology. Moreover, the ratio of reserves to consumption for the two largest world consumers of phosphorus rock, China and the United States, is expected to be not higher than 37 and 40 years, respectively (Daneshgar et al. 2018). Therefore, it is clear that both P and K are finite resources and alternatives must be found soon to sustain food production and prevent fertilizers from becoming unaffordable.

As two of the most important nutrients used in fertilizers are finite resources and their indiscriminate use can cause eutrophication in aquatic ecosystems, leading to the formation of "dead zones" lacking oxygen, the European Commission (2015) has acknowledged the urgent need to apply the Circular Economy (CE) concept to fertilizers as well. CE is usually depicted as a loop comprising four stages, namely production, use, collection, and recycling. Underlying this loop is the need for avoiding losses at all stages. This fact is acutely relevant for nutrients, in particular those that are mined and finite. For example, it has been estimated that the nutrient-use efficiency along the P supply chain is only 5–10% (Geissler et al. 2018). Moreover, large amounts of nutrients are lost in agroindustrial and municipal wastewaters, which need treatment before being released to the surrounding ecosystems and avoid eutrophication. This treatment requires the removal of N and P, a significant part of which is lost in the form settling solids. Even though these solids (sewage sludge) may be used in the fertilizers production, concerns with heavy metal contamination often preclude their wide adoption as sources of nutrients (Frišták et al. 2018). Secondary treatments are usually used to oxidize organic matter through heterotrophic bacteria in the form of biofilms or activated sludge in suspended growth reactors. Further removal of N and P is attained employing tertiary treatments that might involve either physicochemical (e.g., chemical precipitation and reverse osmosis) or biological technologies. The latter can include microorganisms able to carry out processes of nitrification and denitrification, resulting in the loss of

N in the form of atmospheric N_2. However, tertiary treatment is often the most expensive step in a wastewater treatment plant, costing four times as much as the primary treatment (Abdel-Raouf et al. 2012). Moreover, wastewaters containing high levels of NH_3 and antibiotics can be a lethal combination for bacteria used in tertiary treatments, resulting in an inefficient process.

A possible solution to these problems is the use of microorganisms (e.g., microalgae) that are not only resistant to antibiotics and NH_3-rich wastewaters, but are also able to remove N and P from the water as they grow, producing a biomass that can be safely used in different applications, including the production of biofertilizers and biostimulants.

3.2 Biofertilizers, Bioinoculants, and Biostimulants

According to the Food and Agriculture Organization of the United Nations, food production will have to increase at least by 60% until 2050 to support the projected growth of the human population (9.6 billion). The growing scarcity of arable land and resources for agriculture and the need to reduce greenhouse gas emissions are additional reasons for the urgent need to implement more sustainable agriculture. Together with other factors, such as a steady freshwater supply, sustainable agriculture cannot be based on mined chemical fertilizers and oil-derived pesticides with a poor record of biodegradability. Chemical fertilizers are formulations of inorganic nutrients such as PO_4^{3-}, NO_3^-, NH_4^+, and K^+, used to increase crop productivity. However, chemical fertilizers often contain toxic metals like Cd, Hg, As, and Pb and radionuclides like ^{238}U, ^{232}Th, and ^{210}Po (Savci 2012; Tóth et al. 2016). In addition to being a source of these pollutants, chemical fertilizers have also been considered to be the main culprit for the excess introduction of nutrients into the environment and the contamination of surface and groundwater with NO_3^- due to the over-application of N fertilizers (Rahmati et al. 2015; Smith and Siciliano 2015). Furthermore, soil microorganisms transform excess N into N_2O, a greenhouse gas (GHG) with 265 times the capacity of CO_2 to trap heat (Lenka et al. 2016).

Environmentally friendly fertilizers (EFF) aim to avoid some of the above-described environmental problems. EFF are designed to release nutrients at a controlled rate, maintaining their concentration in the soil at levels strictly necessary for optimum productivity (Ni et al. 2011). They can be used alone or in combination with nitrification/urease inhibitors. Different formulations have been developed including coated and non-coated alternatives. In coated fertilizers, nutrients are encapsulated in an environmentally friendly material that degrades in soil, releasing the nutrients slowly. On the other hand, non-coated formulations use compounds with a low degradation rate (e.g., urea-formaldehyde reaction products, isobutylidene diurea, and crotonylidene diurea), delaying nutrient release. Since the nutrient release is slowed down, aiming to last for the entire crop period, nutrients are lost through processes, such as leaching, nitrification, and/or volatilization (Timilsena et al. 2014). The rate of N release from EEF is usually dependent on soil microorganisms, as well as soil moisture, pH, and temperature (Nardi et al. 2018). Although EFF use can have several important advantages over traditional chemical fertilizers, such as decreased nitroxide emissions, increased soil organic matter content, improved soil water content and holding capacity, they are still quite expensive. Additionally, nutrient release from coated EFF is difficult to predict and control since it is largely dependent on soil microorganisms and enzymes, which hampers the timing between the application of EFF and the nutritional requirements of crop plants (Chen et al. 2018). Hence, more cost-effective products and realistic models for nutrients release prediction are necessary for EFF to become a suitable technology (Timilsena et al. 2014).

Recognizing the key role of soil microorganisms in crop productivity, biofertilizers are products containing living organisms (i.e., bioinoculants), nutrients, and other molecules (i.e., biostimulants) that, when applied to soils, seeds, or plants, stimulate seed germination and plant growth (Calvo et al. 2014). The action of bioinoculants and biostimulants is based on the improvement of microbial processes able to increase nutrient bioavailability to plants and consequently crop yield. Upon the application of biofertilizers, soil fertility is increased through enhanced atmospheric N fixation, solubilization of insoluble phosphates and release of plant growth-stimulating compounds (Mahanty et al. 2017). The main advantages of biofertilizers rely on their ability to ensure a steady supply of nutrients throughout plant growth. While more than 50% of the N delivered by chemical fertilizers may be lost to the atmosphere or adjoining waters upon its release to the soil, and a large percentage of the P supplied by chemical fertilizers precipitates in the form of insoluble metal complexes (Adesemoye and Kloepper 2009), the use of bioinoculants as supplements to fertilizers ensures a steadier integrated nutrient supply (Bhardwaj et al. 2014). The radicular microbiome can contain about 10^{11} cells/g of the root, including more than 30,000 different species. Among these, three major groups have been used as bioinoculants: (i) arbuscular mycorrhiza fungi, (ii) N-fixers, and (iii) plant growth-promoting rhizobacteria. Their mechanisms of action, advantages, and disadvantages will be discussed below.

3.2.1 Fungi- and Bacteria-Based Biofertilizers

3.2.1.1 Arbuscular Mycorrhiza Fungi

Mycorrhiza are symbiotic associations between fungi and plants, acting in the root system, in which the fungi meet its carbon requirements from the plant and, in turn, facilitate the plant absorption of nutrients such as P, Ca, Cu, and Zn (Clark and Zeto 2000; Mahdi et al. 2010). These associations are present in around 80% of all terrestrial plant species and are thought to help plants resist different types of abiotic stress such as drought (Gholamhoseini et al. 2013), salinity (Krishnamoorthy et al. 2014), pollution (Upadhyaya et al. 2010), and cold stress (Gai et al. 2009; Lenoir et al. 2016). Moreover, mycorrhiza might also help plants to withstand biotic factors like soil-borne pathogens (Harrier and Watson 2004). Most of the arbuscular mycorrhiza fungi (AMF) belong to the phylum Glomeromycota, although environmental conditions can determine abundances and diversity of around 250 species described so far, which can be found in different continents and climatic zones (Öpik et al. 2013). Nonetheless, most commercial formulas are based on species from the genera *Rhizophagus* and *Funneliformis* (Rouphael et al. 2015).

3.2.1.2 N-fixers

Although half of the N input into agriculture comes from the industrial Haber-Bosch process used to synthesize ammonia for producing fertilizers, biological fixation is responsible for the other half (Smith 2002). The latter occurs via an enzymatic complex called nitrogenase comprising two enzymes: dinitrogenase reductase (DNR) and dinitrogenase (DN). DNR uses Fe as a co-factor to provide electrons to DN, an Fe- and Mo-containing enzyme that is able to reduce N_2 to NH_3 (Santi et al. 2013). In legume nodules, the produced NH_3 diffuses out of the alkaline bacterial protoplasm into the acidic peribacteroid space. In this location, it is protonated to NH_4^+ and is transferred into the cytoplasm of the plant host cells, where it is converted into amino acids (Vessey et al. 2005). In non-leguminous plants, but where bacteria also occur in nodules, the mechanism is suspected to be similar. In other cases, the mechanism of NH_4^+ transfer from the N-fixing bacteria to the plant is not entirely clear (Steenhoudt and Vanderleyden 2000). Biologically, this process is limited to prokaryotic organisms, most of them bacteria, although archaeal methanogens also appear to be important contributors to the process of N fixation (Franche et al. 2009).

3.2.1.3 Plant Growth-Promoting Rhizobacteria (PGPR)

The microorganisms described above that aid plant growth by increasing the availability of nutrients or protecting the plants against abiotic or biotic stress can be considered plant growth-promoting microorganisms. Still, some of these organisms can enhance plant growth through the production of metabolites such as auxins, gibberellins, and cytokinins, which act as growth-promoting phytohormones (Mahanty et al. 2017). These compounds significantly enhance the productivity of crops such as maize, wheat, and rice by increasing plant height, root length, and dry matter production. For example, *Azospirillum*, a Gram-negative, non-nodule forming, nitrogen-fixing bacterium produces indole-3-acetic acid (IAA) that induces changes in the root system of inoculated plants, enhancing mineral uptake (Steenhoudt and Vanderleyden 2000). As observed with some AMF and N-fixing bacteria, flavonoids excreted by plants' roots seem to induce the colonization process (Santi et al. 2013). *Azospirillum* has also been observed to stimulate maize growth by enhancing shoot and root growth and increasing root-hair density in a process mediated by gibberellins (Fulchieri et al. 1993). Other bacteria have been identified as PGPR; these include *Azotobacter*, *Alcaligenes*, *Bacillus*, *Beijerinckia*, *Campylobacter*, *Derxia*, and some Enterobacteriaceae and *Pseudomonas* (Franche et al. 2009). Besides the capacity to produce gibberellins, cytokinins and IAA, *Azotobacter* can also produce vitamins such as thiamine and riboflavin. For example, *A. chroococcum* is able to enhance the germination of seeds of maize and root architecture as well as inhibit the growth of pathogenic microorganisms around the plant root system, thus improving plant growth (Bhardwaj et al. 2014; Gholami et al. 2009). The same inhibiting strategy is used by two *Pseudomonas* species that also colonize maize roots (Kozdrój et al. 2004).

Although the use of the beneficial microbes here described as bioinoculants are becoming increasingly important in sustainable crop production, there are still some limitations in their application. The main restraint arises from the difficulty of scaling up bioinoculant production. Since host free production is currently not possible, problems such as quality control arise from the difficulty in species identification and tracking after release in the field. Furthermore, introduction of unwanted species such as invasive or exotic species may occur through contaminated inocula (Ijdo et al. 2011).

3.2.2 Microalgal Biomass: An Emergent Biofertilizer Feedstock

As biofertilizers play a major role in improving the availability of nutrients necessary for plant growth and are essential for the controlled mineralization and fertilization processes, being also cost-effective, eco-friendly, and a renewable resource, they have become a sustainable alternative for the costly, energy-intensive chemical fertilizers (Kawalekar 2013). The use of biofertilizers produced from microalgae could provide a possible solution not only to enhance agriculture production but also to diminish environmental pollution (Bhalamurugan et al. 2018).

This growing interest in the application of natural products from microalgae to stimulate plant growth, increasing productivity without causing further environmental degradation (e.g., eutrophication, soil infertility, and biodiversity loss) has also been highlighted by Garcia-Gonzalez and Sommerfeld (2016) and Calvo et al. (2014). These natural products are often called "biostimulants." They can promote plant growth even when applied in small quantities, boosting mineral nutrient uptake and expanding the tolerance of plants to abiotic stresses, thus being an alternative to synthetic plant protection products (Chojnacka et al. 2014; Khan et al. 2009). In addition, biostimulants guarantee the growth of food crops, including vegetables and fruit-bearing plants. These natural products provide improved crop quality with full respect for human health and the environment (du Jardin 2015; Povero et al. 2016) with emphasis on its application in sustainable agriculture (Khan et al. 2009). Among biostimulants, a special attention has been given to microalgal extracts (Michalak et al. 2016, 2017) or algal biomass. Both are also known to protect plants from diseases, insects, and abiotic stress (e.g., high salinity, drought, and frost; Khan et al. 2019). Thus, they are alternatives to chemical pesticides. The microalgae-based biostimulants have high efficiency, biodegradability, no phytotoxicity, making them promising products for plant protection (Kim and Chojnacka 2015).

Khan et al. (2019) purported that dried microalgal biomass can substitute commercial chemical fertilizers. Applying dried microalgae biomass to soils might not result in NH_3 volatilization as in the case of manure (Thompson and Meisinger 2002). Moreover, algal nitrogen would present a lower potential for leaching or run-off losses since only about 5% of the algal N would be available as a mineral at the time of application (Mulbry et al. 2005). In fact, according to Edwards (2016), algae-based biofertilizers can improve yields by over 30%, increasing germination rates and speed to maturity by 10–30% and improving soil porosity by 500%.

As the global agricultural market is huge, with a current trade value of USD 1.6 trillion (FAO 2018), microalgae-based biofertilizers/biostimulants show great potential to grow in scale and importance, leading to higher crop productivity and contributing to a more positive impact of agriculture on the environment. This might be particularly true, because microalgae have a great capacity to recycle nutrients from effluents, thus lowering the costs of biomass production. In addition, their potential to increase germination and growth of different seeds and plants make microalgae an important and appealing renewable feedstock for the formulation of innovative biofertilizers and biostimulants in the near future.

3.3 Production of Microalgae for Biofertilizer Production

3.3.1 Microalgae as Feedstocks for Biofertilizers

Microalgae contain valuable compounds useful as plant growth promoters, mainly amino acids and phytohormones, in addition to mineral nutrients such as N as NH_4^+, P as PO_4^{3-}, and organic compounds, K, Mg, and Fe. As proteins are a major component of the microalgal biomass, protein hydrolysis to amino acids might be a major contributor to a biofertilizer efficiency. Conversely, phytohormones (e.g., IAA and salicylic acid, among others) are mainly biostimulants that can act at very low concentrations. For example, eukaryotic microalgae such as *Nannochloropsis*, *Chlorella*, and *Scenedesmus* can produce phytohormones, including IAA, cytokinins, gibberellins, abscisic (ABA), and/or jasmonic acids as well as other molecules with biostimulant activity (Lu and Xu 2015; Table 3.1). Similar results have been described for cyanobacteria like *Arthrospira* (*Spirulina*), *Oscillatoria*, *Chlorogloea*, *Arthronema*, or *Calothrix*, which are also able to biosynthesize IAA, cytokinins, and/or jasmonic acid (Tarakhovskaya et al. 2007; Table 3.1).

Moreover, both cyanobacteria and eukaryotic algae, including macro- and microalgae, are known to be rich sources of bioactive metabolites. Marine algae have

Table 3.1 Biostimulants content of some freshwater and marine microalgae (ng/g) (this study)

	"Phylum"	*Cyanobacteria*	*Chlorophyta*			*Stramenopiles*
Phytohormone class	Species/Molecule	*Arthrospira platensis*	*Tetraselmis suecica*	*Chlorella vulgaris*	*Scenedesmus almeriensis*	*Nannochloropsis gaditana*
Ethylene precursor	ACC[a]	545.55	4.52	11.19	341.42	0.00
Cytokinins	Trans-zeatin	191.37	229.97	813.66	676.30	153.90
	Zeatin riboside	0.00	0.00	0.00	6.82	0.00
	Isopentenyl adenine	2560.73	1647.41	3816.28	**45,561.97**	**28,764.35**
Gibberellins	GA_1[b]	24.97	25.79	7.29	**208.81**	4.96
	GA_3[b]	0.00	0.00	0.02	0.00	0.01
	GA_4[b]	31.27	27.24	26.63	9.09	1.96
Auxin	IAA[c]	10.31	7.07	2.53	**5965.01**	0.89
Others	Abscisic acid	1.03	2.72	2.18	**3718.25**	3.66
	Salicylic acid	0.61	4.76	408.86	**156,713.72**	69.45
	Jasmonic acid	0.84	0.50	0.00	**75.13**	0.26

[a]1-aminocyclopropane-1-carboxylic acid
[b]Gibberellic acid
[c]Indole-3-acetic acid

been shown to have bactericidal (Goud et al. 2007; Srinivasakumar and Rajashekhar 2009), antifungal (Najdenski et al. 2013; Uyisenga et al. 2010), and insecticidal activities (Hernández-Carlos and Gamboa-Angulo 2011; Tringali 1997). In addition, the chlorophytes *Chlorella* and *Scenedesmus* are able to produce antibiotics and siderophores, whereas cyanobacteria belonging to the genera *Nostoc*, *Anabaena*, *Oscillatoria*, and *Tolypothrix* are very effective "biopesticide" agents.

3.3.2 Growth of Microalgae on Wastewaters for Nutrient Recycling

One of the major advantages of microalgae as a source of biofertilizers/biostimula nts is their capacity to be produced on wastewaters such as sewage, manure, or industrial effluents (Acién et al. 2016). In this case, the sustainability of the entire process is largely enhanced, because microalgae are able to recover nutrients, such as N and P, from these wastewaters, avoiding the need to be cultivated using primary resources (atmospheric N_2 and mined phosphorus rock), which require large amounts of energy to be fixated and extracted, respectively. As an example, AQUALIA is the second largest company in Europe in wastewater treatment, each year treating up to 500 Mm^3 of wastewater containing 60 mg/L of NH_4^+-N and 10 mg/L of PO_4^{3-}-P by conventional methods. If microalgae-based methods were used to treat this volume of wastewaters, up to 0.5 Mt per year of microalgae biomass could be produced, recovering up to 25,000 t/year of N and 5000 t/year of P, thus avoiding the need for producing this quite significant amount of fertilizers.

In addition to urban wastewater, there are several examples of the utilization of microalgae for nutrient recovery from centrate (Posadas et al. 2013), manure (García et al. 2017; Godos et al. 2009), agroindustrial (Posadas et al. 2014), and industrial (Muñoz et al. 2009) effluents. Whatever the effluent, to reduce the cost of treatment and to be able to perform these processes at a large scale, the utilization of open reactors is mandatory. Thus, open raceways are the most used containers for microalgal cultivation, although thin-layer reactors have also been proposed (del Mar Morales-Amaral et al. 2015; Posadas et al. 2015).

The utilization of microalgae for nutrient recycling has been previously demonstrated in indifferent scenarios, from laboratory to large-scale units (Craggs et al. 2012). Whatever the scenario, this process must consider that microalgae are not alone, but they form consortia with bacteria. The composition of these consortia depends largely on the effluent composition, location, and environmental conditions (Ferrero et al. 2012; García et al. 2017). Moreover, the predominant microalgal strains in these processes also vary significantly, with the development of pure cultures almost impossible (Olguín 2012). In any case, by producing microalgae from wastewaters, several benefits can be obtained: from the reduction of energy consumption and cost of conventional wastewater treatment systems to the improvement of the sustainability of microalgal biomass production, which becomes almost "cost-free." This last advantage opens a new scenario for microalgae to be used in different applications, especially in the production of biofertilizers/biostimulants. Thus, in a scenario in which clean water, fertilizers, and pure CO_2 were used to cultivate microalgae in raceways, the minimum biomass production cost would be around 5–10 €/kg; however, if wastewater and flue gases replaced the aforementioned nutrients, the production cost could be lower than 0.5 €/kg, without even considering the benefit of generating treated wastewater, which could cover the production costs of the biomass (Acién et al. 2016).

Yet, major drawbacks need to be solved before microalgal biomass production coupled with wastewater treatment can become a reality. One of the major problems is related with the composition of the produced biomass and existing regulations for products from wastewaters. For example, it is possible for microalgae to accumulate heavy metals and emerging pollutants present in the wastewater, becoming contaminated by these compounds. However, there is no evidence of significant contamination of the microalgal biomass when produced on wastewaters, mainly because the heavy metal content of wastewater is typically low and the emerging pollutants are largely removed by the high dissolved oxygen concentration and UV radiation (Matamoros et al. 2015). Concerning regulatory measures, the reuse of materials from wastewater treatment is strongly limited to avoid public health problems, so its direct use for human consumption is totally forbidden. However, water and sludge can be used for agriculture and industrial uses if regulatory guidelines are met. In the case of microalgae, new guidelines must be developed to allow the industrial development of biofertilizers/biostimulants from microalgae, but without compromising any safety criteria.

3.3.3 Large-Scale Production of Microalgae for Biofertilizer Production

The biofertilizers market is a focus in the agriculture industry for its multi-functional benefits and functional properties. The global biofertilizers market is expected to reach USD 1.88 billion by 2020 at a compound annual growth rate of 14.0% from 2015 to 2020. The microalgae/cyanobacteria segment is projected to be the largest market for biofertilizers by 2020 in terms of microorganisms. To be able to produce the huge amount of microalgal biomass required by this market, large-scale reactors/facilities must be developed. Thus, it is expected that the production of microalgae for agricultural use will become one of the most important applications of microalgal biomass in the short term.

Large-scale production of microalgae requires the utilization of open reactors such as raceways, which is currently the technology most often used. Indeed, more than 90% of microalgal biomass commercialized worldwide is produced using this technology (Benemann 2013). Such wide adoption is due to its relatively low cost and scale-up ease; however, the overall performance of this type of reactors is low.

Major drawbacks of this technology are related with low light availability inside the reactors due to the high-water depth, inadequate light integration efficiency, and poor control of culture parameters such as temperature, pH, and dissolved oxygen. However, one of the most important problems are linked to the occurrence of contaminants, which limits the utilization of these reactors to the production of extremophile microorganisms such as *Arthrospira* (*Spirulina*) and *Dunaliella*, and fast-growing species like *Chlorella*. Design and operation of raceways have been revised over the last few years, with relevant improvements being proposed, which are mainly related with the reduction of energy consumption (Low Energy Algae Reactor, patented by Aqualia), optimization of oxygen desorption (Barceló-Villalobos et al. 2019), improvement of CO_2 capture efficiency (Godos et al. 2014), and overall control strategies (Pawlowski et al. 2017), among others.

Despite these improvements, microalgal biomass productivity in raceway reactors is always lower than 50 tons/ha·year, whatever the strain or location of the reactor. The reason for that is because in these reactors, light utilization efficiency can be lower than 1% (Tredici 2010). The high-water depth and low vertical liquid velocity in these reactors lead to infrequent exposure of cells to light. As a result, the cultures become adapted to local irradiance, preventing light integration from taking place (Barceló-Villalobos et al. 2019). As the enhancement of liquid velocity is not feasible due to the high energy consumption needed, decreasing the water depth is the only viable solution to solve the aforementioned problem. Thus, the concept of a thin-layer reactor has been suggested (Masojídek et al. 2015). In these reactors, light utilization efficiencies can be increased up to 4%, resulting in biomass productivities higher than 50 g/m^2·day or 90 tons/ha·year (Masojídek et al. 2015). Even though this technology appears to be suitable to treat wastewater using microalgae (del Mar Morales-Amaral et al. 2015), it must be further optimized and scaled up for producing biofertilizers at industrial scale.

3.4 Microalgal Biofertilizers

3.4.1 Biofertilizer Production

Different technologies are available for preparing biofertilizers from algae, namely water extracts, direct application of algae to the soil (either fresh or dried), or after composting or biodigestion.

Water extracts and direct soil application allow the beneficial effects of the plant growth promoters, mainly amino acids and phytohormones, present in algae to exert their effect on the seed and/or plant. Composting and biodigestion will probably destroy many of those products present in algae, although preserving their mineral content.

Composting is one of the best technical and economical solutions to dispose of and process algal biomass into biofertilizers (Mazé et al. 1993), mainly macroalgae.

However, no clear guidelines are yet available for the full-scale composting of algal biomass (Han et al. 2014).

Composting is a simple biological aerobic process with different technical options to obtain materials—composts—with large and biodiverse microbiological populations. For centuries, composts have been used mainly as a source of good quality organic matter for soil fertilization. In the first half of the twentieth century, it was found that some composts from organic residues not only could increase plant growth, but also protect plants against pests and diseases (Howard 1945). The capacity of controlling or reducing soil diseases—suppressive capacity or suppressivity—was related to chemical and biological properties of the composts, namely the occurrence of some fungal and bacterial species (antagonists). Some of these organisms also showed the capacity to induce systemic acquired resistance in plants (Hoitink et al. 1977; Maurhofer et al. 1994), and to establish symbiosis with the plants through the exchange of nutrients from the soil with organic compost supplied by the plants (Coyne 1999).

Composting technologies can be divided into three main groups: windrow, aerated static pile, and in-vessel composting, with a wide variety of technical solutions to control aeration, moisture, and the emissions of gases (Fig. 3.1).

Windrow is the most common composting method due to its simple operation and lower cost, but an aerated static pile provides better control of the process if the aeration is automatically controlled. Sometimes these methods are used together: after an initial composting phase in an aerated static pile, the process is finalized in

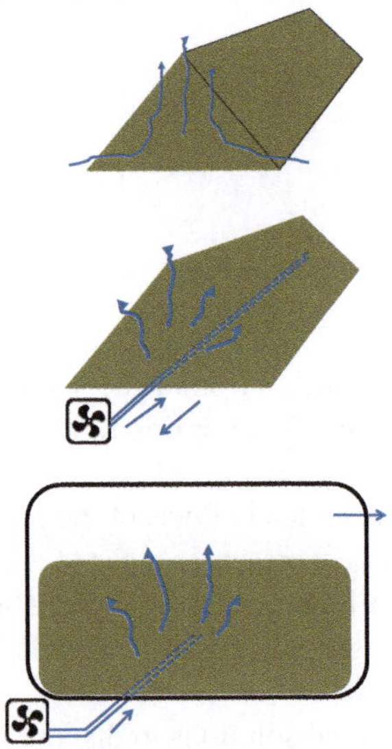

Designation	Protection	Aeration	Odor control	Leachate control
Windrow	Open air or under a shelter	Mechanical turning	No	Possible
Aerated static pile	Open air or under a shelter	Forced ventilation (air suction and/or blowing)	Possible (if air suction)	Possible
In vessel	Closed space	Forced ventilation	Possible	Yes

Fig. 3.1 Basic composting systems and respective operation

a windrow. The higher cost of in-vessel composting decreases its application in most situations, however, is the most appropriate method for bench- or pilot-scale research on composting (Mason and Milke 2005). Composting converts an initial mixture of organic materials into a lower volume of easily handled and stable product, which can be characterized according to its maturity and stability, two very important criteria regarding the utilization of composts. Maturity describes the fitness of the compost for a particular use, while stability refers to the resistance of the compost's organic matter to decomposition (Sullivan and Miller 2001). During composting, easily degradable fractions of organic matter are decomposed. The more resistant fractions of organic matter will remain, however, which increases the number of possible end-uses of the compost (maturity). According to an EU proposal (Procedure 2016/0084/COD 2016), compost must meet at least one of the following stability criteria: a maximum oxygen uptake rate of 25 mmol O_2 kg organic matter^{-1} h^{-1} (method not suitable for material with more than 20% of particle sizes > 10 mm) or a minimum self-heating factor Rottegrad III, which corresponds to a temperature rise by 40–50 °C above ambient temperature in a Dewar's test. A compost with the latter stability level is considered to be a "fresh compost", that is, a half-matured, thermally sanitized compost, showing a relatively high biological activity, which can be used in less sensitive applications such as arable land (Barth et al. 2008).

In addition to the selection of the correct technology, composting of organic materials, namely agricultural residues, require adequate physical and chemical conditions. Algae are also a suitable feedstock for the preparation of composts, provided that they are mixed with the proper complementary materials—co-composting.

The main restrictions of algae as feedstock for composting are high moisture content, low porosity, a relatively low C:N ratio, potential high salinity, as well as high metal and toxin contents (Eyras et al. 1998; Han et al. 2014).

The most important conditions required for composting are temperature, oxygen, water, and nutrients for microorganisms (total amounts and their ratio), and the presence of a strong and biodiverse microbiological population (Golueke 1991).

Typically, temperature increases sharply during the first days of the process up to a maximum, decreasing more or less slowly afterwards, as easily degradable organic matter is consumed. Temperature rise depends on microbial metabolic activity, availability of degradable organic matter, environmental conditions (e.g., O_2, moisture, pH, and C:N ratio), and the rate of heat transfer from the composting material. During composting, temperature usually rises above 45 ° C, entering a thermophilic phase, and in some cases, temperature may reach 75–80 °C (Finstein and Hogan 1993). This thermophilic phase is very important, because high temperature increases the degradation of the most resistant organic matter fractions, eliminate some important human pathogens (e.g., *Salmonella* and *Escherichia coli*) and most plant pathogenic agents (de Bertoldi et al. 1991).

The temperature profile during composting is a suitable control indicator of the process, as it reflects the microbial activity that is influenced by the environmental conditions of the whole process. Oxygen supply to microorganisms can be promoted by choosing an adequate particle size of the raw materials, typically from 3 to 50 mm in diameter for common residues (Biddlestone and Gray 1985; Haug 1993) to reach an air-filled porosity (AFP) above 30% (Haug 1993), ideally 32–36% (Epstein 1997). Nevertheless, the regular turning of the material or forced ventilation is required to keep a continuous and active composting. Aeration of the composting mix not only provides oxygen but is also used to control the temperature of the process by removing excess heat. Temperatures up to 45 °C create conditions for maximum microbial biodiversity and activity. However, temperatures between 55 and 60 °C for a minimum time are required to eliminate most plant pathogens and increase the degradation of the most resistant organic matter fractions (Table 3.2).

More recently, an EU proposal (Procedure 2016/0084/COD 2016) indicates that, during the composting process, all parts of each batch should have one of the following temperature-time profiles:

Table 3.2 Requirements of temperature and time for compost sanitization per country (Hogg et al. 2002)

Country	Minimum temperature (°C)	Time
Australia	55[a]	3 days
Belgium	60	4 days
Canada (CCME)	55	3 days[b], 15 days[c], 3 days[d]
France	60	4 days
Denmark	55	14 days
Germany		
Ordinance for compost:	55	14 days
In-vessel	60	7 days
Not-in-vessel	65	7 days
Ordinance for digestion:	55	1 day (dwelling time: 20 days)
	70[e]	1 h
Italy	55[a]	3 days
Netherlands, The	55	4 days
New Zealand	55[a]	3 days
United Kingdom[f]	55	3 days[g], 15 days[h]
United Kingdom[i]	60[j]	Not dependent on temperature
United States	55	5 days[b], 15 days[c]

[a]Requirement for three turns of windrow with internal temperature reaching 55 °C for 3 days before each turn
[b]In-vessel
[c]In a windrow
[d]Aerated static pile
[e]Higher temperature as pre- or post-treatment
[f]Depends on the compost/digestion plant and risk potential of the material; UK Composting Association
[g]Outdoor aerated static pile or in-vessel
[h]Outdoor turned windrows (mixed 5 times)
[i]UK Soil Association
[j]Target temperature, although not required

- 65 °C or more for at least 5 days
- 60 °C or more for at least 7 days, or
- 55 °C or more for at least 14 days

Above 65 °C, higher pathogen control could be achieved, but microbial activity is strongly decreased (Golueke 1991), almost ceasing at 70 °C (Waksman et al. 1938). An AFP above 70% seems excessive to achieve thermophilic conditions in wastes with low biodegradable organic matter content (Han et al. 2014) due to excessive heat loss. Interestingly, de Guardia et al. (2010a) reported that composting of microalgal biomass is more sensitive to aeration than other organic residues (e.g., food and household wastes, pig slurry, and slaughterhouse sludge). Thus, this fact should be considered when using aeration to control the temperature when microalgae-based feedstocks are used to produce compost.

Moisture content is another essential factor for proper composting. A minimum moisture content of 12% (w/w) is required for microbial activity. In practice, composting will be severely reduced at a moisture content below 30% (w/w; Gray et al. 1971), while high moisture content may reduce AFP (Haug 1993), decreasing airflow, creating anaerobic conditions, leading to odor formation, and lower the degradation rate of the materials. Recommended values lie between 40 and 65% (w/w), but these will depend on the type of residue, particle size, water retention capacity, and the composting method (Gray et al. 1973). Coarser mixes can be kept at higher moisture contents, while fine mixes must be kept at lower moisture values to avoid decreased air circulation. Moisture must be regularly checked to keep it at the highest possible level according to particle size. This control must include all the profile of the mix, especially in composting piles, since the center of the piles may lose too much water, while the outside layers might keep adequate moisture level due to internal water vapor condensation. Since water content in wet microalgal biomass can be higher than 90%, it is preferable to include very dry materials to adjust the moisture in the initial mix at around 50% on a wet basis (de Guardia et al. 2010b).

Mixtures of materials usually do not show nutrient deficiency for composting. The most important situation regarding nutrient composition is the C:N ratio, which must be between around 25 to 30 (Haug 1993). N losses (via NH_3 volatilization) or a slower composting process (due to N deficiency) might take place if the C:N ratio is below or above the recommended interval, respectively. Since the C:N ratios in algae are usually low (Han et al. 2014), algal biomass should be mixed with materials with sufficient content of readily biodegradable carbon to achieve optimal ratios. Lignocellulosic residues such as wood chips, shredded tree barks, dry leaves, straw, sawdust, or grape marc might be used (Eyras et al. 2008), improving also the aeration of the mix (bulk agent). The most degradation-resistant fibers during composting (e.g., cellulose and hemicellulose) will contribute to create a coarser structure of the compost.

Salinity is another important factor concerning using algae as feedstocks for compost to be applied to soil (Table 3.3), particularly if the biomass comes from marine algae. An electrical conductivity (EC) in the mix above 8 dS m^{-1} has been shown to have a negative impact on microorganisms during composting (Santamaría-Romero and Ferrera-Cerrato 2001). Another crucial issue regarding salinity may be

Table 3.3 Maximum application rates (L m^{-2}) of compost according to its electrical conductivity (EC) for plants with different sensitivity to salinity

EC range (dS m^{-1})	Sensitive plants	Tolerant plants
0–1	Unlimited	Unlimited
1–2	<15	<60
2–4	<8	<32
4–8	<4	<16
8–12	<2.5	<10
>12	<2	<8

The rates are for compost mulches or soil incorporation up to 5 cm depth; if incorporated into soil to a depth > 10 cm these rates may be doubled (Hogg et al. 2002)

the lower soil disease suppressed observed in composts from agroindustrial residues with an EC above 3 dS m^{-1} (Suárez-Estrella et al. 2012).

Another potential issue might be the occurrence of high metals and toxins in algae. Heavy metal concentration will depend on the algal feedstock, a factor that should be controlled during the cultivation step. Moreover, if need be, co-composting with other organic residues can lead to lower heavy metal concentration in the final compost (Han et al. 2014). It is also known that compost organic matter can interact with heavy metals, reducing its absorption by plants. This statement suggests that such composts can be used as plant-growing media without affecting horticultural products (Castaldi and Melis 2004). However, in other cases, marine algae-based composts might be recommended for plant cultivation in soil, but not as plant growing media in soilless applications (Greger et al. 2007).

Microalgae co-composting with agroindustrial residues allows for nutrient recycling and organic matter for agriculture, reducing chemical fertilizer needs, and contributing to the circular economy. A mix of raw materials will promote biodiverse populations of microorganisms, which, if need be, can also be inoculated during the process after the thermophilic phase.

Residues from biological energy recovery processes, such as anaerobic digestion, can be incorporated into the composting mixes and transformed into soil fertilizers (Procedure 2016/0084/COD 2016).

3.4.2 *Application of Biofertilizers to Soil*

As they are composed of living organisms, biofertilizers are more sensitive to storage conditions as compared to most chemical fertilizers. It is necessary to evaluate and determine the best storage conditions to increase the shelf life of the biofertilizer. For this purpose, simple methods, before field application, are useful to ensure biofertilizer activity.

Biofertilizers may be applied in different ways, according to their physical state. Microorganisms commercialized as a powder or liquid formulations are easily applied by mixing them in the soil, substrate, or irrigation water.

When the biofertilizers are naturally or artificially present in composts (via inoculation), their application can be achieved by adding the compost to the soil, followed by soil mobilization or as mulch, or even as compost water extracts that can be added to the irrigation water. Compost water extracts are usually prepared under aerobic conditions, possibly removing a large number and variety of beneficial microorganisms and organic compounds such as vitamins, enzymes, and organic chelators (Merill et al. 1998).

3.4.3 *Effect of Microalgae-Based Biofertilizers on Plant Growth*

Examples of commercially available algae-based products for agriculture are presented in the work of Michalak and Chojnacka (2016). Algal extracts were used to stimulate not only plant growth, but also plant mass, chlorophyll and carotenoids and improve root development.

Field trials carried out in 2013 on winter wheat (variety Tacitus) showed that the supercritical *Spirulina* sp. extract (applied in two doses—1.20 L/ha and 1.8 L/ha) slightly increased the crop yield per hectare in comparison with the control group—untreated (4.0%) and the reference product, Asahi SL, with an applied dose of 0.60 L/ha (2.0%). Dose choice was dependent on the content of polyphenols in the applied products. A similar effect was observed with *Spirulina* sp. using doses of 1.0, 1.5, and 1.8 L/ha in the next growing season (i.e., 2015). Among the tested supercritical extracts, the best results in terms of grain yield, number of grains in ear were obtained with *Spirulina* (1.5 L/ha; Michalak et al. 2016).

Other examples have been carried out by Garcia-Gonzalez and Sommerfeld (2016) using *Acutodesmus dimorphus* to promote the growth and production of flowers in Roma tomato plants, with good results. Faheed and Fattah (2008) studied the growth parameters in the lettuce (*Lactuca sativa*) plant using *Chlorella vulgaris*, which significantly enhanced the process of germination. It also increased significantly the amount of pigments (chlorophyll *a*, chlorophyll *b* and carotenoids) in the lettuce seedlings. Agwa et al. (2017) also reported positive results in the development of *Hibiscus esculentus* and enhanced soil fertility upon using *C. vulgaris* as biofertilizer.

Wastewater has increasingly been used to grow microalgae at a lower cost and make the bioavailability of nutrients, and later for enhancing the growth of crops through controlled release of nutrients. A treatment study done by Renuka et al. (2016) compared two formulations (unicellular and filamentous microalgae grown in wastewater) and applied them as biofertilizers to a wheat crop. Interestingly, the best efficacy was achieved with the filamentous one. Dineshkumar et al. (2018)

evaluated the impact of *C. vulgaris* and *Arthrospira (Spirulina) platensis*, both grown in cow manure, on maize crops with various treatments. The authors concluded that maize increased by 51.1% following 60 days after planting in terms of yield and growth.

Mógor et al. (2018) investigated the biostimulant properties of hydrolyzed biomass of *A. platensis*. Bioassays were performed to determine auxin-like and cytokinin-like bioactivity of the hydrolysates on lettuce seedling growth. The hydrolysates had a cytokinin-like effect in the bioassay. However, neither *Scenedesmus almeriensis* or *S. obliquus* showed a cytokinin-like effect, but rather a very positive auxin-like effect on mung bean and cucumber seeds. This is apparently the case in other instances, because microalgal strains displaying an auxin-like effect usually fail to present a significant cytokinin-like activity (Stirk et al. 2013).

Besides these reports in the literature, the authors of the present chapter tested the capacity of different microalgae biomass or extracts as a biostimulant of watercress seeds (*Lepidium sativum*). The Germination Index (GI) was determined as follows:

$$GI\left(\%\right)=\frac{G\times L}{Gw\times Lw}\times 100 \tag{3.1}$$

where *G* and *L* are the number of germinated seeds and their length treated with microalgal extracts and Gw and Lw are the same parameters for the control (distilled water). The values with an increased GI as compared to the control are highlighted on Table 3.4. *Scenedesmus* sp. grown in piggery effluent without pre-treatment showed a GI for watercress seeds of 108 and 95% at microalgal biomass concentrations of 0.5 and 0.1 g/L, respectively. With cell disruption, these values increased to 119 and 109%, for the same concentrations, respectively. *Scenedesmus obliquus* grown in brewery effluent, besides its effect as a biostimulant on watercress seeds (*Lepidium sativum*) (Table 3.4), proved to be effective on mung bean (*Vigna radiata*) and cucumber (*Cucumis sativus* L.; data not shown). The GI for watercress seeds increased by 40% and 20%, compared to the control, when using 0.1 g/L and 0.5 g/L of microalgal biomass, respectively. The auxin-like effect on mung bean root induction increased by around 25% as compared to the control. The best performance of the *Scenedesmus* extracts was achieved without enzymatic hydrolysis or cell disruption and at the lowest concentration (0.1 g/L; data not shown). The auxin-like effect on cucumber root development showed an improvement of around 23% compared to the control (data not shown). For *S. obliquus* biomass grown in brewery wastewater, the need for a minimal pre-treatment was confirmed, as the best results were obtained without cell disruption or enzymatic hydrolysis at the lowest microalgal biomass concentration (0.1 g/L).

Table 3.4 Germination index of watercress (*Lepidium sativum* L.) seeds for different microalgae strains (this study)

	Without cell disruption				With Cell disruption			
Microalgae	Centrifugation		Without Centrifugation		Centrifugation		Without Centrifugation	
Biomass (g/L)	0.5	0.1	0.5	0.1	0.5	0.1	0.5	0.1
Chlorella vulgaris	**110.9**	**106.0**	**109.8**	**105.7**	**129.7**	**123.9**	**134.3**	**131.7**
Nostoc carcicola	**127.9**	**124.6**	**115.4**	**129.6**	91.3	**136.4**	**100.6**	**121.5**
Scenedesmus sp.	99.1	99.5	**108.0**	95.2	**112.6**	95.2	**119.1**	**108.9**
Scenedesmus obliquus	118.9	**139.1**	108.2	118.8	100.2	121.8	113.0	109.2
Arthrospira platensis[a]	84.3	97.6	92.2	**106.9**	81.3	**106.1**	95.6	**106.3**

[a]Usually known as *Spirulina*

3.5 Conclusion

Microalgae-based biofertilizers hold great promise as an innovative way to recycle nutrients by promoting the production of large amounts of biomass that can be upgraded in value and used in several applications. These biofertilizers show a significant potential as biostimulants, due to their phytohormone contents and the occurrence of other plant-growth promoters, such as amino acids and vitamins. However, a major hurdle must yet be tackled, so that the industrial production of biofertilizers from microalgal biomass can become a reality: the consistent production of microalgal biomass in outdoor raceways using wastewaters, free from toxic contaminants, either from chemical or biological origin. Thus, further research is needed to solve this bottleneck and determine the optimal pipelines for the industrial production of microalgae-based biofertilizers/biostimulants shortly.

References

Abdel-Raouf N, Al-Homaidan AA, Ibraheem IBM (2012) Microalgae and wastewater treatment. Saudi J Biol Sci 19:257–275

Acién FG, Gómez-Serrano C, Morales-Amaral MM, Fernández-Sevilla JM, Molina-Grima E (2016) Wastewater treatment using microalgae: how realistic a contribution might it be to significant urban wastewater treatment? Appl Microbiol Biotechnol 100:9013–9022

Adesemoye AO, Kloepper JW (2009) Plant–microbes interactions in enhanced fertilizer-use efficiency. Appl Microbiol Biotechnol 85:1–12

Agwa OK, Ogugbue CJ, Williams EE (2017) Field evidence of *Chlorella vulgaris* potentials as a biofertilizer for *Hibiscus esculentus*. Int J Agric Res 12:181–189

Arioli T, Mattner SW, Winberg PC (2015) Applications of seaweed extracts in Australian agriculture: past, present and future. J Appl Phycol 27:2007–2015

Barceló-Villalobos M, Olmo PF, Guzmán JL, Fernández-Sevilla JM, Fernández FGA (2019) Evaluation of photosynthetic light integration by microalgae in a pilot-scale raceway reactor. Bioresour Technol 280:404–411

Barth J, Amlinger F, Favoino E, Siebert S, Kehres B, Gottschall R, Bieker M, Löbig A, Bidlingmaier W (2008) Compost production and use in the EU. Orbit ECN, Weimar

Benemann J (2013) Microalgae for biofuels and animal feeds. Energies 6:5869–5886

Bhalamurugan GL, Valerie O, Mark L (2018) Valuable bioproducts obtained from microalgal biomass and their commercial applications: a review. Environ Eng Res 23:229–241

Bhardwaj D, Ansari MW, Sahoo RK, Tuteja N (2014) Biofertilizers function as key player in sustainable agriculture by improving soil fertility, plant tolerance and crop productivity. Microb Cell Factories 13:66–76

Biddlestone AJ, Gray KR (1985) Composting. In: Moo-Young M, Robinson C-W, Howell JA (eds) Comprehensive biotechnology, vol 4. Pergamon Press, Oxford, pp 1059–1070

Calvo P, Nelson L, Kloepper JW (2014) Agricultural uses of plant biostimulants. Plant Soil 383:3–41

Castaldi P, Melis P (2004) Growth and yield characteristics and heavy metal content on tomatoes grown in different growing media. Commun Soil Sci Plant Anal 35:85–98

Chen J, Lü S, Zhang Z, Zhao X, Li X, Ning P, Liu M (2018) Environmentally friendly fertilizers: a review of materials used and their effects on the environment. Sci Total Environ 613–614:829–839

Chojnacka K, Michalak I, Dmytryk A, Wilk R, Gorecky H (2014) Innovative natural plant growth biostimulants. In: Sinha S, Pant KK, Govil JN (eds) Advances in fertilizer technology II. Studium Press, Houston, pp 451–489
Clark RB, Zeto SK (2000) Mineral acquisition by arbuscular mycorrhizal plants. J Plant Nutr 23:867–902
Coyne MS (1999) Soil microbiology: an exploratory approach. Delmar Publications, Albany
Craggs R, Sutherland D, Campbell H (2012) Hectare-scale demonstration of high rate algal ponds for enhanced wastewater treatment and biofuel production. J Appl Phycol 24:329–337
Daneshgar S, Callegari A, Capodaglio AG, Vaccari D (2018) The potential phosphorus crisis: resource conservation and possible escape technologies: a review. Resources 7:37. https://doi.org/10.3390/resources7020037
de Bertoldi M, Zucconi F, Civilini M (1991) Temperature, pathogen control and product quality. In: The Staff of BioCycle (ed) The biocycle guide to the art & science of composting. The JG Press, Inc., Emmaus, pp 195–198
de Guardia A, Mallard P, Teglia C, Marin A, Le Pape C, Launay M, Benoist JC, Petiot C (2010a) Comparison of five organic wastes regarding their behaviour during composting: part 1, biodegradability, stabilization kinetics and temperature rise. Waste Manag 30:402–414
de Guardia A, Mallard P, Teglia C, Marin A, Le Pape C, Launay M, Benoist JC, Petiot C (2010b) Comparison of five organic wastes regarding their behaviour during composting: part 2, nitrogen dynamic. Waste Manag 30:415–425
del Mar Morales-Amaral M, Gómez-Serrano C, Acién FG, Fernández-Sevilla JM, Molina-Grima E (2015) Outdoor production of *Scenedesmus* sp. in thin-layer and raceway reactors using centrate from anaerobic digestion as the sole nutrient source. Algal Res 12:99–108
Dineshkumar R, Kumaravel R, Gopalsamy J, Sikder MNA, Sampathkumar P (2018) Microalgae as bio-fertilizers for rice growth and seed yield productivity. Waste Biomass Valori 9:793–800
du Jardin P (2015) Plant biostimulants: definition, concept, main categories and regulation. Sci Hortic (Amsterdam) 196:3–14
Edwards M (2016) Algae secrets – algae food solutions to avoid pesticide exposure. AlgaeIndustryMagazine.com. http://www.algaeindustrymagazine.com/algae-food-solutions-avoid-pesticide-exposure/. Accessed 15 May 2019
Epstein E (1997) The science of composting. Technomic Publishing Inc., Lancaster
European Commission (2015) Closing the loop – an EU action plan for the Circular Economy. COM(2015) 614 final, EC, Brussels
Eyras MC, Rostagno CM, Defossé GE (1998) Biological evaluation of seaweed composting. Compost Sci Util 6:74–81
Eyras MC, Defossé GE, Dellatorre F (2008) Seaweed compost as an amendment for horticultural soils in Patagonia, Argentina. Compost Sci Util 16:119–124
Faheed FA, Fattah ZA-E (2008) Effect of *Chlorella vulgaris* as bio-fertilizer on growth parameters and metabolic aspects of lettuce plant. J Agric Soc Sci 4:165–169
FAO (2018) The state of agricultural commodity markets 2018: agricultural trade, climate change and food security. Food and Agriculture Organization of the United Nations, Rome
Ferrero EM, de Godos I, Rodríguez EM, García-Encina PA, Muñoz R, Bécares E (2012) Molecular characterization of bacterial communities in algal-bacterial photobioreactors treating piggery wastewaters. Ecol Eng 40:121–130
Finstein MS, Hogan JA (1993) Integration of composting process microbiology, facility structure and decision-making. In: Hoitink HAJ, Keener HM (eds) Science and engineering of composting: design, environmental, microbiological and utilization aspects. Renaissance Publications, Worthington, pp 1–23
Franche C, Lindström K, Elmerich C (2009) Nitrogen-fixing bacteria associated with leguminous and non-leguminous plants. Plant Soil 321:35–59
Frišták V, Pipíška M, Soja G (2018) Pyrolysis treatment of sewage sludge: a promising way to produce phosphorus fertilizer. J Clean Prod 172:1772–1778

Fulchieri M, Lucangeli C, Bottini R (1993) Inoculation with *Azospirillum lipoferum* affects growth and gibberellin status of corn seedling roots. Plant Cell Physiol 34:1305–1309

Gai JP, Christie P, Cai XB, Fan JQ, Zhang JL, Feng G, Li XL (2009) Occurrence and distribution of arbuscular mycorrhizal fungal species in three types of grassland community of the Tibetan Plateau. Ecol Res 24:1345–1350

García D, Posadas E, Grajeda C, Blanco S, Martínez-Páramo S, Acién G, García-Encina P, Bolado S, Muñoz R (2017) Comparative evaluation of piggery wastewater treatment in algal-bacterial photobioreactors under indoor and outdoor conditions. Bioresour Technol 245:483–490

Garcia-Gonzalez J, Sommerfeld M (2016) Biofertilizer and biostimulant properties of the microalga *Acutodesmus dimorphus*. J Appl Phycol 28:1051–1061

Geissler B, Hermann L, Mew MC, Steiner G (2018) Striving toward a circular economy for phosphorus: the role of phosphate rock mining. Fortschr Mineral 8:395. https://doi.org/10.3390/min80900395

Gholamhoseini M, Ghalavand A, Dolatabadian A, Jamshidi E, Khodaei-Joghan A (2013) Effects of arbuscular mycorrhizal inoculation on growth, yield, nutrient uptake and irrigation water productivity of sunflowers grown under drought stress. Agric Water Manag 117:106–114

Gholami A, Shahsavani S, Nezarat S (2009) The effect of plant growth promoting rhizobacteria (PGPR) on germination, seedling growth and yield of maize. Int J Agric Biosyst Eng 3:9–14

Godos I, Blanco S, García-Encina PA, Becares E, Muñoz R (2009) Long-term operation of high rate algal ponds for the bioremediation of piggery wastewaters at high loading rates. Bioresour Technol 100:4332–4339

Godos I, Mendoza JL, Acién FG, Molina E, Banks CJ, Heaven S, Rogalla F (2014) Evaluation of carbon dioxide mass transfer in raceway reactors for microalgae culture using flue gases. Bioresour Technol 153:307–314

Golueke CG (1991) Understanding the process. In: The Staff of BioCycle (ed) The biocycle guide to the art & science of composting. The JG Press, Inc., Emmaus, pp 14–27

Goud M, Seshikala D, Charya M (2007) Antibacterial activity and biomolecular composition of certain fresh water micro-algae from River Godavari (India). Sci World J 2:19–23

Gray KR, Sherman K, Biddlestone AJ (1971) A review of composting, Part 2 – the practical process. Process Biochem 6:22–28

Gray KR, Biddlestone AJ, Clark R (1973) A review of composting, Part 3 – processes and products. Process Biochem 8:11–15

Greger M, Malmb T, Kautsky L (2007) Heavy metal transfer from composted macroalgae to crops. Eur J Agron 26:257–265

Han W, Clarke W, Pratt S (2014) Composting of waste algae: a review. Waste Manag 34:1148–1155

Harrier L, Watson CA (2004) The potential role of arbuscular mycorrhizal (AM) fungi in the bioprotection of plants against soil-borne pathogens in organic and/or other sustainable farming systems. Pest Manag Sci 60:149–157

Haug RT (1993) The Practical handbook of compost engineering. Lewis Publishers, Boca Raton

Hernández-Carlos B, Gamboa-Angulo MM (2011) Metabolites from freshwater aquatic microalgae and fungi as potential natural pesticides. Phytochem Rev 10:261–286

Hogg D, Barth J, Favoino E, Centemero M, Caimi V, Amlinger F, Devliegher W, Brinton W, Antler S (2002) Comparison of compost standards within the EU, North America and Australasia – Main Report. The Waste and Resources Action Programme, Banbury, Oxon

Hoitink HAJ, Van Doren DRJ, Schmitthenner AF (1977) Suppression of *Phytophthora cinnamomi* in a composted hardwood bark potting medium. Phytopathology 64:561–565

Howard A (1945) Farming and gardening for health or disease. Faber & Faber, London

Ijdo M, Cranenbrouck S, Declerck S (2011) Methods for large-scale production of AM fungi: past, present, and future. Mycorrhiza 21:1–16

Kawalekar JS (2013) Role of biofertilizers and biopesticides for sustainable agriculture. J Bio Innov 2:73–78

Khan AA, Jilani G, Akhtar MS (2009) Phosphorus solubilising bacteria occurrence, mechanisms and their role in crop production. J Agric Biol Sci 1:48–58

Khan SA, Sharma GK, Malla FA, Kumar A, Gupta RN (2019) Microalgae based biofertilizers: a biorefinery approach to phycoremediate wastewater and harvest biodiesel and manure. J Clean Prod 211:1412–1419

Kim SK, Chojnacka K (2015) In: Se-Kwon K, Chojnacka K (eds) Marine algae extracts: processes, products, and applications. Wiley, New York

Kozdrój J, Trevors JT, van Elsas JD (2004) Influence of introduced potential biocontrol agents on maize seedling growth and bacterial community structure in the rhizosphere. Soil Biol Biochem 36:1775–1784

Krishnamoorthy R, Kim K, Kim C, Sa T (2014) Changes of arbuscular mycorrhizal traits and community structure with respect to soil salinity in a coastal reclamation land. Soil Biol Biochem 72:1–10

Lenka S, Rajendiran S, Coumar MV, Dotaniya ML, Saha JK (2016) Impact of fertilizers use on environmental quality. In: National seminar on Environmental concern for fertilizer use in future. Bidhan Chandra Krishi Viswavidyalaya, Kalyani

Lenoir I, Fontaine J, Sahraoui AL-H (2016) Arbuscular mycorrhizal fungal responses to abiotic stresses: a review. Phytochemistry 123:4–15

Lu Y, Xu J (2015) Phytohormones in microalgae: a new opportunity for microalgal biotechnology? Trends Plant Sci 20:273–282

Mahanty T, Bhattacharjee S, Goswami M, Bhattacharyya P, Das B, Ghosh A, Tribedi P (2017) Biofertilizers: a potential approach for sustainable agriculture development. Environ Sci Pollut Res 24:3315–3335

Mahdi SS, Hassan GI, Samoon SA, Rather HA, Dar SA, Zehra B (2010) Bio-fertilizers in organic agriculture. J Phytol 2:42–54

Masojídek J, Malapascua JR, Kopecký J, Sergejevová M (2015) Thin-layer systems for mass cultivation of microalgae: flat panels and sloping cascades. In: Prokop A, Bajpai R, Zappi M (eds) Algal Biorefineries. Springer, Cham, pp 237–261

Mason IG, Milke MW (2005) Physical modelling of the composting environment: a review. Part 1: reactor systems. Waste Manag 25:481–500

Matamoros V, Gutiérrez R, Ferrer I, García J, Bayona JM (2015) Capability of microalgae-based wastewater treatment systems to remove emerging organic contaminants: a pilot-scale study. J Hazard Mater 288:34–42

Maurhofer M, Hase C, Meuwly P, Metraux JP, Defago G (1994) Induction of systemic resistance of tobacco necrosis virus by the root-colonizing *Pseudomonas fluorescens* strain Chao – influence of the *Gaca* gene and of pyoverdine production. Phytopathology 84:139–146

Mazé J, Morand P, Potoky P (1993) Stabilisation of "green tides" Ulva by a method of composting with a view to pollution limitation. J Appl Phycol 5:183–190

Merill R, Hoberecht K, McKeon J (1998) Organic teas from compost and manures. Organic Farming Research Foundation, Santa Cruz

Michalak I, Chojnacka K (2016) The potential usefulness of a new generation of agro-products based on raw materials of biological origin. Acta Sci Pol Hortorum Cultus 15:97–120

Michalak I, Chojnacka K, Dmytryk A, Wilk R, Gramza M, Rój E (2016) Evaluation of supercritical extracts of algae as biostimulants of plant growth in field trials. Front Plant Sci 7:1591. https://doi.org/10.3389/fpls.2016.01591

Michalak I, Chojnacka K, Saeid A (2017) Plant growth biostimulants, dietary feed supplements and cosmetics formulated with supercritical CO_2 algal extracts. Molecules 22:66. https://doi.org/10.3390/molecules22010066

Mógor ÁF, Ördög V, Lima GPP, Molnár Z, Mógor G (2018) Biostimulant properties of cyanobacterial hydrolysate related to polyamines. J Appl Phycol 30:453–460

Mulbry W, Westhead EK, Pizarro C, Sikora L (2005) Recycling of manure nutrients: use of algal biomass from dairy manure treatment as a slow release fertilizer. Bioresour Technol 96:451–458

Muñoz R, Köllner C, Guieysse B (2009) Biofilm photobioreactors for the treatment of industrial wastewaters. J Hazard Mater 161:29–34

Najdenski HM, Gigova LG, Iliev II, Pilarski PS, Lukavský J, Tsvetkova IV, Ninova MS, Kussovski VK (2013) Antibacterial and antifungal activities of selected microalgae and cyanobacteria. Int J Food Sci Technol 48:1533–1540

Nardi P, Neri U, Di Matteo G, Trinchera A, Napoli R, Farina R, Subbarao GV, Benedetti A (2018) Nitrogen release from slow-release fertilizers in soils with different microbial activities. Pedosphere 28:332–340

Ni B, Liu M, Lü S, Xie L, Wang Y (2011) Environmentally friendly slow-release nitrogen fertilizer. J Agric Food Chem 59:10169–10175

Olguín EJ (2012) Dual purpose microalgae-bacteria-based systems that treat wastewater and produce biodiesel and chemical products within a biorefinery. Biotechnol Adv 30:1031–1046

Öpik M, Zobel M, Cantero JJ, Davison J, Facelli JM, Hiiesalu I, Jairus T, Kalwij JM, Koorem K, Leal ME, Liira J, Metsis M, Neshataeva V, Paal J, Phosri C, Pölme S, Reier Ü, Saks Ü, Schimann H, Thiéry O, Vasar M, Moora M (2013) Global sampling of plant roots expands the described molecular diversity of arbuscular mycorrhizal fungi. Mycorrhiza 23:411–430

Pawlowski A, Guzmán JL, Berenguel M, Acién FG, Dormido S (2017) Event-based control systems for microalgae culture in industrial reactors. In: Tripathi B, Kumar D (eds) Prospects and challenges in algal biotechnology. Springer, Singapore

Pfromm PH (2017) Towards sustainable agriculture: fossil-free ammonia. J Renew Sustain Energy 9:034702. https://doi.org/10.1063/1.4985090

Posadas E, García-Encina P-AA, Soltau A, Domínguez A, Díaz I, Muñoz R (2013) Carbon and nutrient removal from centrates and domestic wastewater using algal-bacterial biofilm bioreactors. Bioresour Technol 139:50–58

Posadas E, Bochon S, Coca M, García-González MC, García-Encina PA, Muñoz R (2014) Microalgae-based agro-industrial wastewater treatment: a preliminary screening of biodegradability. J Appl Phycol 26:2335–2345

Posadas E, Morales MDM, Gomez C, Acién FG, Muñoz R (2015) Influence of pH and CO_2 source on the performance of microalgae-based secondary domestic wastewater treatment in outdoors pilot raceways. Chem Eng J 265:239–248

Povero G, Mejia JF, Di Tommaso D, Piaggesi A, Warrior P (2016) A systematic approach to discover and characterize natural plant biostimulants. Front Plant Sci 7:435. https://doi.org/10.3389/fpls.2016.00435

Procedure 2016/0084/COD.2016.COM (2016) 157: proposal for a REGULATION OF THE EUROPEAN PARLIAMENT AND OF THE COUNCIL laying down rules on the making available on the market of CE marked fertilising products and amending Regulations (EC) n° 1069/2009 and (EC) n° 1107/2009. https://eur-lex.europa.eu/procedure/EN/2016_84. Accessed 14 May 2019

Rahmati O, Samani AN, Mahmoodi N, Mahdavi M (2015) Assessment of the contribution of n-fertilizers to nitrate pollution of groundwater in Western Iran (Case Study: Ghorveh–Dehgelan Aquifer). Water Qual Exposure Health 7:143–151

Renuka N, Prasanna R, Sood A, Ahluwalia AS, Bansal R, Babu S, Singh R, Shivay YS, Nain L (2016) Exploring the efficacy of wastewater-grown microalgal biomass as a biofertilizer for wheat. Environ Sci Pollut Res 23:6608–6620

Rouphael Y, Franken P, Schneider C, Schwarz D, Giovannetti M, Agnolucci M, De Pascale S, Bonini P, Colla G (2015) Arbuscular mycorrhizal fungi act as biostimulants in horticultural crops. Sci Hortic (Amsterdam) 196:91–108

Santamaría-Romero S, Ferrera-Cerrato R (2001) Dynamics and relationships among microorganisms, C-organic and N-total during composting and vermicomposting. Agrociencia 35:377–383

Santi C, Bogusz D, Franche C (2013) Biological nitrogen fixation in non-legume plants. Ann Bot 111:743–767

Savci S (2012) An agricultural pollutant: chemical fertilizer. Int J Environ Sci Dev 3:77–80

Smith BE (2002) Nitrogenase reveals its inner secrets. Science 297:1654–1655

Smith LED, Siciliano G (2015) A comprehensive review of constraints to improved management of fertilizers in China and mitigation of diffuse water pollution from agriculture. Agric Ecosyst Environ 209:15–25

Srinivasakumar KP, Rajashekhar M (2009) In vitro studies on bactericidal activity and sensitivity pattern of isolated marine microalgae against selective human bacterial pathogens. Indian J Sci Technol 2:16–23

Steenhoudt O, Vanderleyden J (2000) *Azospirillum*, a free-living nitrogen-fixing bacterium closely associated with grasses: genetic, biochemical and ecological aspects. FEMS Microbiol Rev 24:487–506

Stirk WA, Ördög V, Novák O, Rolčík J, Strnad M, Bálint P, van Staden J (2013) Auxin and cytokinin relationships in 24 microalgal strains. J Phycol 49:459–467

Suárez-Estrella F, Bustamante MA, Moral M, Vargas-García MC, López MJ, Moreno J (2012) In vitro control of Fusarium wilt using agroindustrial subproduct-based composts. J Plant Pathol 94:59–70

Sullivan DM, Miller RO (2001) Compost quality attributes measurements and variability. In: Stoffela P, Khan B (eds) Compost utilization in horticultural cropping systems. Lewis Publishers, Boca Raton, pp 95–119

Tarakhovskaya ER, Maslov YI, Shishova MF (2007) Phytohormones in algae. Russ J Plant Physiol 54:163–170

Thompson RB, Meisinger JJ (2002) Management factors affecting ammonia volatilization from land-applied cattle slurry in the Mid-Atlantic USA. J Environ Qual Abstr – Waste Manag 31:1329–1338

Timilsena YP, Adhikari R, Casey P, Muster T, Gill H, Adhikari B (2014) Enhanced efficiency fertilisers: a review of formulation and nutrient release patterns. J Sci Food Agr 95:1131–1142

Tóth G, Hermann T, da Silva MR, Montanarella L (2016) Heavy metals in agricultural soils of the European Union with implications for food safety. Environ Int 88:299–309

Tredici MR (2010) Photobiology of microalgae mass cultures: understanding the tools for the next green revolution. Biofuels 1:143–162

Tringali C (1997) Bioactive metabolites from marine algae: recent results. Curr Org Chem 1:375–394

Upadhyaya H, Panda SK, Bhattacharjee MK, Dutta S (2010) Role of arbuscular mycorrhiza in heavy metal tolerance in plants: prospects for phytoremediation. J Phytology 2:16–27

Uyisenga JP, Nzayino P, Seneza R, Hishamunda L, Uwantege K, Gasana N, Bajyana ES (2010) In vitro study of antibacterial and antifungal activity of *Spirulina platensis*. Int J Ecol Dev 16:80–88

Vessey JK, Pawlowski K, Bergman B (2005) Root-based N_2-fixing symbioses: legumes, actinorhizal plants, *Parasponia* sp. and cycads. Plant Soil 266:205–230

Waksman SA, Cordon TC, Hulpoi N (1938) Influence of temperature upon the microbiological population and decomposition processes in composts of stable manure. Agricultural Experiment Station, New Jersey, pp 83–113

World Bank (2019) Commodity market outlook. World Bank Group, Washington, DC

Chapter 4
Harvesting Microalgae: Overcoming the Bottlenecks in Microalgae-Based Wastewater Treatment Through Industrial-Scale Gravity Sedimentation and Thickening

Antonio Ortiz Ruiz, Joan García Serrano, Rubén Díez Montero, Igor Carvalho Fontes Sampaio, and Etiele Greque de Morais

Abstract Microalgae culture is currently receiving considerable attention for its potential in wastewater treatment and production of algal biomass from which high-value bioproducts and bioenergy can be obtained, as well as the consequent carbon dioxide sequestration via the photosynthesis process. However, biomass harvesting is one of the bottlenecks in the culture and microalgae-based wastewater treatment systems. Energy-intensive technologies are required to separate the solid and liquid phases due to the low density of microalgae. Low-cost processes, such as sedimentation, are not efficient enough due to the low settling velocities of the microalgae.

A. O. Ruiz · J. G. Serrano
GEMMA—Group of Environmental Engineering and Microbiology, Department of Civil and Environmental Engineering, Universitat Politècnica de Catalunya-BarcelonaTech, Barcelona, Spain

R. D. Montero
GEMMA—Group of Environmental Engineering and Microbiology, Department of Civil and Environmental Engineering, Universitat Politècnica de Catalunya-BarcelonaTech, Barcelona, Spain

GIA—Group of Environmental Engineering, Department of Water and Environmental Sciences and Technologies, Universidad de Cantabria, Santander, Spain

I. C. F. Sampaio
CPID—Espírito Santo's Center for Research, Innovation and Development, Cariacica, Brazil

E. G. Morais (✉)
GEMMA—Group of Environmental Engineering and Microbiology, Department of Civil and Environmental Engineering, Universitat Politècnica de Catalunya-BarcelonaTech, Barcelona, Spain

DEAB—Department of Food Engineering and Biotechnology, EEABB—Barcelona School of Agricultural and Biosystems Engineering, Universitat Poliècnica de Catalunya-BarcelonaTech, Barcelona, Spain
e-mail: etiele.greque@upc.edu

J. A. V. Costa et al. (eds.), *Microalgal Bioengineering*,
https://doi.org/10.1007/978-3-031-61253-4_4

Sedimentation coupled with coagulation and flocculation has been widely studied on a lab scale in order to increase the microalgae settling velocity. However, only a few studies have addressed the scaling up of these experimental results to industrialize the process. In this chapter, issues of scaling up gravity sedimentation and thickening systems for microalgae culture at demonstrative or full-scale were addressed, including techno-economic analysis.

Keywords Biomass recuperation · Nutrients recovery · Harvesting · Environmental sustainability · Microalgae biorefinery · Pilot-scale harvesting

4.1 Introduction

Microalgae culture is considered a sustainable extensive technology to produce feedstock for bioproducts and biofuels (Wijffels et al. 2013; Zhu 2015). Also, this technology is widely known for its application in wastewater treatment since it is possible to uptake nutrients for microalgae growth, and the produced biomass can be harvested and valorized (Gudin and Chaumont 1991). Moreover, arable land is not required for microalgae cultivation, avoiding direct competition with the farming industry (Benedetti et al. 2018). Thus, microalgae can be a solution for sustainable food and energy production, as well as recovery of resources from wastewaters. However, there are several issues which limit this technology from being scaled up and established in the industry (Vanthoor-Koopmans et al. 2013). One of the most important hurdles for biomass production is the microalgae harvesting (Roy and Mohanty 2019) process, usually representing 20–30% of the total microalgal biomass production cost (Najjar and Abu-Shamleh 2020).

There are many studies on this issue testing different harvesting techniques as filtration (Zhao et al. 2022), centrifugation (Ghazvini et al. 2022), flotation (Demir-Yilmaz et al. 2023), and sedimentation (Singh and Patidar 2018; Roy and Mohanty 2019; Ortiz et al. 2021). Mechanical-based technologies, such as centrifugation and filtration, yield concentrated biomass (Lu et al. 2022) and are suitable for obtaining high-value products, such as food and cosmetics since they do not require the use of chemical coagulants that can affect the products' composition (Gerardo et al. 2015). These harvesting processes have a great capability to concentrate the biomass but require expensive mechanical equipment and have a high energy consumption and operation costs. On the other hand, sedimentation is a low-energy consumption process because gravity is the force used to separate the solid and liquid phases of the culture broth. In addition, sedimentation can be used as a first step to reduce the volume of water in the harvested biomass (Lam et al. 2019), making subsequent downstream processes easier and reducing the operation cost (Hapońska et al. 2018).

Flotation and sedimentation are the most used techniques for effluent clarification and biomass harvesting in microalgae-based wastewater treatment systems, as it is in conventional wastewater treatment processes. Sedimentation in microalgae cultures is a low-efficiency process since the cells have a very low settling velocity (in the order of 1 cm/h) due to the small size of cells (5–20 μm) and a negatively charged surface cell wall (Smith and Davis 2013; Smith et al. 2013; Gutiérrez et al. 2016b). For flotation processes, the mechanical injection of dissolved air is necessary to favor the rise of the biomass to be collected, which increases the operational cost due to high energy consumption. The efficiency of these processes strongly depends on the speciation of the culture, the growth of which depends entirely on the location and climatic conditions (García et al. 2018; Díez-Montero et al. 2020a). The use of coagulants and flocculants are applied to improve the performance of microalgal biomass harvesting (Uggetti et al. 2018; García et al. 2018).

A lot of lab-scale studies have been carried out to increase the terminal settling velocity, mainly using coagulants and flocculants, which stabilize the negatively charged surface cell wall and enhance the flocculation, increasing the size of the flocs and, therefore, the terminal settling velocity (by around ten times) (Mubarak et al. 2019). However, scaling up from batch laboratory studies (Gutiérrez et al. 2015a, b) to full-scale continuous operation is a big challenge (Ortiz 2021; Ortiz et al. 2022a).

In this chapter, issues of scaling up gravity sedimentation and thickening systems for microalgae culture at demonstrative or full scale were addressed and analyzed, including techno-economic analysis. Some studies have taken up the challenge with pilot-scale sedimentation studies (Ndikubwimana et al. 2016; Hom-Diaz et al. 2017), but very few have exceeded the volume of the harvesting unit upper than 200 L, and even fewer have studied a settler in continuous working conditions (Ortiz et al. 2021). This chapter focuses on microalgae harvesting techniques and the design of settling tanks for biomass collection discussing primary and secondary concentration methods, including physical and chemical processes, such as sedimentation, filtration, and centrifugation. It emphasizes the combination of sedimentation with coagulation and flocculation as a cost-effective method for microalgae harvesting. The design principles and performance analysis of settling tanks are explored, considering different settling theories and their application in industrial-scale systems.

4.2 Microalgae Harvesting

The objectives of microalgae harvesting systems are both to produce clarified effluent water and to collect the biomass produced in the culture system, which can be closed (or semi-closed) photobioreactors (PBR) or open high-rate algal ponds (HRAP). Microalgal harvesting involves separating the solid and liquid phases and increasing the biomass concentration from the culture, which is in the range of 0.02–0.06% (Deconinck et al. 2018). This low concentration and the small size (5–20 μm) make microalgae harvesting challenging, and most of the techniques need high energy demand (Najjar and Abu-Shamleh 2020). Microalgae harvesting typically involves primary and secondary concentration steps. Primary concentration methods thicken the microalgal biomass slurry up to 1–3% to aid in separating it from the culture medium. After this, further dewatering is necessary, and this is referred to as secondary concentration. Secondary concentration techniques can produce microalgal sludge with an average concentration of up to 200 g/L (Deconinck et al., 2018). These concentration techniques are based on physical and chemical processes, with some physical techniques using mechanical or electrical forces, such as filtration and centrifugation to concentrate the microalgal biomass.

Sedimentation techniques rely on gravity to separate microalgae from the culture medium, while chemical techniques use inorganic or organic additives to improve coagulation neutralizing the negative charge of the microalgae, which favors coagulum formation. Typically, coagulation is followed by flocculation processes that involve using other chemical products to facilitate the joining of the coagulum into larger floccules. Some researchers have suggested that combining sedimentation with coagulation and flocculation represents the most cost-effective microalgae harvesting method (Deconinck et al., 2018). As part of conventional wastewater treatment, the harvested biomass usually undergoes a multi-stage process that includes a second sedimentation step to further thicken it. An example of the thickened biomass was reported to have a concentration of 7 g VSS/m^3 d in winter and 43 g VSS/m^3 d (García et al. 2018).

To address the challenge of scaling up harvesting processes, it is necessary to change the perspective from the studies carried out at the lab scale, in which the different species' behavior and coagulants are the main focus (Gutiérrez et al. 2016b). For the design and operation of a microalgae harvesting process based on large-scale sedimentation, the physical principles of sedimentation must be retaken and applied.

According to Navier-Stokes, which is based on Newton's laws, the terminal settling velocity of a particle is determined by its size and the density difference between the particle and the medium. Later, Hazen Cam's postulates outlined a method to determine the hydraulic load of a settler working continuously. This method indicates the minimum sedimentation rate based on the inflow and the useful surface of the settler, providing a way to estimate the theoretical requirements of a settler. Attempts have been made to expand this theory for microalgae sedimentation processes by considering flocculent and hindered settling. However, the significant variability in microalgae species, and therefore in their size and shape, has resulted in sedimentation values being based mainly on experimental data (Crittenden et al. 2012). Therefore, it is necessary to conduct full-scale studies to obtain experimental results that can provide useful information for scaling up the process (Ortiz 2021).

To meet the requirements of downstream biomass processes and enhance the feasibility of valorization, it is necessary to increase the concentration of the harvested biomass obtained from the settler. This can be achieved through a thickening step performed after the biomass is harvested, which increases the total solids concentration by compression of the sludge (Ortiz et al. 2022b). The resulting slurry-like consistency is ideal for subsequent processes, such as feeding an anaerobic digester for biogas and biomethane production (Passos et al. 2013). Sedimentation-based thickening is a process where harvested solids are compressed through sedimentation (Bustos et al. 1999). During this process, particles at the bottom of the thickener are compressed due to their own weight, causing the water to move up through the voids.

Although gravity-based thickening is commonly used in conventional wastewater treatment, it is not typically used in biorefineries for the harvesting of microalgae that are cultivated in wastewater. The low density of microalgae is not considered sufficient to induce the compression process (Muylaert et al. 2017). Furthermore, a recent study of microalgae biomass harvesting in gravity thickeners has shown a significant improvement in the performance of the system (Ortiz et al. 2022b). This study explored different configurations and operational strategies to optimize the system and achieve an adequate total solids (TS) concentration of harvested biomass. The authors found a concentration factor of 1.9 and a recovery efficiency of 28% after using a single thickener. The addition of a second thickener and further modification of its operation led to the achievement of TS of 26.5 g/L and a concentration factor of 3.6, which represents a successful performance of the gravity-thickening system in a pilot-scale facility (Ortiz 2021). This approach allowed low investment, energy requirements, and operation costs, making it a viable option for microalgae biomass harvesting in the context of wastewater treatment (Ortiz et al. 2022b).

4.2.1 Gravity Separation Theory and Design of Settling Tanks

The gravity separation process involves the use of appropriately sized settling tanks or settlers, which mainly depend on the inlet flow rate and the terminal sedimentation velocity of the suspended solids present in the inlet (Bajcar et al. 2011). The terminal settling velocity refers to the maximum velocity that particles can attain during sedimentation. Initially, particles accelerate due to the force of gravity, and as their velocity increases, so does the frictional force with the liquid culture medium. When the force of gravity on the particle equals the frictional force, the velocity becomes constant, and this is known as the terminal settling velocity (Yin and Song 2021).

Microalgae harvesting using sedimentation coupled with coagulation and flocculation is a conceptually simple process, but it poses challenges for designing proper settlers on an industrial scale (Ortiz et al. 2021). The performance of settlers can be described using particle-settling theories, but practical aspects must also be considered when operating large-scale settlers continuously. Settling theories can be categorized into four types that typically occur in different zones of a settling tank.

The first type is discrete particle settling, which occurs at the top of the settler where spherical particles do not interact with each other. The terminal sedimentation velocity depends on the size of each particle (Vahidifar et al. 2018) and is constant. The second type is flocculent settling, which occurs in the central zone of the settler, where the interaction between different particles can form large particle aggregates with faster settling velocities that tend to sweep the smaller particles (De 2017). In this case, there is an increasing terminal velocity, related to the increasing particle size. The third type is hindered settling, which occurs in the lower zone of the settler where the particle concentration is high enough to restrict the fluid velocity field around individual particles, forming a blanket of particles with a different interface with the clarified liquid (Dankers and Winterwerp 2007). The interface descends with a velocity which decreases according to the increasing particle concentration. The fourth type is compression settling, which occurs at the bottom of the settler where the particle blanket, with a high concentration of particles, compresses by itself due to the weight of the particles (Crittenden et al. 2012).

It is important to note that typologies (2), (3), and (4) can only be analyzed through isoperformance curves, which are obtained from experimental column tests. However, Hazen Cam's theory can be used to estimate settler performance by considering three parameters: the critical sedimentation velocity of the particles (Vc (m/h)), the flow rate (Q (m^3/h)), and the useful surface area of the settler where particles can settle (S (m^2)). By combining these three parameters, the hydraulic loading rate can be defined as shown in Eq. 4.1.

$$v_c \geq \frac{Q}{S}(\mathrm{m/h}) \tag{4.1}$$

Particles with a sedimentation velocity lower than Vc are unable to reach the bottom of the settler and remain unharvested. However, particles with a terminal sedimentation velocity higher than Vc are able to reach the bottom and can be harvested. Thus, it is important to know the sedimentation velocity of particles when sizing and designing settlers. The tank's working flow rate, volume, and dimensions should allow for sufficient hydraulic retention time (HRT) and low enough hydraulic loading rate to let the particles to settle to the bottom. The primary purpose of settlers is to produce large volumes of clear water and small quantities of concentrated biomass (Bajcar et al. 2011).

The low sedimentation velocity of microalgae and the requirement for laminar flow at a low speed necessitate a larger surface area to satisfy Eq. 4.1. To achieve a larger surface area within a reduced volume, lamella settling tanks can be utilized. These tanks are modified with a set of plates within the tank to increase the surface area available for particle settling.

The efficiency of settlers can be increased by reducing the depth of the settler, which reduces the particle trajectory and the time it takes for a particle to settle to the bottom to be harvested (Šikić 2017). This allows for the clarification of larger volumes of water in a shorter period of time. To achieve this, lamella settling tanks were developed based on the idea that the sedimentation process depends more on the sedimentation area than on the hydraulic retention time (Tchobanoglous et al. 2014). The lamellar settler operates on the principle that the hydraulic loading rate or surface load (m^3/m^2 d) of a settling tank is independent of its depth. This means that the capacity of a settling tank can be increased by using inclined plates at a specific angle. Lamellar settlers are intended for the continuous separation of solid particles from water and have two primary objectives: to increase the settling area and to achieve a laminar flow (Robescu et al. 2010).

This system is not frequently used for microalgal harvesting. However, it was proposed by García et al. (2018), as the best solution in a pilot plant for harvest the microalgal biomass of three photobioreactors at demonstrative scale, which treated 7 m^3/d of agricultural runoff and urban wastewater with a nitrogen removal efficiency of 65% (ranging from 90% in summer to 50% in winter). The removal efficiency for phosphorus was outstanding, usually greater than 95%, with no clear seasonal trends observed. Volumetric biomass production varied greatly between seasons, with much lower values in winter than in summer. Some of the applications of the biomass produced under these circumstances include the production of biogas and biofertilizers (Uggetti et al. 2018).

4.2.2 *Sedimentation Velocity of Microalgae and Physical Properties of Microalgal Flocs*

The settling velocity of a particle in a fluid is described by Stokes' Law, which states that the velocity is proportional to the density and square diameter of the particle (Shaikh et al. 2017). As discussed in Sect. 2.1, the settling velocity of particles also affects the hydraulic loading equation, which determines the working flow rate. In terms of microalgae harvesting by settling, a lower sedimentation rate results in a longer settling time, making gravity settling less competitive compared to other methods like filtration, centrifugation, and flotation (Fasaei et al. 2018). Therefore, to successfully scale up microalgae biomass harvesting based on sedimentation and achieve large volumes of clarified water and small volumes of concentrated biomass, it is important to enhance the settling velocity of the microalgal biomass (Ortiz et al. 2021).

Many studies have been conducted to address the issue of increasing the settling velocities of microalgae. Two approaches have been explored: increasing the size of flocs using chemical and natural coagulants and flocculants (Gutiérrez et al. 2015b, 2016b; Ogbonna and Nwoba 2021; Niemi and Gentili 2021) and using flocculant carriers with large density to increase the flocs' density and weight (Li et al. 2019; Wei et al. 2020).

In a study by Gutiérrez et al. (2015b), settling velocities of microalgal biomass were investigated in column tests with different column depths (12.5, 20, 30, and 40 cm). The tests were performed with and without the use of flocculants, Ecotan and Tanfloc (Fig. 4.1). The results of the study demonstrated an increase in settling velocity with an increase in column depth. In the absence of a flocculant, the settling velocity increased only from the top to 20 cm and remained constant until the

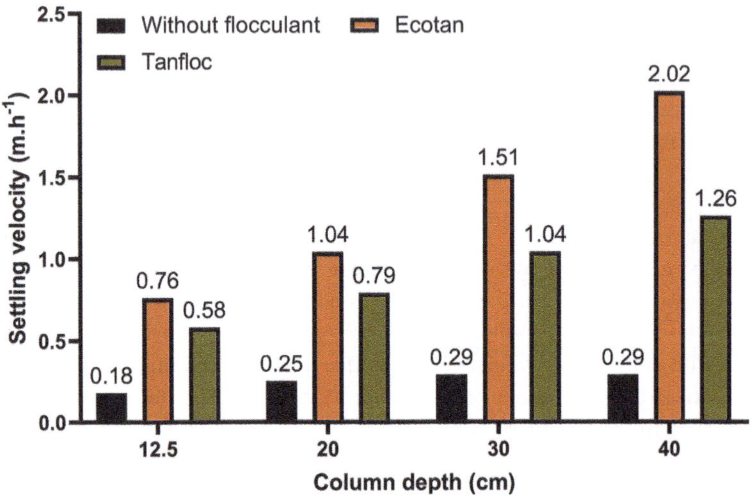

Fig. 4.1 Velocities to attain 90% biomass recovery at different column depth. (Adapted from Gutiérrez et al. 2015b)

bottom. However, with the addition of a flocculant, the settling velocity continued to increase throughout the particles' descent.

The settling velocities without flocculant ranged from 0.18 to 0.29 m/h, while the tests with Ecotan and Tanfloc achieved 2.02 and 1.26 m/h, respectively. These findings indicate that the use of flocculants significantly increases the settling velocity and that having a sufficient settling height can promote the aggregation of particles in larger flocs and further increase settling velocity. Thus, it is recommended to include a high settling zone in settler design to encourage flocculent settling.

This contradicts the concept of the sloped settling system used in lamella settling tanks, where closely spaced plates are used to reduce settling time. Thus, it is necessary to segregate the coagulation, flocculation, and sedimentation processes into different compartments and customize their dimensions for each specific process. A strongly agitated chamber is required for proper mixing of the coagulant with the culture medium, followed by an area without agitation and with sufficient height for the flocculation process to form large flocs (Sun et al. 2019). Then, an area of plates or lamellae, inclined at approximately 60° to the horizontal, is used to facilitate the biomass reaches a surface where it settles and descent to an accumulation and purge zone. Ortiz et al. (2021) presented a 0.7 m^3 lamella settler designed to perform coagulation, flocculation, and sedimentation processes separately.

The settler was operated under a continuous flow of 0.4 L/h, and the researchers found that a coagulant dosage of 1–10 mg/L of PAX-18 (Polyaluminum Chloride) was necessary. The results demonstrated a remarkable removal and harvesting efficiency of 95% of total solids and a turbidity lower than 5 NTU in the clarified water outflow. These findings offer valuable insights into the possibility of adopting sedimentation as an economical option for microalgae biomass harvesting at an industrial scale.

On the other hand, some studies have focused on increasing the effective density of microalgae flocs to enhance their settling velocity. The settling velocity is heavily influenced by the initial concentration and microalgae species in the culture broth. However, the use of biopolymers and cationic starch as flocculants to increase the floc size, and clay powder to improve the effective density of microalgae flocs, has shown promising results (Table 4.1). For instance, Gutiérrez et al. (2015b) were able to increase the settling velocity from around 0.29 m/h for microalgae without using any flocculant to 1.26 m/h using Tanfloc as bioflocculant.

Moorthy et al. (2017) did not achieve any improvement in the value, even with the use of chitosan and natural bentonite clay, which resulted in velocities of up to 1.04 m/h. In contrast, Vandamme et al. (2013) had achieved a promising sedimentation rate of 2.16 m/h by using chitosan cationic starch on an industrial scale. However, Gutiérrez et al. (2016a) achieved a significant qualitative advancement in the flocculation process by using Tanfloc SG and obtaining values of up to 6.52 m/h.

Wei et al. (2020) used montmorillonite clay as a flocculant carrier to increase the effective density of microalgae flocs, and the results showed a settling velocity of 11.05 m/h. Finally, Ortiz et al. (2021) obtained an average velocity of 13.75 m/h using a chemical coagulant, PAX-18. Nevertheless, the use of metal salt-based coagulants may not be suitable for photosynthesis and other biological processes, resulting in microalgae growth inhibition and inhibiting methane formation in anaerobic digestion (Rossi et al. 2021).

The mentioned studies indicate a growing interest in harvesting technology for microalgae grown in wastewater. Additionally, they have sparked interest in sedimentation combined with coagulation and flocculation as the most cost-effective and environmentally friendly method for harvesting algal biomass. Moreover, the use of bioflocculants has opened up opportunities for the valorization of biomass into various high-value products, such as biofertilizers, animal feed, and bioenergy. Furthermore, the clays used to enhance effective density can be recovered and recycled, adding to their economic and environmental benefits (Moorthy et al. 2017).

Table 4.1 Comparison of microalgae settling velocities of different lab-studies

Flocculant	Microalgae species	Settling velocity (m/h)	References
None	Mixed culture	0.29	–
CNBC	*Scenedesmus abundans*	1.04	Moorthy et al. (2017)
Tanfloc	Bacterial form	1.26	Gutiérrez et al. (2015b)
CS	*Chlorella*	2.16	Vandamme et al. (2013)
Tanfloc SG	*Chlorella*	6.52	Gutiérrez et al. (2016b)
CMCS	*Chlorella* FACB-31	11.05	Wei et al. (2020)
PAX-18	Mixed culture	13.75	Ortiz et al. (2021)

CNBC Chitosan and natural bentonite clay, *CS* Chitosan cationic starch; *PAX-18* Polyaluminum chloride; *CMCS* Clay modified cationic starch

4.2.3 Thickening of Microalgal Biomass

Thickening is the process where the suspended solids of an already concentrated bioprocess are further condensed, resulting in a much more concentrated biomass product and a relatively solids-free supernatant. Gravity thickening can achieve a total solids (TS) concentration of 20–30 g/L, which is lower than that obtained by centrifugal and filtration processes (EPA 2003; Pahl et al. 2013) but is suitable for feeding an anaerobic digestion process. Regarding it, there is a lack of reporting on the application of gravity thickening in full-scale microalgae-based plants (Lundquist et al. 2018). Although some microalgae collection studies have referred to a thickening process, the processes studied, in general, are sedimentation or working in batch and extending the harvesting time (Pahl et al. 2013; Vandamme 2017).

Gravity thickening is typically carried out in circular tanks with a sloped conical bottom and scrapers that aid in the transport and collection of settled solids at the tank bottom. Thickened solids are periodically removed from the tank by pumping from the bottom. The primary operational challenges or difficulties in gravity thickeners involve the formation of gas bubbles that adhere to solid particles, increasing their buoyancy and reducing settling. To mitigate this issue, scrapers or pickets attached to the thickener arm can be used to release gas from the sludge blanket. To prevent significant turbulence and resuspension, these devices move at a low speed.

Another phenomenon that must be avoided is ratholing, which occurs when there is a short-circuiting of the draw-off resulting in the removal of poorly consolidated biomass, while a high concentration of solids remains in the tank (Ortiz et al. 2022b). Ratholing can happen if the thickened biomass is drawn off too quickly or too slowly, resulting in a cone of depression at the draw-off point. To prevent ratholing, the pumping speed for solids removal should be carefully adjusted (EPA 2003).

Lundquist et al. (2018) implemented and operated a pilot plant which consisted of 9 HRAPs (3.5 m^2 and 1 m^3 total), three cone-bottom settling tanks (settlers of 200 L each), and a sludge settling tank for harvested biomass thickening (thickener of 1.9 m^3). The ponds were fed with wastewater previously clarified in a primary sedimentation tank, and the HRT of the ponds ranged between two and five days. The effluent left the ponds through vertical ramped standpipes, which controlled the pond depth at 30 cm, and flowed by gravity to the settlers with an HRT ranging between 2 and 3 h. The settled biomass was pumped to the thickener tank, but the study did not report any results regarding the achieved concentration and thickener performance.

Ortiz et al. (2022b) optimized and operated a gravity-harvesting system consisting of three tubular photobioreactors (30 m^3 total) followed by a lamella settler (700 L) and two gravity thickeners of 200 L each for the treatment of 7 m^3/d of urban wastewater and agricultural runoff. The objective was to concentrate the algal biomass from the lamella settler (TS of 0.1–0.8%) into a biomass with 2% TS for subsequent anaerobic digestion. Two thickeners were connected in series, and between 180 and 240 L of settled biomass were pumped every day to the first thickener at 400 L/h from the purge of a previous lamella settler in nine equal events every two hours. Both thickeners were equipped with scrapers to avoid the ratholing phenomenon, which appeared one minute after starting the purge and prevented complete biomass recovery. Different scraper time strategies were performed in order to achieve an effective in detachment of the biomass stuck on the inner walls to avoid the ratholing phenomenon. Eventually, a concentration of 26.5 g/L TS was achieved.

Sedimentation-based microalgae thickening is an underutilized technique in microalgae harvesting due to the low density and weight of microalgae, which makes it difficult to achieve compression sedimentation and obtain a higher concentration of solids. However, recent studies have shown that it is possible to achieve these processes to some extent with the proper timing of operating elements, such as scrapers and purges (Ortiz et al. 2022b). This opens up the possibility of obtaining a product of a certain concentration through a low-cost process, which may be suitable for further processes, such as anaerobic digestion (Díez-Montero et al. 2020b) or pre-products that can lower the costs of the final production of high-value products, such as bioplastics and pigments (Rueda et al. 2020; Senatore et al. 2023).

4.3 Techno-Economic Assessment of Low-Cost Microalgae Harvesting

One of the main costs associated to microalgal biomass production is related to the harvesting process. Recent studies report that it usually accounts for about 20–30% of the total production cost (Zhu et al. 2020; Suparmaniam et al. 2022). Although one of the earliest instances of cost estimation came from a paper published in the 1980s (Gudin and Thepenier 1986), recent estimations have reported similar results, highlighting the disparities between microalgae harvesting costs in open/closed systems (Fasaei et al. 2018). An update of this value can be estimated from more recent studies to provide a more reliable and current value. Despite the existence of a theoretical low-cost approach for microalgae harvesting in an industrial environment (Wang et al. 2019), there are very few examples of low-cost and feasible approaches to harvest this biomass when cultivated in wastewater.

Microalgae harvesting based on sedimentation has been very poorly studied on a large scale, so it has not been able to be compared with other more widely used techniques, such as dissolved air flotation (DAF) and centrifugation. To reduce the costs related to microalgae harvesting, it is common to perform a two-step separation process consisting of initial sedimentation and further dewatering, which results in a high solids content biomass. The selection of adequate harvesting methods depends on the characteristics of the target microorganism and also on the type and value of the end product (Barros et al. 2015).

However, biomass concentration as a first step can be directly used to feed an anaerobic digestion process to produce biogas. This was the case of the demonstration site 1 of the INCOVER project (Fig. 4.2), funded by the European program

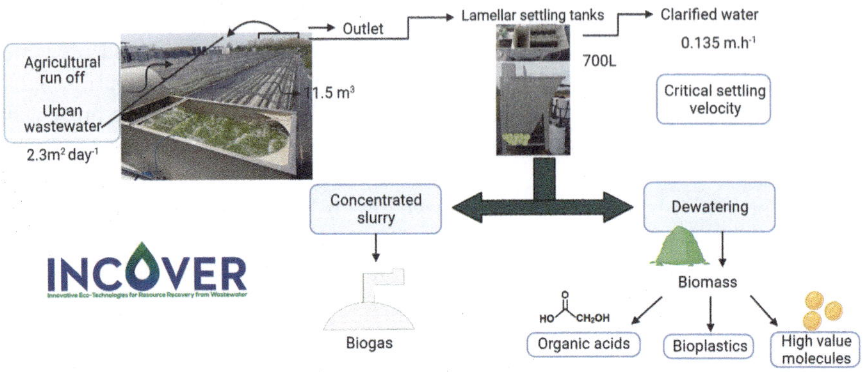

Fig. 4.2 Illustration summary of the INCOVER project

Horizon 2020 (https://incover-project.eu/case-study/case-study-1; GA 689242) (Uggetti et al. 2018; García et al. 2018; Marín et al. 2019; Díez-Montero et al. 2020b).

In a different project, PAVITR (http://www.pavitr.net; GA 821410), also funded by the Horizon 2020 program, it has been designed a microalgae-based system for wastewater treatment in India. In this case, the microalgal biomass is harvested by DAF (Ortiz et al. 2022a) and used as a biofertilizer in agriculture. The design of the PAVITR microalgae-based treatment system was based on the findings published by Ortiz et al. (2022b).

Ortiz et al. (2022) performed a technical and economic evaluation of various methods of microalgae harvesting at full or demonstrative scale. To achieve this, the experiences of the INCOVER and PAVITR projects have been gathered, and a specific techno-economic assessment has been conducted. In particular, the operational costs of the INCOVER and PAVITR systems were estimated and compared to determine the percentage of the harvesting process' contribution to the overall operation costs (Ortiz 2021) (Table 4.2).

The microalgal biomass in the INCOVER system was produced in three semi-closed tubular photobioreactors, with a volume of 11.5 m^3 each. These were fed with a daily mix of 2.3 m^3 of agricultural runoff and urban wastewater. The harvesting process was carried out in a lamella settler with a capacity of 700 L and an operational critical settling velocity of 0.135 m/h (Ortiz et al. 2021) (configuration 1 (C1)). In the PAVITR project (Fig. 4.3), two High Rate Algal Ponds (HRAP) with a volume of 100 m^3 each were designed to treat 50 m^3/d of urban wastewater in Aligarh, India. The biomass was harvested using a DAF system (configuration 2 (C2)). To compare both harvesting systems under the same conditions, a third hypothetical configuration with the same HRAPs as the PAVITR project and a lamella settler to harvest the biomass was evaluated (configuration 3 (C3)).

Labor costs were not factored into the analysis as they were considered proportional to both biomass production and harvesting processes. Pumping water was assumed to have a specific energy consumption of 0.1 kWh/m^3 based on the average power consumption of several commercially available centrifugal pumps (ZUWA

Table 4.2 Operational cost of biomass production and sedimentation-based harvesting process

Configuration	*1—Photobioreactors–Settler*						*2—HRAP–DAF*						*3—HRAP—Settler*					
Culture	m³/d	m³/h	h	W	kWh/d	€/d	m³/d	m³/h	h	W	kWh/d	€/d	m³/d	m³/h	h	W	kWh/d	€/d
Influent	7	3.3	2.1	300	0.64	0.13	50	4.2	12	400	4.8	0.96	50	4.2	12	400	4.8	0.96
Effluent	7	3.3	2.1	300	0.64	0.13	50	4.2	12	400	4.8	0.96	50	4.2	12	400	4.8	0.96
Paddlewheels	(6 uts.)		24	250[a]	18	3.60	(2 uts.)		24	500	24	4.80	(2 uts.)		24	500	24	4.80
						3.85						**6.72**						**6.72**
Harvesting																		
Influent	7	0.4	17.5	50	0.35	0.070												
Purge pump	0.15	0.4	0.4	50	0.0075	0.002	50	4.2	12	1000	12.0	2.4	1	3	0.3	300	0.30	0.060
Stirrer	–	–	17.5	90[a]	0.788	0.16	–	–	–	–	–	–	–	–	12	90	1.08	0.22
Dosing pump	–	–	17.5	30	0.525	0.11	–	–	–	–	–	–	–	–	12.0	50	0.6	0.12
Coagulant		7.6 mg/L		0.4 €/kg		0.07		7.6 mg/L		0.4 €/kg		0.51	–	7.6 mg/L		0.4 €/kg		0.51
						0.40						**2.91**						**0.90**

[a]Engine working at 50% of full capacity

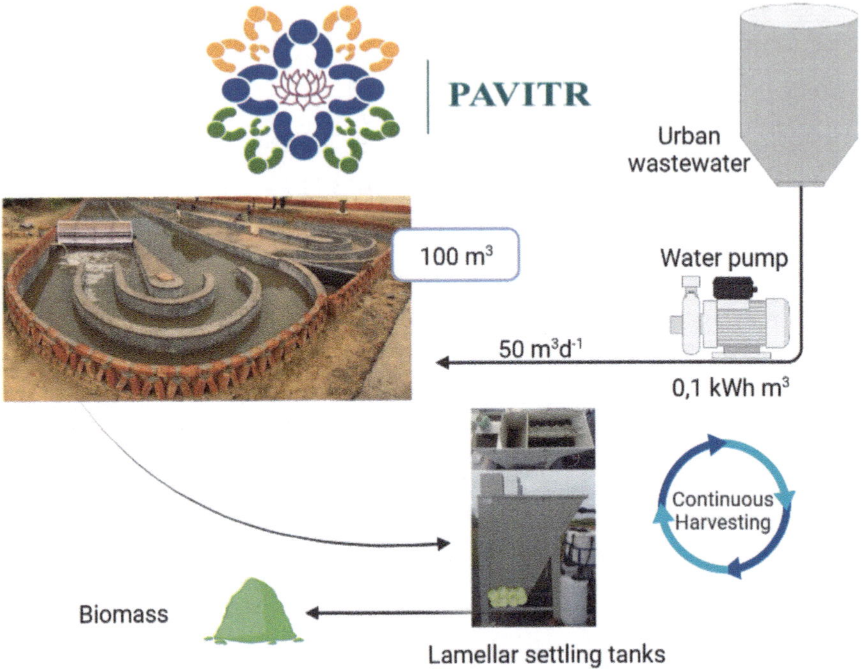

Fig. 4.3 Illustration summary of the PAVITR project

P80; Pedrollo CPm610; Cattini CPM146–1; Hidromundo AB13; Damova DM 10). The price of electricity was assumed to be 0.2 €/kWh, which is the current average cost in Europe (Ortiz 2021).

The three semi-batch INCOVER photobioreactors were fed daily with 2.3 m^3 each, and the same volume was emptied every day for 0.7 h into three tanks, from which the settler was fed at a flow rate of 400 L/h for 17.5 h/day. As a result, an influent pump was needed for the harvesting unit.

The PAVITR HRAPs were designed to operate continuously, with 50 m^3 of urban wastewater fed every day for 12 h. Consequently, the harvesting unit was fed directly from the HRAP, eliminating the need for an additional harvesting influent pump. For configuration 2 (C2) with a DAF harvesting unit, a compact unit with a power consumption of 1 kW was considered. For configuration 3 (C3), a lamella settler with double the size of C1 was evaluated. This size and the settler inflow rate of 4.2 m^3/h resulted in a critical settling velocity of 0.40 m/h, which falls within the range of microalgae settling velocities reported in the literature (Table 4.2). The coagulant dosage was estimated using a jar-test carried out in (Ortiz et al. 2021), and the coagulant cost was calculated as the average of coagulant chemical prices from Henan GO Biotech, Anqing Haida Chemical, and AquaChem. The liquid solution was excluded from consideration due to its higher price and the need to avoid unnecessary water transport, which is not economically or environmentally suitable from an industrial-scale perspective. The volume of biomass produced in the HRAP to be harvested in the harvesting units was estimated using the biokinetic model BIO_ALGAE (Ortiz 2021; Ortiz et al. 2021).

The results reveal notable variations among the three configurations. In C1, the cost of harvesting accounts for only 9.5% of the total operational cost (Fig. 4.4), which is significantly lower than the widely cited range of 20–30% in harvesting review studies. However, it is important to note that the biomass production system in C1 was oversized. The total influent volume (7 m^3) was split into three photobioreactors, leading to an increased number of paddlewheels and energy consumption. This fact resulted in a low percentage value of harvesting cost despite the oversized harvesting unit in this configuration due to the influent pump required to feed the settler.

For C2, the compact DAF unit in the harvesting process constituted 30.2% of the total operational cost (Fig. 4.4), which aligns with the values cited in the literature. This outcome was attributed to the optimized design of the HRAP and lower paddlewheel energy consumption. However, the high energy consumption of the harvesting unit (DAF) compared to gravity-settling methods also contributed to this high value.

Furthermore, C1 exhibited a low harvesting percentage cost of the total operational cost (9.5%), but the production configuration of C1 was oversized, leading to an unusually low value that cannot be compared to the notion of an optimized industrial process. On the other hand, C2 and C3 were comparable since both adopted the same production configuration (HRAP). In this context, C3 showed that a sedimentation-based harvesting process can significantly reduce the harvesting cost percentage and consequently the total production costs (Fig. 4.4). This result was mainly due to the lower consumption of the settler unit compared to the DAF since the sedimentation process uses gravity forces to separate the liquid-solid phase, while the DAF requires additional energy consumption to inject dissolved air to

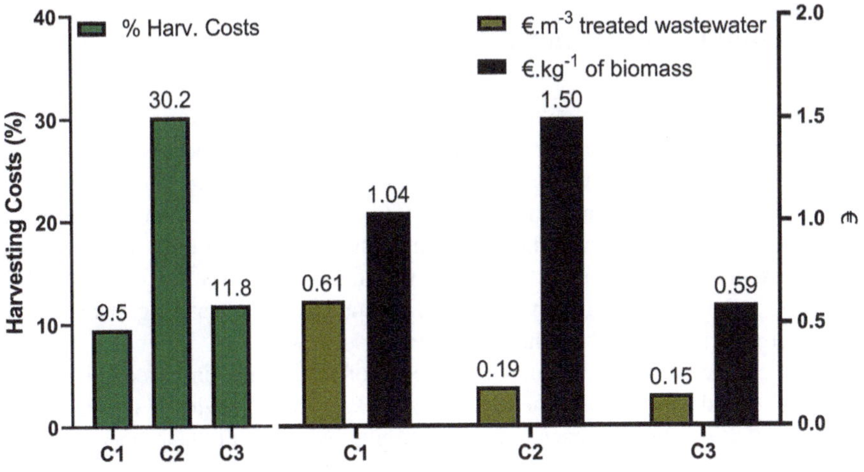

Fig. 4.4 Techno-economic performance indicators

induce biomass flotation. Additionally, the continuous operation of the HRAP enables direct feeding of the settler from the HRAP effluent, without requiring an extra pump.

The current study estimated the costs of biomass production and treated wastewater. Two sources were used for biomass production: experimental results from the INCOVER plant (C1) and estimated results from the BIO_ALGAE model of the HRAP of PAVITR project (C2 and C3). The cost of wastewater treatment ranged from 0.61 €/m^3 for C1 to 0.15 €/m^3 for C3. The high cost of C1 was attributed to a research-focused design that led to excessive treatment costs. The treatment costs of C2 and C3 were 0.19 €/m^3 and 0.15 €/m^3, respectively, which align with the treatment-focused design of HRAP and fall within the operating cost range of large-scale conventional wastewater treatment (0.1–0.3 €/m^3) (Ortiz et al. 2022b).

The production-focused design of the tubular semi-close system resulted in a lower production cost of 1.04 €/kg of biomass compared to C2 (1.50 €/kg), mainly due to the high consumption of the DAF for biomass harvesting. C3 had a significantly lower biomass production cost of 0.59 €/kg compared to C2, which reflected the use of gravity sedimentation for microalgal biomass harvesting. The results highlight the potential of gravity sedimentation as a viable option for microalgal biomass harvesting and the use of lamella settlers to increase the critical settling velocity and reduce the required space for the process. Moreover, lamella settlers can reduce coagulant consumption, as shown by Ortiz et al. (2021), by an average of 7.5 mg/L compared to conventional settlers (20 g/L). This is particularly relevant since energy consumption per m^3 of treated water decreases with scale up, while coagulant consumption remains proportional. However, sedimentation-based harvesting is not yet fully consolidated at full scale, and further studies are necessary to validate recent findings and optimize the operation of microalgae harvesting by large-scale sedimentation with lamella settlers.

4.4 Final Considerations

In conclusion, microalgae culture is a promising sustainable technology that can produce feedstock for a variety of applications, including bioproducts, biofuels, and wastewater treatment. However, the microalgae harvesting process has long been a major obstacle to achieving efficient biomass production. While mechanical-based technologies offer high separation efficiency, their costs often limit their practical use in real-world scenarios, particularly in the context of wastewater treatment. The two-step separation process involving sedimentation and dewatering has emerged as a cost-effective solution to this problem.

This chapter has highlighted the benefits of sedimentation-based microalgae thickening and the use of lamella settling tanks, which provide a more energy-efficient and scalable approach to harvesting microalgae. With the proper timing of operating elements setting-up, microalgae gravity thickening can be achieved, and the resulting biomass can be recovered by sudden pulses in harvesting flow. This approach can contribute to the techno-economic feasibility of microalgae sedimentation-based harvesting and opens up opportunities for further downstream processes, such as anaerobic digestion, which can lead to the production of high-value products like bioplastics and pigments.

Overall, the principles and advancements demonstrated in this chapter have the potential to improve the performance of microalgae harvesting units in biorefineries and treatment plants at full scale. By reducing operation costs, limiting the use of coagulants and flocculants, and improving biomass thickening, the approach presented here provides a valuable solution to overcoming the limitations of the microalgae WWT.

Acknowledgments Authors would like to thank the Government of Catalonia [Consolidated Research Group 2017 SGR 1029] for their financial support. Igor C.F. Sampaio would like to thank the Brazilian National Council for Scientific and Technological Development (CNPq) for the financial support and his post-doctoral scholarship (Call 26/2021). The authors also thank Biorender online program for figures design.

References

Bajcar T, Steinman F, Širok B et al (2011) Sedimentation efficiency of two continuously operating circular settling tanks with different inlet- and outlet arrangements. Chem Eng J 178:217–224. https://doi.org/10.1016/j.cej.2011.10.054

Barros AI, Gonçalves AL, Simões M et al (2015) Harvesting techniques applied to microalgae: a review. Renew Sust Energ Rev 41:1489–1500

Benedetti M, Vecchi V, Barera S et al (2018) Biomass from microalgae: the potential of domestication towards sustainable biofactories. Microb Cell Factories 17(1):173. https://doi.org/10.1186/s12934-018-1019-3

Bustos MC, Concha F, Bürger R et al (1999) Sedimentation and thickening. Springer, Dordrecht

Crittenden JC, Trussell RR, Hand DW et al (2012) Gravity separation. MWH's Water Treat (Type I):641–725. https://doi.org/10.1002/9781118131473.ch10

Dankers PJT, Winterwerp JC (2007) Hindered settling of mud flocs: theory and validation. Cont Shelf Res 27(14):1893–1907. https://doi.org/10.1016/j.csr.2007.03.005

De A (2017) Flocculant settling. In: Sedimentation process and design of settling systems. Springer Transactions in Civil and Environmental Engineering. Springer, New Delhi. https://doi.org/10.1007/978-81-322-3634-4_6

Deconinck N, Muylaert K, Ivens W et al (2018) Innovative harvesting processes for microalgae biomass production: a perspective from patent literature. Algal Res 31:469–477. https://doi.org/10.1016/j.algal.2018.01.016

Demir-Yilmaz I, Ftouhi MS, Balayssac S et al (2023) Bubble functionalization in flotation process improve microalgae harvesting. Chem Eng J 452:139349. https://doi.org/10.1016/j.cej.2022.139349

Díez-Montero R, Belohlav V, Ortiz A et al (2020a) Evaluation of daily and seasonal variations in a semi-closed photobioreactor for microalgae-based bioremediation of agricultural runoff at full-scale. Algal Res 47(February):101859. https://doi.org/10.1016/j.algal.2020.101859

Díez-Montero R, Vassalle L, Passos F et al (2020b) Scaling-up the anaerobic digestion of pretreated microalgal biomass within a water resource recovery facility. Energies 13(20):5484. https://doi.org/10.3390/en13205484

EPA (2003) US Environmental Protection Agency, September 2003

Fasaei F, Bitter JHH, Slegers PMM et al (2018) Techno-economic evaluation of microalgae harvesting and dewatering systems. Algal Res 31(June 2017):347–362. https://doi.org/10.1016/j.algal.2017.11.038

García J, Ortiz A, Álvarez E et al (2018) Nutrient removal from agricultural run-off in demonstrative full scale tubular photobioreactors for microalgae growth. Ecol Eng 120:513–521. https://doi.org/10.1016/j.ecoleng.2018.07.002

Gerardo ML, Van Den Hende S, Vervaeren H et al (2015) Harvesting of microalgae within a biorefinery approach: a review of the developments and case studies from pilot-plants. Algal Res 11:248–262. https://doi.org/10.1016/j.algal.2015.06.019

Ghazvini M, Kavosi M, Sharma R et al (2022) A review on mechanical-based microalgae harvesting methods for biofuel production. Biomass Bioenergy 158:106348. https://doi.org/10.1016/j.biombioe.2022.106348

Gudin C, Chaumont D (1991) Cell fragility – the key problem of microalgae mass production in closed photobioreactors. Bioresour Technol 38(2–3):145–151. https://doi.org/10.1016/0960-8524(91)90146-B

Gudin C, Thepenier C (1986) Bioconversion of solar energy into organic chemicals by microalgae. Adv Biotechnol Process 6:73–110

Gutiérrez R, Ferrer I, García J et al (2015a) Influence of starch on microalgal biomass recovery, settleability and biogas production. Bioresour Technol 185:341–345. https://doi.org/10.1016/j.biortech.2015.03.003

Gutiérrez R, Passos F, Ferrer I et al (2015b) Harvesting microalgae from wastewater treatment systems with natural flocculants: effect on biomass settling and biogas production. Algal Res 9:204–211. https://doi.org/10.1016/j.algal.2015.03.010

Gutiérrez R, Ferrer I, González-Molina A et al (2016a) Microalgae recycling improves biomass recovery from wastewater treatment high rate algal ponds. Water Res 106:539–549. https://doi.org/10.1016/j.watres.2016.10.039

Gutiérrez R, Ferrer I, Uggetti E et al (2016b) Settling velocity distribution of microalgal biomass from urban wastewater treatment high rate algal ponds. Algal Res 16:409–417. https://doi.org/10.1016/j.algal.2016.03.037

Hapońska M, Clavero E, Salvadó J et al (2018) Pilot scale dewatering of Chlorella sorokiniana and Dunaliella tertiolecta by sedimentation followed by dynamic filtration. Algal Res 33(February):118–124. https://doi.org/10.1016/j.algal.2018.05.007

Hom-Diaz A, Jaén-Gil A, Bello-Laserna I et al (2017) Performance of a microalgal photobioreactor treating toilet wastewater: pharmaceutically active compound removal and biomass harvesting. Sci Total Environ 592:1–11. https://doi.org/10.1016/j.scitotenv.2017.02.224

Lam MK, Khoo CG, Lee KT (2019) Scale-up and commercialization of algal cultivation and biofuels production. In: Biofuels from algae. Elsevier, pp 475–506

Li K, Liu Q, Fang F et al (2019) Microalgae-based wastewater treatment for nutrients recovery: a review. Bioresour Technol 291:121934. https://doi.org/10.1016/j.biortech.2019.121934

Lu Z, Beal CM, Johnson ZI (2022) Comparative performance and technoeconomic analyses of two microalgae harvesting systems evaluated at a commercially relevant scale. Algal Res 64:102667. https://doi.org/10.1016/j.algal.2022.102667

Lundquist T, Spierling R, Crowe B et al (2018) Final report: scale-up of algal biofuel production using waste nutrients. Golden

Marín D, Ortíz A, Díez-Montero R et al (2019) Influence of liquid-to-biogas ratio and alkalinity on the biogas upgrading performance in a demo scale algal-bacterial photobioreactor. Bioresour Technol 280(December 2018):112–117. https://doi.org/10.1016/j.biortech.2019.02.029

Moorthy RK, Premalatha M, Arumugam M (2017) Batch sedimentation studies for freshwater green alga Scenedesmus abundans using combination of flocculants. Front Chem 5(June):1–13. https://doi.org/10.3389/fchem.2017.00037

Mubarak M, Shaija A, Suchithra TV (2019) Flocculation: an effective way to harvest microalgae for biodiesel production. J Environ Chem Eng 7(4):103221. https://doi.org/10.1016/j.jece.2019.103221

Muylaert K, Bastiaens L, Vandamme D et al (2017) Harvesting of microalgae: overview of process options and their strengths and drawbacks. In: Microalgae-based biofuels and bioproducts. Elsevier, pp 113–132

Najjar YSH, Abu-Shamleh A (2020) Harvesting of microalgae by centrifugation for biodiesel production: a review. Algal Res 51:102046. https://doi.org/10.1016/j.algal.2020.102046

Ndikubwimana T, Zeng X, Murwanashyaka T et al (2016) Harvesting of freshwater microalgae with microbial bioflocculant: a pilot-scale study. Biotechnol Biofuels 9(1):1–11. https://doi.org/10.1186/s13068-016-0458-5

Niemi C, Gentili FG (2021) The use of natural organic flocculants for harvesting microalgae grown in municipal wastewater at different culture densities. Physiol Plant 173(2):536–542. https://doi.org/10.1111/ppl.13409

Ogbonna CN, Nwoba EG (2021) Bio-based flocculants for sustainable harvesting of microalgae for biofuel production. A review. Renew Sustain Energy Rev 139:110690. https://doi.org/10.1016/j.rser.2020.110690

Ortiz A (2021) Microalgae-based wastewater treatment systems at demonstrative scale: gravity harvesting and thickening of biomass, and advanced design of bioreactors. PhD thesis, Universitat Politècnica de Catalunya

Ortiz A, García-Galán MJ, García J et al (2021) Optimization and operation of a demonstrative full scale microalgae harvesting unit based on coagulation, flocculation and sedimentation. Sep Purif Technol 259:118171. https://doi.org/10.1016/j.seppur.2020.118171

Ortiz A, Díez-Montero R, García J et al (2022a) Advanced biokinetic and hydrodynamic modelling to support and optimize the design of full-scale high rate algal ponds. Comput Struct Biotechnol J 20:386–398. https://doi.org/10.1016/j.csbj.2021.12.034

Ortiz A, García J, Uggetti E et al (2022b) Optimization of multi-stage thickening of biomass in a demonstrative full–scale microalgae-based wastewater treatment system. Sep Purif Technol 281:119830. https://doi.org/10.1016/j.seppur.2021.119830

Pahl SL, Lee AK, Kalaitzidis T et al (2013) Harvesting, thickening and dewatering microalgae biomass. In: Borowitzka MA, Moheimani NR (eds) Algae biofuels energy, Developments in Applied Phycology 5, pp 1–288. https://doi.org/10.1007/978-94-007-5479-9

Passos F, García J, Ferrer I (2013) Impact of low temperature pretreatment on the anaerobic digestion of microalgal biomass. Bioresour Technol 138:79–86. https://doi.org/10.1016/j.biortech.2013.03.114

Robescu D, Mandiş C, Robescu D (2010) Design lamellar secondary settling tank using numerical modeling. UPB Sci Bull Ser D Mech Eng 72(4):211–216

Rossi S, Visigalli S, Castillo Cascino F et al (2021) Metal-based flocculation to harvest microalgae: a look beyond separation efficiency. Sci Total Environ 799:149395. https://doi.org/10.1016/j.scitotenv.2021.149395

Roy M, Mohanty K (2019) A comprehensive review on microalgal harvesting strategies: current status and future prospects. Algal Res 44(September):101683. https://doi.org/10.1016/j.algal.2019.101683

Rueda E, García-Galán MJ, Ortiz A et al (2020) Bioremediation of agricultural runoff and biopolymers production from cyanobacteria cultured in demonstrative full-scale photobioreactors. Process Saf Environ Prot 139:241–250. https://doi.org/10.1016/j.psep.2020.03.035

Senatore V, Rueda E, Bellver M et al (2023) Production of phycobiliproteins, bioplastics and lipids by the cyanobacteria Synechocystis sp. treating secondary effluent in a biorefinery approach. Sci Total Environ 857:159343. https://doi.org/10.1016/j.scitotenv.2022.159343

Shaikh SMR, Nasser MS, Hussein IA et al (2017) Investigation of the effect of polyelectrolyte structure and type on the electrokinetics and flocculation behavior of bentonite dispersions. Chem Eng J 311:265–276. https://doi.org/10.1016/j.cej.2016.11.098

Šikić T (2017) Five methods for secondary settler design. Filtr + Sep 54(4):28–31. https://doi.org/10.1016/S0015-1882(18)30043-0

Singh G, Patidar SK (2018) Microalgae harvesting techniques: a review. J Environ Manag 217:499–508. https://doi.org/10.1016/j.jenvman.2018.04.010

Smith BT, Davis RH (2013) Particle concentration using inclined sedimentation via sludge accumulation and removal for algae harvesting. Chem Eng Sci 91:79–85. https://doi.org/10.1016/j.ces.2013.01.007

Smith BT, Halperin J, Darzins A et al (2013) Enhanced sediment flow in inclined settlers via surface modification or applied vibration for harvesting microalgae. Algal Res 2(4):369–377. https://doi.org/10.1016/j.algal.2013.05.004

Sun Y, Zhou S, Chiang P-C et al (2019) Evaluation and optimization of enhanced coagulation process: water and energy nexus. Water-Energy Nexus 2(1):25–36. https://doi.org/10.1016/j.wen.2020.01.001

Suparmaniam U, Shaik NB, Lam MK et al (2022) Valorization of fish bone waste as novel bioflocculant for rapid microalgae harvesting: experimental evaluation and modelling using back propagation artificial neural network. J Water Process Eng 47:102808. https://doi.org/10.1016/j.jwpe.2022.102808

Tchobanoglous G, Burton FI, Stensel HD (2014) Wastewater engineering: treatment and resource recovery, 5th edn. McGraw-Hill, New York

Uggetti E, García J, Álvarez JA et al (2018) Start-up of a microalgae-based treatment system within the biorefinery concept: from wastewater to bioproducts. Water Sci Technol 78(1):114–124. https://doi.org/10.2166/wst.2018.195

Vahidifar S, Saffarian MR, Hajidavalloo E (2018) Introducing the theory of successful settling in order to evaluate and optimize the sedimentation tanks. Meccanica 53(14):3477–3493. https://doi.org/10.1007/s11012-018-0907-2

Vandamme D (2017) Harvesting, thickening and dewatering processes, pp 202–223. https://doi.org/10.2174/9781681085227117010010

Vandamme D, Foubert I, Muylaert K (2013) Flocculation as a low-cost method for harvesting microalgae for bulk biomass production. Trends Biotechnol 31(4):233–239. https://doi.org/10.1016/j.tibtech.2012.12.005

Vanthoor-Koopmans M, Wijffels RH, Barbosa MJ et al (2013) Biorefinery of microalgae for food and fuel. Bioresour Technol 135:142–149. https://doi.org/10.1016/j.biortech.2012.10.135

Wang L, Pan B, Gao Y et al (2019) Efficient membrane microalgal harvesting: pilot-scale performance and techno-economic analysis. J Clean Prod 218:83–95. https://doi.org/10.1016/j.jclepro.2019.01.321

Wei C, Liao Q, Huang Y et al (2020) Simultaneous enhancing the sedimentation and adsorption performance of Chlorella vulgaris with montmorillonite modified cationic starch. Biochem Eng J 164(September). https://doi.org/10.1016/j.bej.2020.107785

Wijffels RH, Kruse O, Hellingwerf KJ (2013) Potential of industrial biotechnology with cyanobacteria and eukaryotic microalgae. Curr Opin Biotechnol 24(3):405–413. https://doi.org/10.1016/j.copbio.2013.04.004

Yin Q, Song C (2021) A high precision model for the terminal settling velocity of drops in fluid medium. Front Earth Sci 15(4):947–955. https://doi.org/10.1007/s11707-020-0835-z

Zhao Z, Blockx J, Muylaert K et al (2022) Exploiting flocculation and membrane filtration synergies for highly energy-efficient, high-yield microalgae harvesting. Sep Purif Technol 296:121386. https://doi.org/10.1016/j.seppur.2022.121386

Zhu L (2015) Biorefinery as a promising approach to promote microalgae industry: an innovative framework. Renew Sust Energ Rev 41:1376–1384. https://doi.org/10.1016/j.rser.2014.09.040

Zhu L, Hu T, Li S et al (2020) Effects of operating parameters on algae Chlorella vulgaris biomass harvesting and lipid extraction using metal sulfates as flocculants. Biomass Bioenergy 132:105433. https://doi.org/10.1016/j.biombioe.2019.105433

Chapter 5
Mass Balance of the Major Cellular Carbon Pools for the Marine Diatom *Thalassiosira pseudonana* as Regulated by Irradiance, CO_2, and Growth Phase

Frank Shang, Laura T. Carney, Elliot L. Weiss, Patricia Abelin, Maria Vernet, Dominick Mendola, Wilson G. Mendoza, and Brian Gregory Mitchell

Abstract Salt tolerant strains of photosynthetic microalgae are promising as a potential new commodity-scale source of biomass for renewable fuels, biomaterials, animal feeds, and human nutraceuticals. Understanding the dynamics of carbon allocation among the cellular pools as a function of the algae cultivation environment is essential for predicting production yields of the biomass components of economic interest. We investigated the effect of irradiance and CO_2 on the cellular accumulation of fatty acids, pigments, carbohydrates, and proteins in the marine diatom *Thalassiosira pseudonana.* Triplicate batch cultures were grown at 18 °C using saturating and non-saturating irradiance levels in combination with ambient and elevated CO_2 concentrations. Carbon allocation was quantified during mid-log (balanced) growth, and stationary phase when metabolism is unbalanced and allocation between carbon pools shifts. Dynamic changes were observed in total carbon partitioning as well as fatty acid and pigment profiles. The percentage of cellular carbon allocated to lipid storage reached a maximum of 45.19 ± 8.20% during stress-induced stationary phase but were significantly lower during mid-log phase. Pigments accounted for a maximum of 4.30 ± 0.46% of cellular carbon during mid-log growth and were significantly lower during stationary phase. Significant trends in fatty acid and pigment profiles based on differences in environmental parameters and growth phase were also observed. We accounted for an average of 85% of cellular carbon based on quantifying the carbon quota in the lipid, protein, carbohydrate, and pigment pools. In this study, we quantify the changes in carbon allocation of the bulk metabolome for *Thalassiosira pseudonana* at sub-saturating and saturating irradiance during balanced exponential growth with adequate nutrients and during unbalanced metabolism with the onset of stationary phase induced by nutrient depletion.

F. Shang · L. T. Carney · E. L. Weiss · P. Abelin · M. Vernet
D. Mendola · W. G. Mendoza · B. G. Mitchell (✉)
Integrative Oceanography Division, Scripps Institution of Oceanography, University of California San Diego, La Jolla, CA, USA
e-mail: gmitchell@ucsd.edu

J. A. V. Costa et al. (eds.), *Microalgal Bioengineering*,
https://doi.org/10.1007/978-3-031-61253-4_5

Keywords *Thalassiosira pseudonana* · Microalgae · Lipids · Pigments · Carbohydrates · Carbon partitioning

5.1 Introduction

Global consumption of food and fuel is projected to increase exponentially over the next century, largely driven by the socioeconomic progress in developing markets and population growth. There will be significant geopolitical pressures on the utilization of water, arable land and biodiversity resources, technological challenges in meeting the demand for protein using terrestrial crops, and risks relative to the allocation of resources if terrestrial crops for biofuel are expanded (FAO 2022; Elias et al. 2023). The adoption of policies and development of techniques to abate these imminent global challenges are increasing in importance as fuel and food consumption grows and natural resource depletion accelerates.

To generate commodities from more sustainable photosynthetic feedstocks requires high yield crops that have minimal competition for arable land, fresh water, and nutrients. Microalgae have significant potential as a feedstock of photosynthetic biomass since yields and resource use (land, top soil, freshwater, nutrients) are superior to any terrestrial crop (Georgianna and Mayfield 2012). In the last four decades, studies on microalgae have shown that biomass can be generated efficiently for animal feed, wastewater treatment, nutraceuticals, and fuel with minimal use of freshwater, arable land, and nutrients (Rosenberg et al. 2008; Sander and Murthy 2010; Scott et al. 2010). However, there remains a large challenge to bring the costs of microalgae cultivation down so that the biomass can compete economically in the commodity markets for fuel, feed, and materials.

Efforts have mostly focused on utilizing lipids from microalgae for biodiesel; either continuous or semi-continuous cultures of algae cultures are metabolically stressed to induce the accumulation of lipids, which are then extracted and used to produce biodiesel (Utting 1985; Sukenik et al. 1993; Yu et al. 2009). Via hydro processing, biodiesel can be converted to a renewable diesel or aviation fuel compatible with the current fuel infrastructure (Doliente et al. 2020). Although life cycle analysis for fuel production from algae has been shown to have a low greenhouse gas burden (Woertz et al. 2014), large fractions of the cellular biomass are not utilized for fuel including proteins, carbohydrates, and pigments. Alternatively, a multi-component economic strategy for the co-production of fuels, feed, materials, and nutraceuticals from the lipids, proteins, carbohydrates, and pigments could enhance the value of algal biomass as a feedstock for multiple downstream uses (Brand 1981; Chaumont 1993; Brown et al. 1997; Poulsen et al. 2007). Effective implementation of this strategy will require an understanding of how abiotic growth conditions control the partitioning of cellular carbon to the major carbon pools (i.e. lipids, proteins, carbohydrates, pigments). The change in proportion and profiles of each pool impact the valuation of the biomass for animal feeds, biofuels, and nutraceutical components including omega-3 fatty acids and pigment antioxidants (Seto et al. 1984; Fabregas et al. 1986; Fernandez-Reiriz et al. 1989). Therefore, quantitative assessment of carbon allocation is essential for a robust understanding of the metabolic regulation of carbon partitioning and to predict the valuation of the biomass as input in techno-economic models to evaluate algae biomass commercialization (Beal et al. 2015, 2018).

Of the numerous abiotic variables that regulate algae growth and cellular carbon partitioning, CO_2 availability and irradiance (as photosynthetically available radiation; PAR) are important from an industrial economics standpoint (Ryther and Guillard 1962; Goldman and Ryther 1976). Assuming sufficient solar energy and a reasonable temperature climate, the most expensive system inputs for photoautotrophic microalgae production are CO_2 and macronutrients (Greenleaf and SenGupta 2009). There is a large literature on inorganic nutrient regulation of microalgae growth, and there are inexpensive primary sources of agricultural nutrients for terrestrial crops that are satisfactory for algae (FAO 2022). However, knowledge of the impact of CO_2 and the optimization of this essential nutrient source is less studied for microalgae. The interactions between CO_2 and microalgae have been examined on a laboratory scale (Hu et al. 2012; Solovchenko and Khozin-goldberg 2013), an ecological scale (Johnson et al. 2013), and an industrial scale (Ou et al. 2021); however, the cost of supplying CO_2 to an aqueous culture of algae is currently too high for industrial production of commodity products from algae and further study is needed to optimize CO_2 delivery and utilization for commercial scale (Acién-Fernández et al. 2012).

For commercial production, management of microalgae crop density, system depth and mixing rates, in combination with solar flux, defines the light intensity experienced by individual cells. Diatom have very high growth rates over a wide range of irradiances and achieve this by significantly regulating their ratio of cellular carbon to pigments (Geider 1987) and saving energy associated with the Si frustule that minimizes the cellular carbon quota (Inomura et al. 2023). A growth rate of a doubling per day can be achieved by most microalgae at a mean irradiance range of 5–10% irradiance at the Earth's surface (~100–200 μmol quanta m^{-2} s^{-1}). Irradiance flux to the microalgal cells has a direct effect on the balance between energy production and expenditure, and thus modifies the biochemical allocation of carbon into cellular pools. Investigating the effects of co-variation in irradiance and CO_2 availability on diatom physiology is an important step in optimizing system design and operation to balance crop density, depth, mixing rate, and flux of CO_2 to achieve the target carbon allocation to optimize crop value.

There are numerous studies employing established analytical techniques to evaluate yields of specific target metabolites in algae (Armbrust et al. 2004; Jiang et al. 2012; Wahidin et al. 2013). However, few studies have applied analytical techniques towards the study of a quantitative mass balance closure of the sub-cellular carbon pools including lipids, carbohydrates, proteins, and pigments (Renaud et al. 1991; Brown et al. 1996; Hu et al. 2008; Pruvost et al. 2011; Courtois de Viçose et al. 2012), and to the best of our knowledge, none of these studies have included the combined co-variation in both irradiance and CO_2. Various studies have used techniques, such as gravimetric analysis, gas and liquid chromatography, or whole-cell

FTIR scans of components but limitations in experimental design did not allow for the assessment of multiple irradiances or CO_2 on mass balance nor were pigment profiles reported together with all the other major cellular carbon pools (Jiang et al. 2012; Wahidin et al. 2013). To evaluate mass balance closure most of the literature reports ash-free dry weights determined gravimetrically as the metric of total biomass yield. However, ash free dry weight has relatively large variations in the ratio of biomass to cellular carbon. This variability is due to differences in desalting procedures and ash content between species of algae (Zhu et al. 1997), and errors related to weighing small mass of sample, changes in humidity and sample contamination by fibers and hair. Direct determination of particulate organic carbon POC using gas chromatographic methods is a far more accurate estimate of cellular carbon than ash-free dry weight. For physiological studies of sub-cellular carbon partitioning, normalizing each fraction to POC rather than ash-free dry weight is a superior representation of carbon allocation and the percentage contribution of the different carbon pools to the total organic carbon of the biomass.

In this study, we investigated the quantitative changes in the allocation of carbon into lipids, protein, carbohydrates, and pigments for the model diatom *Thalassiosira pseudonana* as a function of irradiance and CO_2 availability during log phase (exponential growth) and within 30 h after the onset of stationary phase. While our study is not a comprehensive matrix of all possible nutrient, light and CO_2 co-limitation scenarios, our controlled experiments, and use of modern methods for total carbon and estimates of carbon in the different biomass pools, provide insight that can guide future efforts to identify commodity-scale production methods for microalgae to optimize the carbon allocation into the biomass fractions to maximize value of the crop.

5.2 Materials and Methods

5.2.1 Cell Culturing and Growth Monitoring

Thalassiosira pseudonana (strain CCMP 1335) was obtained from the Provasoli-Guillard National Center for Marine Algae and Microbiota (formerly the CCMP), Bigelow Laboratory for Ocean Sciences (East Boothbay Harbor, ME, USA). Cultures were maintained at 18 °C in sterilized Enhanced Artificial Seawater Media (ESAW/HESNW) as defined by the North East Pacific Culture Collection (Harrison et al. 1980). Irradiance was supplied to the stock cultures at an intensity of 20–30 μmol photons $m^{-2}·s^{-1}$ under a 12 h:12 h light:dark cycle with cool-white fluorescent tubes (General Electric, Fairfield, CT, USA); irradiance from the source lamps was attenuated by distance and neutral-density screens.

Stock cultures were used to inoculate triplicate experimental cultures grown at 18 °C in vertically oriented 1 L tubular glass photobioreactors filled to ~800 mL. We used three irradiance levels; one below and two above the known saturating irradiance for *T. pseudonana* (Strzepek and Price 2000): low: 83 ± 11, medium: 237 ± 11, and high: 420 ± 11 μmol photon $\cdot m^{-2} \cdot s^{-1}$. Irradiance was measured for each culture condition at the time of inoculation and during each of the two sampling intervals using a Biospherical Instruments QSL-100 scalar quantum irradiance sensor (Biospherical Instruments, Inc., San Diego, CA). Variability in the irradiance of cultures was caused largely by growth of the culture leading to light attenuation within the photobioreactors as biomass increased. Experiments were carried-out over a period of 12–13 days using a 12:12 light:dark cycle, with sampling points at mid-log growth phase and 24–30 h after the cultures entered stationary phase, defined by a leveling off of the growth curve. Ambient air (as control) and 2% CO_2 in air (as a treatment variable) were delivered to the bottom of each bioreactor by a thin glass tube with the bubbles providing mixing. Eighteen 1-liter tubes were inoculated with stock cultures to achieve an initial seeding density of 1×10^4 cells mL^{-1} and grown until the cultures entered stationary phase caused by nitrate depletion.

A 5 mL sample was collected daily from each photobioreactor during the middle of the light period to quantify cell density and pH. Cell density was quantified with a Varian Cary 100 UV–VIS Spectrophotometer (Varian Inc., Santa Clara, CA, USA) using optical density for a 1 cm pathlength at 750 nm, and microscopically using a hemocytometer. The pH was monitored using a Thermo Orion Star LogR Meter (Thermo Fisher Scientific Inc., Beverly, MA, USA). During mid-exponential phase and stationary phase (e.g., between 24 and 36 h after the cessation of cell accumulation), 200 mL were collected from each replicate tube and concentrated by centrifugation (10 min at 3000 g), then lyophilized and stored at −80 °C for compositional analysis. Lipid content based on total fatty acid methyl ester analysis (FAME), protein, carbohydrates, pigments, and particulate organic carbon and nitrogen (POC and PON) were quantified from aliquots of each sample using methods described below. At the time of experimental sampling, cell size was determined using a Coulter particle counter (Beckman Coulter Inc., Fullerton, CA) equipped with a 30 μm aperture.

5.2.2 Cellular Organic Carbon and Nitrogen

Particulate organic carbon (POC) and particulate organic nitrogen (PON) measurements were determined using analytically precise elemental analysis methods (Holmes 1963). 200 μg of lyophilized sample was combusted with tungsten trioxide catalyst using an ECS 4010 CHNSO analyzer (Costech Inc., Valencia, CA, USA,) calibrated with acetyl nitrile. Samples were corrected by subtracting values for a blank that included only the catalyst.

5.2.3 Lipid Analysis

By definition, lipids are comprised of fatty acids and derivatives (Christie 2005). Determination of lipids based on total FAME is regarded as the most accurate measurement for evaluating lipids in biomass as compared to older methods that include extraction, evaporation of solvent and gravimetric estimates (Laurens 2013; Laurens et al. 2012a, b; Patel et al. 2019). Fatty acid extraction and methyl ester derivatization was performed according to methods as described by Bligh and Dyer (1959) with modifications as described below for fatty acid methyl ester (FAME) quantification using gas chromatography mass spectrometry (GCMS; AOCS 1989). To standardize the esterification efficiency, 200 μL of 1.098 $mg \cdot mL^{-1}$ pentadecanoic acid (C15 FA) in methanol was added to the lyophilized algae pellet before extraction. The pellet was then resuspended in 2 mL of 2:1 chloroform: methanol (v/v) containing 0.01% butylated hydroxytoulene (BHT) (w/v) (Ackman 1998). The chloroform fraction was separated and dried under N2 (100%). The residual lipid film was then derivatized with potassium hydroxide in methanol in accordance with the American Oil Chemists' Society method Ce 1b-89 (AOCS 1989). A total of 450 μL from the organic phase was added to a GC autosampler vial along with 50 μL of C13/C19 FAME as the quantification standard. The assay of FAME was carried out with an HP 5890 gas chromatograph system (Hewlett-Packard, Wilmington, DE, USA) equipped with a fixed outlet capillary column splitter to utilize both mass selective and flame ionization split detection on one column. Separation was achieved in a DB-Wax column (30 m, 0.25 mm, and 0.25 μm) from Agilent Technologies, Inc. (Santa Clara, CA, USA) with a helium mobile phase at a flow rate of 1.4 mL min^{-1} and a temperature ramp from 140 to 260 °C with incremental holds at 140, 160, 210, and 260 ° C.

Quantification of C14 to C18 and C20 to C24 fatty acids was performed relative to known concentrations and Flame Ionization Detection (FID) peak areas of C13 and C19 FAME internal standards (Nu Chek Prep, Waterville, MN, USA), respectively. The response factor for each standard compound was applied to the respective sample peak area to calculate the final concentration of each peak of interest. The fatty acid methylation reaction (i.e., trans-esterification) efficiency was determined from the C15 FA internal standard added to each lyophilized algal pellet. Final mass of carbon in lipids was determined based on carbon chain length.

5.2.4 Carbohydrate Analysis

Mono- and polysaccharides were quantified for each sample using the 2,4,6-tripyridyl-s-triazine method (TPTZ) (Myklestad et al. 1997). This method was designed to analyze carbohydrates in seawater and is extremely sensitive with an optimal reading range between 0.5 and 5 mg/L, thus we had to optimize the method for analysis of dense cultures. The reliability and accuracy of this method was assessed by performing analyses on three sets of reference solutions including glucose, mixed monosaccharides (i.e., galactose, mannose, ribose, xylose, arabinose), and a mixture of glucose and mixed monosaccharides. These reference solutions were analyzed for monosaccharides prior to hydrolysis, and total carbohydrate post-hydrolysis. Results showed that with or without hydrolysis, the measured carbohydrate percentage error ranged from 1% to 10%. The method had very good accuracy in total carbohydrate analysis (error 1–3%) prior to and after hydrolysis, using glucose and mixed monosaccharides as a reference.

All glassware used for these reactions was acid washed and pre-combusted (400 ° C, 1 h), and samples were kept in dark vials after harvesting. Six mL of culture was pelleted by centrifugation (10 min, 1207 × g), and the pellet was diluted to the initial volume with Milli-Q water and stored in liquid N_2 until analysis. The TPTZ method involves two reactions (Witter and Luther 2002). First, monosaccharides (non-reducing sugars) or polysaccharides (reducing sugars freed after hydrolysis) are oxidized at alkaline pH, driving the reduction of Fe^{3+} to Fe^{2+}. Next, Fe^{2+} was measured colorimetrically at 595 nm on a Varian Cary 100 UV-VIS Spectrophotometer (Varian Inc., Santa Clara, CA, USA), using TPTZ. Carbohydrate was quantified using a calibration curve made from D-glucose (Arcos Organics, New Jersey, USA) in Milli-Q water ranging from 0 to 5 mg L^{-1}. Monosaccharide content was analyzed directly from the thawed sample, and an equivalent fraction was hydrolyzed and used to determine total carbohydrates. Hydrolysis was achieved by adding 0.2 mL of 1 M HCl to 2 mL of sample in an ampoule, which was then flame-sealed and heated for 20 h at 100° C. The samples were then cooled and neutralized with 0.2 mL 1 M NaOH to pH 7, and an appropriate aliquot was taken for analysis (100–500 μL sample). Polysaccharides were estimated as the difference between total carbohydrates and monosaccharides.

The TPTZ method was compared with the method of Dubois et al. (1956), which is commonly used for carbohydrate analysis in algal cultures (Ogbonna and Tanaka 1996; Choix et al. 2012; Laurens et al. 2012a, b). Given that our testing of the TPTZ method used a relatively dense culture, the hydrolyzed microalgal culture was diluted so the carbohydrate concentration was within the standard curve of our calibration. Total (cellular) carbohydrate was found to be, on average, 7% higher using the TPTZ method than with the Dubois method, indicating that the TPTZ method is a viable alternative for total carbohydrate analysis. In addition to total carbohydrates, the TPTZ method provided estimates of both cellular and extra-cellular carbohydrate monosaccharide and polysaccharide contents in dense cultures, which cannot be determined using the Dubois method.

5.2.5 *Protein*

Crude protein was calculated from particulate organic nitrogen (see PON method above) measurements using the nitrogen to protein factor of 4.78 ± 0.62 for *T. pseudonana* derived by Lourenco et al. (2004). The percent carbon of the protein was then calculated using a factor of 0.47 (Rivkin 1989).

5.2.6 Pigment Analysis

Photosynthetic pigments were analyzed using high performance liquid chromatography (HPLC) to separate and quantify specific chlorophylls and carotenoids (Jeffrey et al. 1999). The HPLC method is based on a reverse-phase monomeric octylsilica (OS, C8) column and pyridine-containing mobile phases (Zapata et al. 2000). The HPLC was calibrated using an 18-pigment suite (DHI Laboratory Products, Horsholm, Denmark) with a 5-point concentration gradient for individual pigment calibration. This method provided good resolution in terms of both pigment composition and quantification for *T. pseudonana.* The assay of pigments was performed using a Waters Alliance HPLC (Waters Corp, Milford, MA, USA), including a Waters 996 diode-array detector (1.2 nm optical resolution) interfaced with a Waters 474 scanning fluorescence detector and Waters variable wavelength detector. Analytical separations were performed using a Waters Symmetry C8 column (WAT200630; 150 mm x· 4.6 mm, 3.5 μm particle size, 100 Å pore size) with solvents detailed by Zapata (Zapata and Garrido 1991). All solvents used were HPLC-grade, with flow rate fixed at 1 mL min^{-1}. Pigments were identified based on time of elution from the column relative to standards in combination and validated also with diode-array spectroscopy (350–750 nm). Chlorophyll signatures were also detected by fluorescence (excitation: 440 nm, emission: 650 nm). Pigment types were identified by comparison of spectra to a reference library previously created and co-chromatography with pure standards to confirm pigment identity and elution time.

5.2.7 Carbon Partitioning Among Cellular Pools

The cellular carbon fraction for each component was determined as the ratio of carbon in each component, determined with the methods above, divided by the total particulate organic carbon (POC). We also determined the total carbon as the sum of all individual carbon pools analyzed relative to total POC as an index of accuracy in our goal to achieve closure for the carbon pool allocation. Our analytical protocols do not quantify sterols or gylcerols resulting from tricylglyceride hydrolysis, hydrocarbons as wax esters, nucleic acids, or refractive cell wall components. Combined, these pools make up 5–10% of cellular carbon (Brown et al. 1996) and are considered the main fractions of the carbon pool that we did not quantify.

5.2.8 Statistical Analysis

All data were analyzed using Systat version 13 (Systat Software, Inc., Chicago, Illinois, USA). Prior to analyses, data were examined for homogeneity of variances using Levene's test and for normality by graphical examination of residuals. Data that did not meet these assumptions were corrected by transformation and re-tested. Differences among treatments were analyzed using 1-, 2-, or 3-way analyses of variance (ANOVA) followed by Tukey's HSD post hoc comparisons (hereafter, THSD).

5.3 Results

5.3.1 *Culture Growth Monitoring*

Cultures were allowed to grow for 12–13 days and final optical densities (OD_{750}) were in the range of 0.41 ± 0.07 to 0.43 ± 0.07. With the addition of CO_2 to the air supply, slightly higher ODs were achieved at both low and medium irradiance (Table 5.1). Minimal light attenuation due to culture opacity was observed throughout this experiment; irradiances remained at least 88% of the initial values at the conclusion of the experiment. Specific growth rates (μ, natural log as defined by Sosik and Mitchell 1991) ranged from 1.34 to 1.44 day^{-1} with an average for all treatments of 1.39 ± 0.19 day^{-1}. Despite the wide range in irradiance and pH, there was no statistical difference in growth among the treatments (Table 5.1). The pH of the culture bubbled with air and supplemented with CO_2 was 6.7 (± 0.2). The pH of the culture bubbled with ambient air only was 9.1 (± 0.1) during log growth but dropped to 7.8 (± 0.03) in stationary phase consistent with expected changes in CO_2

Table 5.1 Culturing parameters of each treatment (L, M, H, L + CO_2, M + CO_2; see methods for definitions) and culture parameters for each treatment at the time of mid-log and stationary phase sampling

		L	M	H	L + CO_2	M + CO_2
	Irradiance	83 ± 11	237 ± 11	420 ± 11	83 ± 11	237 ± 11
	Temperature (°C)	18 ± 1.0	18 ± 1.0	18 ± 1.0	18 ± 1.0	18 ± 1.0
	Specific Growth (μ, day^{-1})	1.34 ± 0.22	1.39 ± 0.07	1.44 ± 0.14	1.39 ± 0.26	1.37 ± 0.12
Mid-log	OD	0.14 ± 0.02	0.19 ± 0.01	0.20 ± 0.01	0.21 ± 0.03	0.24 ± 0.02
	Cell Count (10^6 cells mL^{-1})	2.7 ± 0.3	3.7 ± 0.3	3.9 ± 1.5	4.1 ± 0.5	4.7 ± 0.5
	Average Cell Diameter (μm)	4.8	4.7	4.8	4.9	4.9
	Light Attenuation (% source)	5.2	5.8	6.1	6.2	6.2
	pH	9.15 ± 0.05	9.02 ± 0.08	9.03 ± 0.02	6.89 ± 0.09	6.79 ± 0.09
Stationary	OD	0.41 ± 0.07	0.43 ± 0.09	0.43 ± 0.07	0.47 ± 0.04	0.47 ± 0.03
	Cell Count (10^6 cells mL^{-1})	8.1 ± 1.5	8.4 ± 1.8	8.5 ± 1.4	9.2 ± 0.7	9.3 ± 0.6
	Average Cell Diameter (μm)	4.8	5.5	5.6	4.9	5.6
	Light Attenuation (% source)	10.2	10.5	10.4	10.8	11.1
	pH	7.78 ± 0.06	7.78 ± 0.20	7.83 ± 0.06	6.68 ± 0.52	6.59 ± 0.15

uptake during rapid growth, but a shift to higher ratio of respiration to photosynthesis in stationary phase. Average cell diameter was measured at 4.9 ± 0.1 μm during mid-log sampling for all treatments and increased 15% during stationary phase to 5.5 ± 0.01 μm at medium and high irradiance. No significant change in cell diameter was observed at low irradiance between the growth phases (Table 5.1).

5.3.2 Partitioning of Major Carbon Pools

The primary goal of this study was to quantify mass balance closure of the major carbon pools relative to the total cellular carbon as regulated by light, CO_2, and growth phase. We accounted for an average of 85% of the total particulate organic carbon per sample when the major pools analyzed were summed (Fig. 5.1). Since we did not quantify sterols, gylcerols resulting from tricylglyceride hydrolysis, hydrocarbons as wax esters, nucleic acids, or refractive cell wall components, we assume these components, and some process losses, account for the approximately 15% of total carbon we did not account for.

Our analysis showed significant variations in the major carbon pools correlating to growth phase and environmental changes in CO_2 (pH) and irradiance. Cellular concentrations of lipids and carbohydrates were significantly higher during stationary phase compared to mid-log phase (Fig. 5.1; Table 5.1), while the concentration of both pigments and protein were significantly lower in all stationary phase samples compared to mid-log phase (Tables 5.2 and 5.3, Fig. 5.1). At stationary phase, higher concentrations of lipids and carbohydrates were detected in cultures exposed to higher irradiances compared to cultures grown at low irradiance. In addition to the bulk components shown in Fig. 5.1, in the following subsections we provide details of the response of the lipid and pigment profiles to the light and CO_2 regulation.

5.3.3 Lipids

The percentage of total lipids, quantified as fatty acids, as a fraction of POC increased significantly from mid-log to stationary phase across all treatments (Fig. 5.2; 2-way ANOVA: growth phase – $F_{1,12} = 45.39$, $P < 0.001$; irradiance – $F_{2,12} = 6.35$, $P = 0.013$). During mid-log and stationary phase, the percentage of lipids in cultures exposed to medium and high irradiance levels were consistently higher than those for low irradiance conditions regardless of CO_2 treatment (Fig. 5.2; THSD: $P < 0.045$). Lipids as a fraction of total carbon were not significantly different between medium and high irradiance treatments (Fig. 5.1; THSD: $P = 0.865$). The highest accumulation of lipids relative to total carbon was 45.19 ± 8.27% (mean ± 1SD), measured during stationary phase, at medium irradiance with the elevated CO_2 treatment (Fig. 5.2). For 2% CO_2 treatments during mid-log phase, the

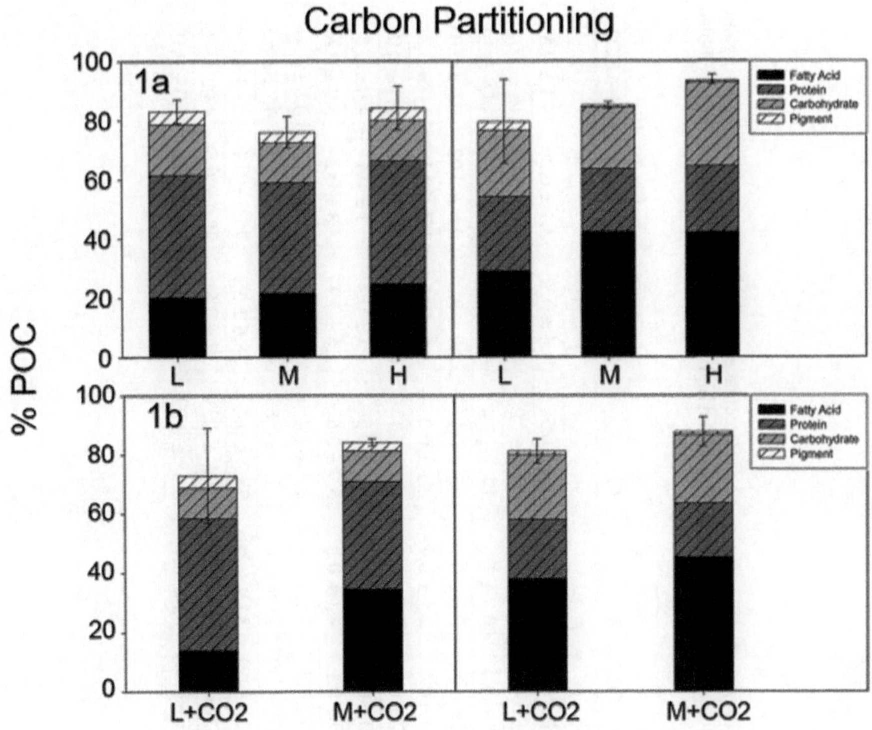

Fig. 5.1 Biochemical composition of cultures of *Thalassiosira pseudonana* as a function of irradiance and CO_2 expressed as a percentage of total particulate carbon (POC) during mid-log phase (left) and stationary phase (right). (**a**) Cultures treated with three irradiances low: 83 ± 11, medium: 237 ± 11 and high: 420 ± 11 μmol photon $m^{-2} \cdot s^{-1}$ bubbled with ambient air. (**b**) Cultures treated with the same irradiances and bubbled with 2% CO_2. Error bars represent the standard deviation of triplicate cultures

high irradiance treatment had 1.5× greater total lipid concentration compared to the low irradiance treatment (Fig. 5.2; 2-way ANOVA: CO_2 – $F_{1,7} = 1.68$, $P = 0.236$; irradiance – $F_{1,7} = 25.52$, $P = 0.001$; CO_2 × irradiance – $F_{1,7} = 19.23$, $P = 0.003$; THSD: $P = 0.002$).

Significant amounts of myristic (C14), palmitic (C16), and palmitoleic (C16:1) acids in addition to the poly-unsaturated stearidonic and eicosapentaenoic acids (C18:4 and C20:5, respectively) were detected in all samples. The relative concentration of poly-unsaturated fatty acids (PUFAs) and saturated fatty acids (SFAs) were inversely correlated (Pearson correlation: $r = -0.76$; Fig. 5.3). PUFAs were consistently higher during mid-log phase compared to stationary phase, with the highest percentage of PUFAs relative to total fatty acids, 54 ± 5% (mean ± 1SD, $n = 3$), detected at the mid-log phase in the low irradiance treatment (2-way ANOVA: irradiance – $F_{2,12} = 16.72$, $P < 0.001$; growth phase – $F_{1,12} = 566.14$, $P < 0.001$; irradiance × growth phase – $F_{2,12} = 0.22$, $P = 0.807$; THSD irradiance $P < 0.001$). Conversely, SFAs were highest during stationary phase in the two higher irradiance

Table 5.2 Carbon composition in the sub-cellular lipid pools for *Thalassiosira pseudonana*, expressed as percentages of particulate organic carbon organic carbon (% POC) for mid-log and stationary phase as a function of irradiances low: 83 ± 11, medium: 237 ± 11, and high: 420 ± 11 μmol photon m^{-2} s^{-1} bubbled with ambient air or 2% CO_2 (+CO_2)

	Mid-Log					Stationary				
	L	M	H	L + CO2	M + CO2	L	M	H	L + CO2	M + CO2
Fatty Acids										
C14	2.37 ± 0.4	3.01 ± 0.6	3.8 .0 ± 0.8	1.8 ± 0.3	4.1 ± 0.2	3.0 ± 0.5	4.4 ± 0.5	4.4 ± 0.7	3.4 ± 0.2	4.1 ± 0.5
C16	1.94 ± 0.2	2.42 ± 0.6	2.61 ± 0.4	1.9 ± 1.0	6 ± 0.4	5.5 ± 1.7	10.0 ± 1.2	10.4 ± 1.7	11.0 ± 0.2	15 ± 3.0
C16:1	3.94 ± 1.4	3.95 ± 0.9	4.08 ± 0.9	2.4 ± 0.4	9.8 ± 0.4	9.7 ± 1.6	14.9 ± 2.4	14.1 ± 2.7	13.1 ± 0.5	14.8 ± 2.1
C16:2	0.83 ± 0.2	1.03 ± 0.2	1.3 ± 0.2	ND	0.98 ± 0.1	0.8 ± 0.2	0.9 ± 0.1	0.9 ± 0.2	0.6 ± 0.02	0.6 ± 0.03
C16:3	3.71 ± 1.3	3.11 ± 0.8	3.1 ± 0.6	1.9 ± 0.4	2.9 ± 0.1	2.1 ± 0.4	2.1 ± 0.2	1.6 ± 0.4	1.7 ± 0.1	1.6 ± 0.2
C18	0.72 ± 0.2	1.22 ± 0.3	1.88 ± 0.6	1.8 ± 0.3	1.2 ± 0.1	0.4 ± 0.1	0.7 ± 0.1	0.74 ± 0.2	0.3 ± 0.03	0.4 ± 0.04
C18:4	2.01 ± 0.4	1.89 ± 0.4	2.05 ± 0.3	1.4 ± 0.5	3.0 ± 0.1	1.32 ± 0.24	2.0 ± 0.5	3.1 ± 0.4	3.1 ± 0.1	4.0 ± 0.6
C20:5	3.6 ± 0.8	3.72 ± 0.8	4.38 ± 0.8	2.6 ± 0.6	4.9 ± 0.1	4.1 ± 0.8	5.9 ± 0.6	5.6 ± 1.1	3.8 ± 0.2	3.9 ± 0.4
C22:6	0.8 ± 1.4	1.09 ± 1.4	1.34 ± 1.4	ND	1.6 ± 1.4	1.9 ± 1.4	1.3 ± 1.4	1.2 ± 1.4	0.8 ± 1.4	0.8 ± 1.4
% POC FA	19.92 ± 6.1	21.4 ± 6.1	24.5 ± 6	13.8 ± 3.5	34.4 ± 2.9	28.9 ± 6.8	42.1 ± 6.8	42 ± 8.8	37.9 ± 2.3	45.2 ± 8.3
% PUFA	54.35 ± 4.9	49 ± 1.0	49 ± 2.9	42.1 ± 9.4	37.6 ± 1.0	34.9 ± 2.1	28.2 ± 0.4	28.6 ± 3.1	25.8 ± 0.4	23.3 ± 0.4
% SFA	25.57 ± 2.5	30.98 ± 1.0	33.4 ± 2.5	36 ± 4.9	33.0 ± 0.5	30.4 ± 2.7	35.8 ± 3.3	37 ± 0.8	38.7 ± 0.6	43.2 ± 1.0
C16 / C16:1	0.49 ± 0.01	0.61 ± 0.03	0.64 ± 0.02	0.8 ± 0.1	0.6 ± 0.01	0.6 ± 0.1	0.68 ± 0.02	0.7 ± 0.02	0.8 ± 0.04	1.0 ± 0.1

Standard deviations represent those of triplicate cultures

Table 5.3 Carbon composition sub-cellular pigment and carbohydrate pools for *Thalassiosira pseudonana*, expressed as percentages of particulate organic carbon (% POC) for mid-log and stationary phase as a function of irradiances low: 83 ± 11, medium: 237 ± 11, and high: 420 ± 11 μmol photon m^{-2} s^{-1} bubbled with ambient air or 2% CO_2

	Mid-Log					Stationary				
	L	M	H	L + CO2	M + CO2	L	M	H	L + CO2	M + CO2
Pigments										
Chl-c2	0.09 ± 0.04	0.1 ± 0.04	0.1 ± 0.05	0 ± 0	0 ± 0.005	0 ± 0	0 ± 0	0 ± 0.002	0.03 ± 0.004	0.017 ± 0.01
Chl-c1	0.13 ± 0.04	0.1 ± 0.03	0.1 ± 0.03	0.01 ± 0.003	0.1 ± 0.01	0 ± 0	0 ± 0.001	0.01 ± 0.005	0.02 ± 0.002	0.017 ± 0.01
Fucox	0.97 ± 0.11	0.8 ± 0.27	0.8 ± 0.18	1.2 ± 0.02	0.5 ± 0.05	0.83 ± 0.11	0.18 ± 0.03	0.14 ± 0.07	0.33 ± 0.039	0.26 ± 0.09
Diadinox	0.2 ± 0.02	0.28 ± 0.06	0.5 ± 0.1	0.19 ± 0.002	0.1 ± 0.02	0.23 ± 0.10	0.07 ± 0.03	0.07 ± 0.05	0.08 ± 0.01	0.1 ± 0.03
Diatox	0.06 ± 0.01	0.05 ± 0.01	0.1 ± 0.01	0.02 ± 0.01	0.2 ± 0.02	0.07 ± 0.05	0.03 ± 0.004	0.04 ± 0.01	0.039 ± 0	0.04 ± 0.02
Chl-a	2.79 ± 0.22	2.2 ± 0.02	2.5 ± 0.32	2.6 ± 0.02	1.9 ± 0.14	1.78 ± 0.15	0.4 ± 0.04	0.23 ± 0.12	0.66 ± 0.05	0.5 ± 0.16
B-carotene	0.08 ± 0.05	0.07 ± 0.01	0.1 ± 0.01	0.05 ± 0.01	0.2 ± 0.22	0.02 ± 0.002	0.01 ± 0	0 ± 0.002	0.013 ± 0	0.01 ± 0.001
% POC TP	4.33 ± 0.46	3.6 ± 0.34	4.2 ± 0.69	4.1 ± 0.003	2.9 ± 0.04	2.93 ± 0.2	0.6 ± 0.08	0.49 ± 0.25	1.17 ± 0.1	1.0 ± 0.31
Chl / Carotenoids	2.28 ± 0.06	1.7 ± 0.08	1.8 ± 0.06	1.8 ± 0.3	2.2 ± 0.01	1.5 ± 0.07	1.36 ± 0.06	0.93 ± 0.02	1.54 ± 0.03	1.4 ± 0.03
Carbohydrates										
Polysaccharide	3.76 ± 0.77	5.5 ± 2.3	3.6 ± 2.8	5.38 ± 1.7	6.6 ± 0.2	6.49 ± 1.1	12.8 ± 5.6	13.6 ± 1.7	17.0 ± 1.87	11.3 ± 0.2
Monosaccharide	13.4 ± 1.2	7.9 ± 1.1	12 ± 2.5	5.1 ± 1.2	3.7 ± 0.7	15.9 ± 1.1	8.02 ± 0.7	14.2 ± 0.8	5.03 ± 0.2	16.7 ± 5.6
% POC Carbs	17.2 ± 2.0	13.5 ± 1.3	14 ± 0.8	10.5 ± 2.4	10 ± 0.8	22.3 ± 7.1	20.8 ± 5.4	28.3 ± 1.4	22.0 ± 1.9	23.2 ± 1.1

Standard deviations represent those of triplicate cultures

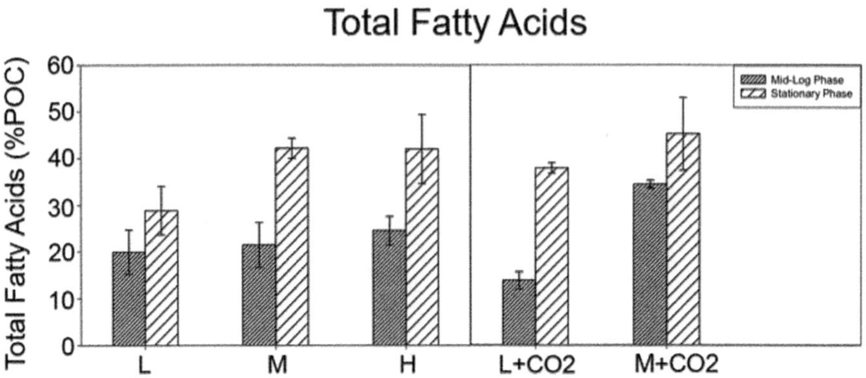

Fig. 5.2 Total fatty acids as a percentage of particulate organic carbon (%POC) at mid-log and stationary phase under the three irradiance treatments as in Fig. 5.1 (L, M, H), bubbled with air and with 2% CO_2

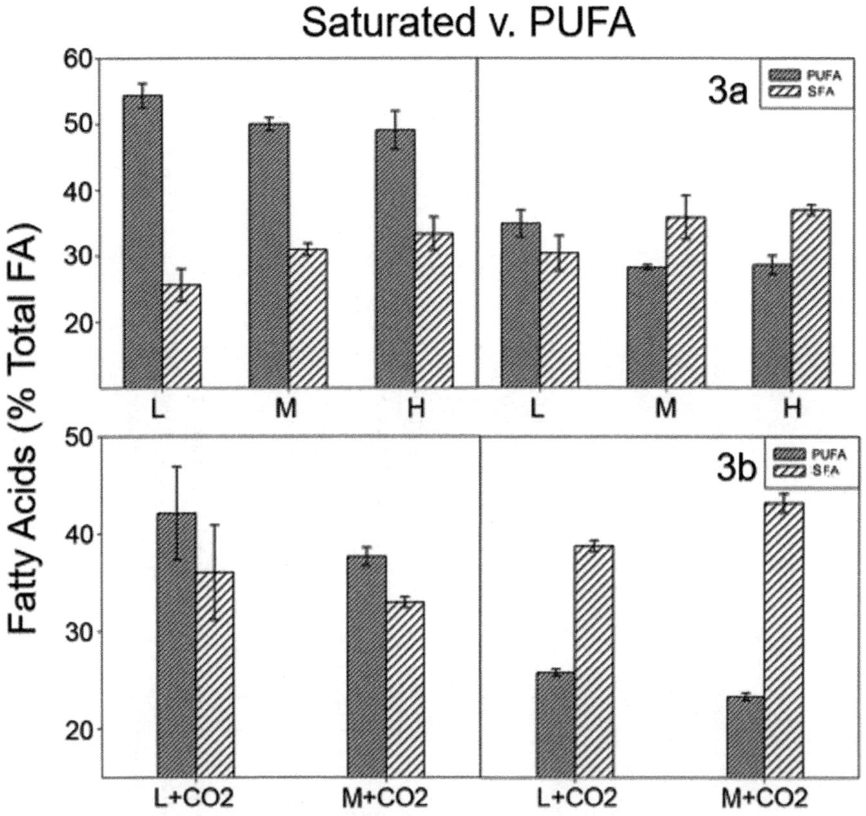

Fig. 5.3 Changes in the relative concentration of saturated fatty acids (SAT) to poly unsaturated fatty acids (PUFA) during mid-log (left) and stationary (right) phase. (**a**) Cultures treated with three irradiances (L M H) bubbled with ambient air. (**b**) Cultures treated with three irradiances (L M H) bubbled with 2% CO_2

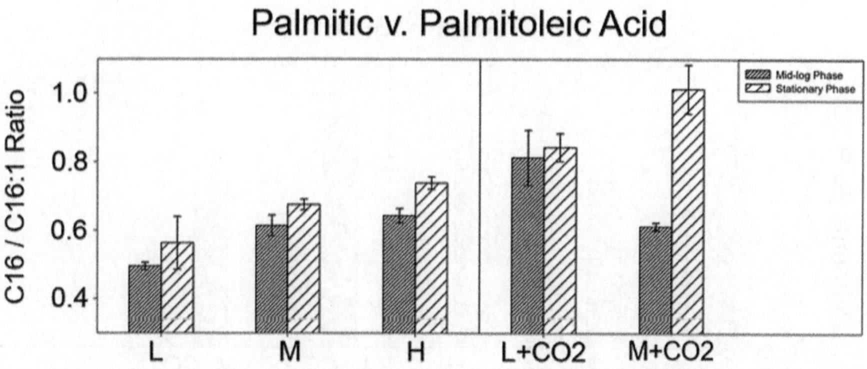

Fig. 5.4 Changes in the ratio of palmitic acid (C16) to palmitoleic acid (C16:1) during mid-log and stationary phase under three irradiance treatments (L, M, H), bubbled with air and with 2% CO_2

treatments (2-way ANOVA: irradiance – $F_{2,12} = 14.83$, $P = 0.001$; growth phase – $F_{1,12} = 15.03$, $P = 0.002$; irradiance × growth phase – $F_{2,12} = 0.19$, $P = 0.824$; THSD irradiance $P < 0.006$). Elevated CO_2 levels exaggerated this trend causing a larger decrease in PUFAs compared to SFAs during stationary phase (separate 2-way ANOVAs by phase: irradiance – $F_{1,7} > 17$, $P < 0.005$; CO_2 – $F_{1,7} > 68$, $P < 0.001$; irradiance × CO_2 – $F_{1,7} > 0.6$, $P > 0.08$). The monounsaturated palmitoleic acid was detected at significant concentrations and accounted for the remaining fatty acid fraction.

There were significant trends in the ratio of palmitic to palmitoleic acid (C16–C16:1) from mid-log phase to stationary phase. The C16–C16:1 ratio increased consistently across all conditions from mid-log to stationary phase, and from low to high irradiance (Fig. 5.4). This ratio also significantly increased during stationary phase with elevated CO_2 (2-way ANOVA: irradiance – $F_{1,7} = 7.05$, $P = 0.033$; CO_2 – $F_{1,7} = 97.81$, $P < 0.001$; irradiance × CO_2 – $F_{1,7} = 3.69$, $P = 0.096$). The highest ratio of C:16–C16:1, 1.01 ± 0.07 (mean ± 1SD; $n = 3$), was measured during stationary phase for the M + CO_2 treatment (Fig. 5.4).

5.3.4 *Pigments*

Chlorophyll *a* and the carotenoids fucoxanthin and diadinoxanthin were the major pigments detected; Chlorophylls *c1* and *c2*, *ß*–carotene and diatoxanthin were also detected at lower concentrations. Pigment composition and concentration showed changes related to both growth phase and experimental treatment. Total pigments as a fraction of total POC ranged from 0.6% to 4.5% (Fig. 5.5) and were at significantly higher concentrations during mid-log phase compared to stationary phase in all treatments. Furthermore, decreased irradiance and increased CO_2 resulted in higher pigment accumulation for mid-log phase, under low irradiance and with CO_2 supplementation, total pigments accounted for ~4.5% of the total POC (Fig. 5.5).

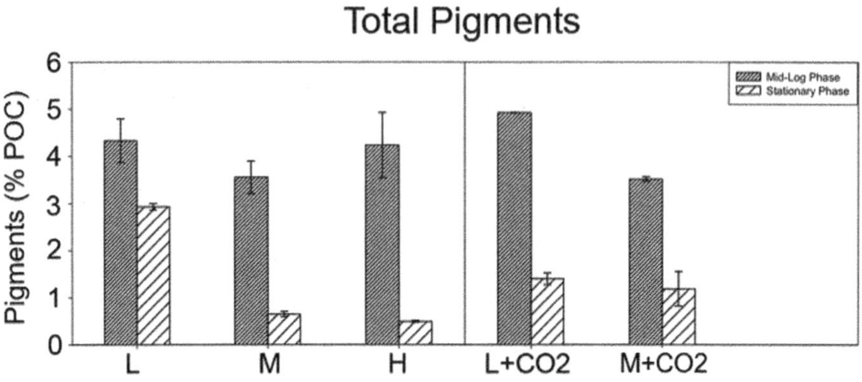

Fig. 5.5 Total pigments as a percentage of particulate organic carbon (%POC) at mid-log and stationary phase under three irradiance treatments (L, M, H), bubbled with air and with 2% CO_2

Significant decreases in total pigment concentration were observed at stationary phase in all treatments, more so at the medium and high irradiance levels (Fig. 5.5; 2-way ANOVA: irradiance – $F_{2,11} = 20.56$, $P < 0.001$; growth phase – $F_{1,11} = 100.79$, $P < 0.001$; irradiance × growth phase – $F_{2,12} = 5.64$, $P = 0.021$; THSD irradiance $P = 0.001$). The largest decrease in total pigments from mid-log phase to stationary phase was observed at high irradiance (Fig. 5.5). Elevated CO_2 appeared to reduce the inhibitory effect of higher irradiance on total pigments during stationary phase (Fig. 5.5; 2-way ANOVA: irradiance – $F_{1,8} = 62.42$, $P < 0.001$; CO_2 – $F_{1,8} = 10.12$, $P = 0.013$; irradiance × CO_2 – $F_{1,8} = 39.87$, $P < 0.001$).

Both total chlorophylls and total carotenoids decreased from mid-log to stationary phase; however, chlorophylls showed a more significant decrease than carotenoids. The ratio of chlorophylls to carotenoids was lower in stationary phase than the corresponding treatment in mid-log phase (separate 1-way ANOVAs by irradiance level: growth phase – $F_{1,3} > 7$, $P < 0.05$) and was significantly lower at medium and high irradiance compared to low irradiance (2-way ANOVA: irradiance – $F_{1,7} = 7.67$, $P = 0.028$; CO_2 – $F_{1,7} = 1.47$, $P = 0.265$; irradiance × CO_2 – $F_{1,7} = 0.48$, $P = 0.511$; Fig. 5.6). Chlorophyll *a* and fucoxanthin were the major chlorophyll and carotenoid components, respectively, and the changes in total pigment concentration and pigment as a fraction of total carbon can be attributed mainly to the decreases in these two pigments in stationary growth phase.

5.4 Discussion

To improve the quantitative understanding of the allocation of carbon in algae feedstock into the major cellular pools of economic interest, we investigated the dynamics of carbon partitioning in the model diatom *T. pseudonana* as regulated by growth phase (mid-log, stationary) and specific environmental treatments (irradiance and CO_2 level). We quantified the cellular concentrations of lipids, protein,

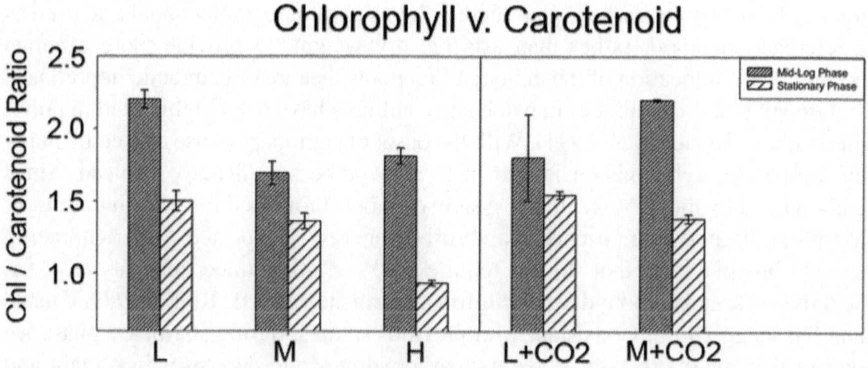

Fig. 5.6 Changes in the ratio of chlorophyll pigments (Chl) to carotenoid pigments during mid-log and stationary phase under three irradiance treatments (L, M, H), bubbled with air and with 2% CO_2

carbohydrate, and pigments, while presenting a detailed characterization of the profile and trends for fatty acids and pigments since these two allocations are of particular interest commercially. We demonstrated that abiotic factors and growth phase significantly affect carbon allocation into the major pools and, within the limits of our analytical methods, we achieved mass balance closure of the major pools of carbon relative to total organic carbon for each treatment.

The sum of the four carbon pools studied accounted for an average of 85% of total particulate organic carbon across all treatments. Approximately 5–10% of the carbon is expected in undetected sterols, glycerol, refractive cell wall components, and nucleic acids that we did not quantify. Accounting for 85% of the carbon, given the components we did not analyze, indicates we achieved a carbon mass balance closure for the allocation between the major pools we did analyze (Brown et al. 1996). One cannot expect to achieve 100% accounting for all cellular carbon due to analytical process losses for each of the four major pools we quantified and those components we did not analyze. Our estimate of protein is based on a constant conversion of total organic nitrogen to protein and C:N for protein; this ratio has variability depending on organism and growth conditions. Considering routine analytical uncertainty including a few percentage of process loss for pools we quantified, compounds we did not analyze, and a range of values for the conversion of total N to protein carbon available in literature, our accounting for ~85% of total carbon in the four major pools gives us confidence that the relative allocation to these pools is very robust.

The methods we developed improved upon previous methods (Brown et al. 1996; Brondz 2002; Courtois de Viçose et al. 2012) by normalizing our major carbon pools to independent measurements of total particulate organic carbon (POC) determined by gas chromatography, a more accurate method to estimate total organic carbon compared to ash-free dry weight. Attempts to achieve mass balance closure of total carbon is rare in the literature, yet it is important for determining the absolute masses of economically relevant carbon pools. We recommend that our approach of using total cellular organic carbon quantified with gas chromatography should be used as a reference standard, rather than ash-free dry weight, to provide more accurate accounting of allocation of the major carbon pools that are of economic importance.

Growth phase differences in batch algae cultures have been highlighted in previous studies (Renaud et al. 1991). With the onset of stationary phase caused by nutrient limitation, cell division is known to stop or be significantly reduced. Since carbon fixed by the photosynthetic system cannot be allocated for growth in stationary phase limited by nutrients, the absorbed energy is allocated to C-rich energy storage biochemicals that do not require N, P, or other trace nutrients (e.g. triacylglycerols and carbohydrates) (Shifrin and Chisholm 1981; Rivkin 1989; Gilstad and Sakshaug 1990). Consistent with previous studies, during stationary phase we observed increases in fatty acids and carbohydrates and decreases in protein and pigments as a percentage of total organic carbon, especially at higher irradiance.

The effect of irradiance on the carbon pool allocation of photosynthetic microalgae is largely regulated by photoacclimation; algal cells undergo biochemical and physiological changes to minimize excess photon absorption above saturating irradiance and to increase pigment absorption at sub-saturating irradiances (Sosik and Mitchell 1991; Strzepek and Price 2000). Surprisingly, for the range of irradiances and changes in CO_2 supplementation that we tested, the growth rates in mid-log phase were not significantly different within the uncertainty of the data, but growth rates did vary from 1.2 to 1.6 day^{-1}. The lipid profile we detected for *T. pseudonana* was consistent with previous lipid studies on diatoms (Class Bacillariophyceae; Brown et al. 1996; Yu et al. 2009). Saturated and monoenic fatty acids from microalgae are major derivative components in triacylglycerols (Hodgson et al. 1991; Sukenik et al. 1993), therefore, increases in the fatty acids, C14, C16, and C16:1 can be correlated to triacylglycerol accumulation. While high irradiance samples had the highest total fatty acid concentration during stationary phase, the relative fraction of polyunsaturated fatty acid (PUFA) for these samples were the lowest in stationary phase. This trend is consistent with previous findings by Sukenik (1991) who showed that cells of the heterokont algae, *Nannochloropsis*, were characterized by high proportions of the PUFA eicosapentaenoic acid at low irradiances. Because PUFAs are a major constituent of the thylakoid membranes in algae, acclimation to high irradiance leading to reduction in pigments is generally coupled to decreases in total thylakoid membranes (Hu and Schulten 1997).

The pigment profile of *T. pseudonana* was characteristic of planktonic diatoms (Brown 1988). Chlorophyll *a* and fucoxanthin were the predominant chlorophyll and carotenoid, respectively. The ratio of chlorophylls to carotenoids decreased from mid-log to stationary phase in all samples. The additional decreases in this ratio from low irradiance levels to medium and high irradiance levels during stationary phase suggest that the antioxidative properties of carotenoids are favored over the photosynthetic properties of chlorophylls under the stress of nutrient limitation and saturating levels of irradiance. Carotenoids derive photoprotective properties from accumulating in the plastoglobuli of plastids and preventing excess light energy from reaching the photosynthetic machinery (Ben-Amotz et al. 1982; Vechtel et al. 1992; Balamurugan et al. 2017).

For commercial purposes there is a range of fuel, feed, biomaterial, and nutraceutical components in microalgae. An understanding of the biochemical responses to cultivation conditions will contribute to the management of commercial production systems; for fuel there is a desire for high yields in monosaturated fatty acids, minimal polyunsaturated acids, and controlled proportions of saturated acids; C18:1 and C16:1 are the most suitable fatty acids in terms of oxidative stability and cold weather behavior (Pinzi et al. 2009). On the other hand, PUFAs are of higher value for feed applications (Santin et al. 2021). Applying the data from this study, *T. pseudonana* should be cultured at saturating irradiance and CO_2 supplementation for the purpose of fuel lipid feed stock. Polyunsaturated fatty acids should be avoided for fuel production due to their contribution to autoxidation of biodiesel (Pinzi et al. 2009). However, given the importance of the polyunsaturated fatty acids and especially the omega-3 PUFAs for nutraceutical and feed applications, low irradiance and minimal CO_2 should be utilized to optimize this fraction (Langdon and Waldock 1981; Wall et al. 2010). The trends we report regarding allocation of total carbon into lipid and pigments as a function of irradiance and CO_2 co-limitation have not been previously reported for microalgae and are important for defining optimal cultivation strategies depending on the commercial goals that require knowledge of total carbon allocation into major carbon pools in the crop and desired valorization targets for the crop.

5.4.1 *Economic Significance*

The capital and operation costs relative to time-dependence of grow-out and optimization for biomolecules of interest are important factors contributing to any techno economic model. Thus, the optimization of yields per unit time and area should be taken into consideration. Applying our data to this concept, we were able to achieve levels of fatty acids up to 45% of the total particulate organic carbon, but only after 287 h. However, at 119 h, 34% fatty acids were already detected in the same cultures during log phase growth. From a commercial standpoint, 75% of the eventual yield in less than half the time with continuous harvest of the crop might be a favorable economic strategy. Furthermore, harvesting during log phase growth minimizes the area needed to implement a two-step process for lipid induction during stationary growth phase. Developing a basic understanding of the plasticity of an organism can contribute significantly to commercializing specific outputs and target compounds. We believe there may be a major economic advantage for harvesting under continuous culture rather than two-stage culture where the lipids are induced in a stationary phase with nutrient limitation. Also, since PUFA concentrations are a larger fraction during log-phase growth, further economic benefit might be attained by harvesting during log phase. The valorization of the carbon pools, yield per area, and regulation of CO_2 and light should all be explored in the context of life cycle analysis and techno economic models.

5.5 Conclusion

The phenotypic variations resulting from physiological acclimation of *T. pseudonana* in response to irradiance and CO_2 demonstrate the importance of these factors in determining the allocation of the sub-cellular carbon pools that are relevant to economic models of microalgae yields and valuations. These two abiotic growth-regulating factors can be controlled by depth and concentration in the production system for light and by control of CO_2 delivery. By studying the major carbon pools, we demonstrate phenotypic plasticity of *T. pseudonana* in response to these abiotic factors and also to the phase of growth (mid-log phase, N-limited stationary phase). Our study is important as it is one of the few where the major carbon pools and total particulate organic carbon are all quantified and the only study that references these pools to total organic carbon for mass balance closure. For example, Brown et al. (1996) conducted an analytically thorough experiment, including HPLC analysis of amino acid makeup of protein; however, it lacked an analysis of pigment compositions and included a narrow range of irradiance levels and CO_2 variability. An experiment conducted by Jiang et al. (2012) implemented FTIR to assess biochemical composition changes during nitrogen starvation; however, this experiment did not utilize a summation of carbon nor a variance in irradiance. Using POC instead of dry weight or ash-free dry weight is superior as a reference for carbon portioning. We find it surprising that others have not routinely used this highly regarded and simple method to determine total organic carbon. We recommend that reference to POC should become a community standard when the abiotic regulation of carbon pools and biomass production are considered for economic valuation and commercial applications of algal biomass. Here we have demonstrated the use of an alternative method to AFDW for estimating carbon partitioning in algae using POC. Similar investigations are needed on the diverse group of microalga strains being evaluated for commercial relevance in order to define the production systems and processes that can maximize economic value and minimize production costs of algal biomass, for example, by harvesting continuously in log-phase exponential growth, rather than batch harvest in stationary phase.

Acknowledgments Support is acknowledged from U.S. Department of Energy grant DE-EE0003373 to the University of California, San Diego, Cal-CAB Consortium and from the laboratory of Professor Lihini Aluwihare at SIO/UCSD.

References

Acién-Fernández FG, González-López CV, Fernández Sevilla JM et al (2012) Conversion of CO_2 into biomass by microalgae: how realistic a contribution may it be to significant CO_2 removal? Appl Microbiol Biotechnol 96:577–586. https://doi.org/10.1007/s00253-012-4362-z

Ackman RG (1998) Remarks on official methods employing boron trifluoride in the preparation of methyl esters of the fatty acids of fish oils. J Am Oil Chem Soc 75:541–545

American Oil Chemists' Society (1989) Official methods and recommended practices of the American Oil Chemists' Society, 6th edn. AOCS Press, Urbana

Armbrust EV, Berges JA, Bowler C et al (2004) The genome of the diatom *Thalassiosira pseudonana*: ecology, evolution, and metabolism. Science 306:79–86

Balamurugan S et al (2017) Occurrence of plastidial triacylglycerol synthesis and the potential regulatory role of AGPAT in the model diatom *Phaeodactylum tricornutum*. Biotechnol Biofuels 10:97. https://doi.org/10.1186/s13068-017-0786-0

Beal CM, Gerber LN, Sills DL et al (2015) Algal biofuel production for fuels and feed in a 100-ha facility: a comprehensive techno-economic analysis and life cycle assessment. Algal Res 10(2015):266–279. https://doi.org/10.1016/j.algal.2015.04.017

Beal CM, Gerber LN, Thongrod S, Phromkunthong W, Kiron V et al (2018) Marine microalgae commercial production improves sustainability of global fisheries and aquaculture. Sci Rep 8:15064. https://doi.org/10.1038/s41598-018-33504-w

Ben-Amotz A, Katz A, Avron M (1982) Accumulation of ß-carotene in halotolerant algae: purification and characterization of ß-carotene-rich globules from *Dunaliella bardawil* (Chlorophyceae). J Phycol:529–537. https://doi.org/10.1111/j.1529-8817.1982.tb03219.x

Bligh EG, Dyer WJ (1959) A rapid method of total lipid extraction and purification. Can J Biochem Physiol 37:911–917

Brand LE (1981) Genetic variability in reproduction rates in marine phytoplankton populations. Evolution 35:1117–1127

Brondz I (2002) Development of fatty acid analysis by high-performance liquid chromatography, gas chromatography, and related techniques. Anal Chim Acta 465:1–37

Brown JS (1988) Photosynthetic pigment organization in diatoms (Bacillariophyceae). J Phycol 24:96–102

Brown MR, Dunstan GA, Norwood SJ, Miller KA (1996) Effects of harvest stage and light on the biochemical composition of the diatom *Thalassiosira pseudonana*. J Phycol 32:64–73

Brown MR, Jeffrey SW, Volkman JK, Dunstan GA (1997) Nutritional properties of microalgae for mariculture. Aquaculture 151:315–331

Chaumont D (1993) Biotechnology of algal biomass production: a review of systems for outdoor mass culture. J Appl Phycol 5:593–604

Choix FJ, de-Bashan LE, Bashan Y (2012) Enhanced accumulation of starch and total carbohydrates in alginate-immobilized *Chlorella spp.* induced by *Azospirillum brasilense*: I. Autotrophic conditions. Enzyme Microb Technol 51:294–299

Christie WW (2005) Lipids analysis: isolation, separation, identification and structural analysis of lipids, 3rd edn. The Oily Press, Bridgwater

Courtois de Viçose G, Porta A, Viera MP et al (2012) Effects of density on growth rates of four benthic diatoms and variations in biochemical composition associated with growth phase. J Appl Phycol 24:1427–1437

Doliente SS, Narayan A, Tapia JFD et al (2020) Bio-aviation fuel: a comprehensive review and analysis of the supply chain components. Front Energy Res Sec Process Energy Syst Eng 8. https://doi.org/10.3389/fenrg.2020.00110

Dubois M, Gilles KA, Hamilton JK, Rebers PA, Smith F (1956) Colorimetric method for determination of sugars and related substances. Anal Chem 28:350–356

Elias E, Tsegaye T, Hapeman C et al (2023) A vision for integrated, collaborative solutions to critical water and food challenges. J Soil Water Conserv 78(3):63A–68A. https://doi.org/10.2489/jswc.2023.1220A

Fabregas J, Herrero C, Abalde J et al (1986) Biomass production and biochemical variability of the marine microalga *Dunaliella tertiolecta* (Butcher) with high nutrient concentrations. Aquaculture 53:187–199

FAO (2022) World food and agriculture – statistical yearbook 2022, Rome. https://www.fao.org/documents/card/en/c/cc2211en

Fernandez-Reiriz MJ, Perezcamacho A, Ferreiro MJ et al (1989) Biomass production and variation in the biochemical profile (total protein, carbohydrates, RNA, lipids and fatty-acids) of 7 species of marine microalgae. Aquaculture 83:17–37

Geider R (1987) Light and temperature dependence of the carbon to chlorophyll a ratio in microalgae and cyanobacteria: implications for physiology and growth of phytoplankton. New Phytol 106(1):1–34. http://www.jstor.org/stable/2434683

Georgianna DR, Mayfield SP (2012) Exploiting diversity and synthetic biology for the production of algal biofuels. Nature 488:329–335

Gilstad M, Sakshaug E (1990) Growth rates of ten diatom species from the Barents Sea at different irradiances and day lengths. Mar Ecol Prog Ser 64:169–173

Goldman JC, Ryther J (1976) Temperature-influenced species competition in mass cultures of marine phytoplankton. Biotechnol Bioeng 18:1125–1144. https://doi.org/10.1002/bit.260180809

Greenleaf JE, SenGupta AK (2009) Flue gas carbon dioxide sequestration during water softening with ion-exchange fibers. J Environ Eng 135:386–396

Harrison PJ, Waters RE, Taylor FJR (1980) A broad spectrum artificial medium for coastal and open ocean phytoplankton. J Phycol 16:28–23

Hodgson PA, Henderson RJ, Sargent JR, Leftley JW (1991) Patterns of variation in the lipid class and fatty-acid composition of *Nannochloropsis oculata* (Eustigmatophyceae) during batch culture. J Appl Phycol 3:169–181

Holmes FL (1963) Elementary analysis and the origins of physiological chemistry. Isis 54:50–81

Hu XC, Schulten K (1997) Pigment organization of bacterial photosynthetic membrane and dynamics of energy transfer. Biophys J 72:1

Hu Q, Sommerfeld M, Jarvis E et al (2008) Microalgal triacylglycerols as feedstocks for biofuel production: perspectives and advances. Plant J 54:621–639

Hu B, Min M, Zhou W, Li Y, Mohr M, Cheng Y et al (2012) Influence of exogenous CO2 on biomass and lipid accumulation of microalgae auxenochlorella protothecoides cultivated in concentrated municipal wastewater. Appl Biochem Biotechnol 166(7):1661–1673. https://doi.org/10.1007/s12010-012-9566-2

Inomura K, Karlusich JJP, Dutkiewicz S et al (2023) High growth rate of diatoms explained by reduced carbon requirement and low energy cost of silica deposition. Microbiol. Spectr. https://doi.org/10.1128/spectrum.03311-22

Jeffrey SW, Wright SW, Zapata M (1999) Recent advances in HPLC pigment analysis of phytoplankton. Mar Freshw Res 50:879–896

Jiang YL, Yoshida T, Quigg A (2012) Photosynthetic performance, lipid production and biomass composition in response to nitrogen limitation in marine microalgae. Plant Physiol Biochem 54:70–77

Johnson VR, Brownlee C, Rickaby RE, Graziano M, Milazzo M, Hall-spencer J (2013) Responses of marine benthic microalgae to elevated CO2. Mar Biol 160(8):1813–1824. https://doi.org/10.1007/s00227-011-1840-2

Langdon CJ, Waldock MJ (1981) The effect of algal and artificial diets on the growth and fatty-acid composition of *Crassostrea gigas* spat. J Mar Biol Assoc UK 61:431–448

Laurens LML (2013) Summative mass analysis of algal biomass-integration of analytical procedures. NREL Technical Report NREL/TP-5100-60943, Revised December 2015

Laurens LML, Dempster TA, Jones HDT, Wolfrum EJ, Van Wychen S, McAllister JSP, Rencenberger M, Parcher KJ, Gloe LM (2012a) Algal biomass constituent analysis: method uncertainties and investigation of the underlying measuring chemistries. Anal Chem 84:1879–1887

Laurens LML, Quinn M, Van Wychen S et al (2012b) Accurate and reliable quantification of total microalgal fuel potential as fatty acid methyl esters by *in situ* transesterification. Anal Bioanal Chem. https://doi.org/10.1007/s00216-012-5814-0

Lourenco SO, Barbarino E, Lavin PL, Marque UML, Aidar E (2004) Distribution of intracellular nitrogen in marine microalgae: calculation of new nitrogen-to-protein conversion factors. Eur J Phycol 39:17–32

Myklestad SM, Skanoy E, Hestmann S (1997) A sensitive and rapid method for analysis of dissolved mono- and polysaccharides in seawater. Mar Chem 56:279–286

Ogbonna JC, Tanaka H (1996) Night biomass loss and changes in biochemical composition of cells during light/dark cyclic culture of *Chlorella pyrenoidosa*. J Ferment Bioeng 82:558–564

Ou L, Banerjee S, Xu H, Coleman AM et al (2021) Utilizing high-purity carbon dioxide sources for algae cultivation and biofuel production in the United States: opportunities and challenges. J Clean Prod 321:128779. https://doi.org/10.1016/j.jclepro.2021.128779

Patel A, Antonopoulou I, Enman J et al (2019) Lipids detection and quantification in oleaginous microorganisms: an overview of the current state of the art. BMC Chem Eng 1:13. https://doi.org/10.1186/s42480-019-0013-

Pinzi S, Garcia IL, Lopez-Gimenez FJ, de Castro MDL, Dorado G, Dorado MP (2009) The ideal vegetable oil-based biodiesel composition: a review of social, economical and technical implications. Energy Fuel 23:2325–2341

Poulsen N, Berne C, Spain J, Kroger N (2007) Silica immobilization of an enzyme through genetic engineering of the diatom *Thalassiosira pseudonana*. Angew Chem Int Ed 46:1843–1846

Pruvost J, Van Vooren G, Le Gouic B et al (2011) Systematic investigation of biomass and lipid productivity by microalgae in photobioreactors for biodiesel application. Bioresour Technol 102:150–158

Renaud SM, Parry DL, Thinh LV et al (1991) Effect of light-intensity on the proximate biochemical and fatty-acid composition of *Isochrysis sp.* and *Nannochloropsis oculata* for use in tropical aquaculture. J Appl Phycol 3:43–53

Rivkin RB (1989) Influence of irradiance and spectral quality on the carbon metabolism of phytoplankton. 1. Photosynthesis, chemical-composition and growth. Mar Ecol Prog Ser 55:291–304

Rosenberg JN, Oyler GA, Wilkinson L, Betenbaugh MJ (2008) A green light for engineered algae: redirecting metabolism to fuel a biotechnology revolution. Curr Opin Biotechnol 19:430–436

Ryther JH, Guillard RR (1962) Studies of marine planktonic diatoms. 2. Use of *Cyclotella nana* Hustedt for assays of vitamin b12 in seawater. Can J Microbiol 8:437

Sander K, Murthy GS (2010) Life cycle analysis of algae biodiesel. Int J Life Cycle Assess 15:704–714

Santin A, Russo MT, Ferrante MI et al (2021) Polyunsaturated fatty acids from microalgae: strategies to improve their yields and their potential exploitation in aquaculture. Molecules 26(24):7697. https://doi.org/10.3390/molecules26247697. PMID: 34946780; PMCID: PMC8707597

Scott SA, Davey MP, Dennis JS et al (2010) Biodiesel from algae: challenges and prospects. Curr Opin Biotechnol 21:277–286

Seto A, Wang HL, Hesseltine CW (1984) Culture conditions affect eicosapentaenoic acid content of *Chlorella minutissima*. J Am Oil Chem Soc 61:892–894

Shifrin NS, Chisholm SW (1981) Phytoplankton lipids: interspecific differences and effects of nitrate, silicate and light-dark cycles. J Phycol 17:374–384

Solovchenko A, Khozin-goldberg I (2013) High-CO2 tolerance in microalgae: possible mechanisms and implications for biotechnology and bioremediation. Biotechnol Lett 35(11):1745–1752. https://doi.org/10.1007/s10529-013-1274-7

Sosik HM, Mitchell BG (1991) Absorption, fluorescence, and quantum yield for growth in nitrogen-limited Dunaliella tertiolecta. Limnol Oceanogr 36:910–921

Strzepek RF, Price NM (2000) Influence of irradiance and temperature on the iron content of the marine diatom *Thalassiosira weissflogii* (Bacillariophyceae). Mar Ecol Prog Ser 206:107–117

Sukenik A (1991) Ecophysiological considerations in the optimization of eicosapentaenoic acid production by *Nannochloropsis sp* (Eustigmatophyceae). Bioresour Technol 35:263–269

Sukenik A, Zmora O, Carmeli Y (1993) Biochemical quality of marine unicellular algae with special emphasis on lipid-composition. 2. *Nannochloropsis sp*. Aquaculture 117:313–326

Utting SD (1985) Influence of nitrogen availability on the biochemical-composition of 3 unicellular marine-algae of commercial importance. Aquac Eng 4:175–190

Vechtel B, Eichenberger W, Ruppel HG (1992) Lipid bodies in *Eremosphaera viridis* DeBary (Chlorophyceae). Plant Cell Physiol 33:41–48

Wahidin S, Idris A, Shaleh SRM (2013) The influence of light intensity and photoperiod on the growth and lipid content of microalgae *Nannochloropsis sp*. Bioresour Technol 129:7

Wall R, Ross RP, Fitzgerald GF, Stanton C (2010) Fatty acids from fish: the anti-inflammatory potential of long-chain omega-3 fatty acids. Nutr Rev 68:280–289

Witter AE, Luther GW (2002) Spectrophotometric measurement of seawater carbohydrate concentrations in neritic and oceanic waters from them US Middle Atlantic Bight and the Delaware Estuary. Mar Chem 77:143–156

Woertz IC, Benemann JR, Du N, Unnasch S, Mendola D, Mitchell BG, Lundquist TJ (2014) Life cycle GHG emissions from microalgal biodiesel – a CA-GREET model. Environ Sci Technol 48(11):6060–6068

Yu ET, Zendejas FJ, Lane PD, Gaucher S, Simmons BA, Lane TW (2009) Triacylglycerol accumulation and profiling in the model diatoms *Thalassiosira pseudonana* and *Phaeodactylum tricornutum* (Baccilariophyceae) during starvation. J Appl Phycol 21:669

Zapata M, Garrido JL (1991) Influence of injection conditions in reversed-phase high-performance liquid-chromatography of chlorophylls and carotenoids. Chromatographia 31:589–594

Zapata M, Rodriguez F, Garrido JL (2000) Separation of chlorophylls and carotenoids from marine phytoplankton: a new HPLC method using a reversed phase C_8 column and pyridine-containing mobile phases. Mar Ecol Prog Ser 195:29–45

Zhu C, Lee Y, Chao T (1997) Effects of temperature and growth phase on lipid and biochemical composition of *Isochrysis galbana* TK1. J Appl Phycol 9:451–457

Chapter 6
Microalgal Products from Agro-industry Wastes

Júlio C. de Carvalho, Eduardo B. Sydney, Paulo C. S. Kirnev, Denisse T. M. Aulestia, and Carlos R. Soccol

Abstract Industrial production of microalgae requires large volumes of water. Agro-industry wastewaters are generated in large volumes and must be adequately treated. The connection of both processes is only natural—conventional wastewater treatment processes end with a stabilizing pond that has a self-established microbiota, which includes microalgae. This chapter discusses how to use agro-industry wastewaters as culture media, describing the composition of the most common residues, presenting selected examples of microalgal production in effluents, and evaluating how the wastewater composition and the intended microalgal biomass use affects process development. Robust strains or effluent-specific microalgae, screened from the microbiota that develops in wastewaters, can be used in processes that are integrable into agro-industries, preferably after an anaerobic digestion step for high-COD residues.

Keywords Pome · Vinasse · Cassava · Chlorella · Scenedesmus · Waste treatment

6.1 Introduction

Agro-industry processes large volumes of agricultural raw materials, use copious amounts of water and generate liquid wastes with generally low toxicity but high eutrophication potential. Aerobic or anaerobic digestion treatments are the most common processes for reducing the organic load in wastes, and the final steps of wastewater stabilization naturally present wild microalgae.

J. C. de Carvalho (✉) · P. C. S. Kirnev · D. T. M. Aulestia · C. R. Soccol
University of Paraná, Department of Bioprocess Engineering and Biotechnology, Curitiba, Parana, Brazil
e-mail: jccarvalho@ufpr.br

E. B. Sydney
Federal University of Technology—Paraná, Department of Bioprocess Engineering and Biotechnology, Ponta Grossa, Parana, Brazil

J. A. V. Costa et al. (eds.), *Microalgal Bioengineering*,
https://doi.org/10.1007/978-3-031-61253-4_6

To fully integrate streams in a circular biorefinery process, it is essential to treat and reuse nutrients from these wastewaters for both regulatory and economic reasons: regulatory, because wastewaters must be returned to water bodies in suitable quality, and economical, because these wastewaters have considerable energy and nutrient content that can be recycled through microalgae production.

The production of microalgae using wastes is a natural evolution of agro-industrial processing and microalgal biotechnology. The volumes and areas for cultivating microalgae in wastewaters are significant but comparable to traditional secondary treatments. This chapter discusses the main aspects and potential of microalgal production in agro-industry wastes and its secondary effluents—those where part of the carbon was already digested.

6.2 Main Agro-industry Wastes

6.2.1 Main Processed Crops and Their Residues

The global crop production in 2017 was estimated at 9.2 billion tons (Fig. 6.1). This includes food, feed, and energy crops. Part of these crops is processed in large industries—for example, a typical sugarcane or maize ethanol distillery, or an oil palm industry, processes thousands of tons of raw material per day. This large volume of biomass generates large volumes of residues, which are considered, in a biorefinery, as raw materials for other processes.

Residues from the agro-industry can be divided into solid (straws, husks, bagasse, etc.) and liquid (washing waters, pressing juices, and vinasses). Compared to liquid

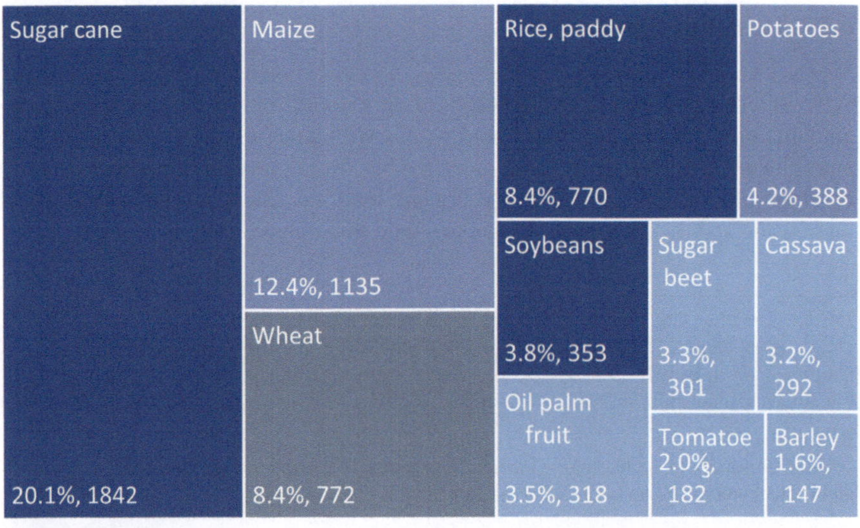

Fig. 6.1 Main crops in the world. These 12 agricultural products accounted for 70% of all crop production in 2017. Values are the percentage of the product among 161 other crops in the database, and its production in million tons (10^9 kg). (Data compiled from the FAO database (FAO 2019))

residues, the "solid" residues from agro-industry have much higher solids concentration, lower water activity, and higher stability. Besides, they can be used for burning, for composting, as soil conditioners, a base for feeds, and for more sophisticated uses such as depolymerization into sugars that can be further converted into a multitude of bioproducts (Woiciechowski et al. 2013). For this conversion, there are several fast-growing microorganisms that can be used, but microalgae are usually *not* the first-choice microorganisms (Pelizer et al. 2015).

Liquid agro-industry wastes, conversely, are too dilute for direct uses such as burning, hydrolysis, or fermentation because of the meager energy-to-volume ratio. And yet, these residues are too concentrated to be discarded, requiring treatment in situ. However, microalgae can be easily cultivated in residues (Pittman et al. 2011; Perez-Garcia et al. 2011; Cai et al. 2013). There are two kinds of liquid wastes that can be used for microalgae cultivation: the primary effluents, with high COD (chemical oxygen demand) and the secondary effluents, already subjected to a (usually) anaerobic digestion step and having a lowered COD (but generally rich in nitrogen and phosphorus). Both residues have variable concentrations of several macro- and micronutrients.

6.2.2 Agro-Industry Residues Composition and Suitability for Microalgae Cultivation

There is significant variability in agro-industry wastewaters. A common feature is that wastewaters have a high COD, around 10–60 g/L. Microalgae require nutrients, similar to bacteria, yeasts, and fungi—and all these microorganisms can grow in residues. The main differences in microalgae media composition are (i) the independence of available organic carbon sources due to its ability to use atmospheric carbon (CO_2) through photosynthesis; (ii) the significant negative impact of turbidity, which impairs the passage of light; (iii) reduced ability to hydrolyze macromolecules such as lipids, proteins, and complex carbohydrates, and (iv) higher demand of magnesium.

The photosynthetic apparatus of microalgae works similarly to those in superior plants, but microalgae can use CO_2 much faster. Many microalgae also have the cellular machinery to use organic carbon, which is much more efficient for energy production within cells. In the presence of organic carbon sources, which is very common in residual wastewaters, the photosynthetic pathway can be entirely or partially inhibit (heterotrophic and mixotrophic growth, respectively). In both cases, part of the dissolved organic carbon in the residues is used. Both at heterotrophic and mixotrophic conditions, microalgae growth is slower than bacteria and similar to yeasts. Thus, large-scale cultures tend to be *cocultures* of algae and bacteria or even *polycultures*, leading to biomass with a dominant, but not axenic, microalga (Molina et al. 2019). In that case, part of the COD is usually converted by heterotrophs (bacteria), while microalgae grow auto- or mixotrophically (Mitra et al. 2012; Wang et al. 2014). Table 6.1 shows selected primary liquid effluents from large-scale agro-industrial processing.

Table 6.1 Average composition of important agro-industrial liquid residues

Components	Palm oil mixed effluents (POME)	Sugarcane Vinasse	Cassava	Spent Wash Sugarcane Molasses	Spent Wash Sugarcane Juice	Spent Wash Corn	Spent Wash Beet Molasses	Wine Vinasse
pH		4.07	5.40	4.10	4.04	3.65	5.35	2.90
BOD (mg/L)	22,411.55	13,879.33		71,050.00	16,700.00	35,000.00	44,900.00	18,900.00
COD (mg/L)	50,140.39	32,673.00	20,600.00	11,9200.00	30,400.00	61,950.00	91,100.00	
Soluble COD (mg O2/L)			13,800.00					
Dissolved Solids (DS) (mg/L)	50,000.00	12,764.00	13,500.00					
Suspended Solid (SS) (mg/L)	25,710.20		15,500.00					
Total Volatile Solid (TVS) (mg/L)	57,000.00							
Total Volatile Solid Soluble (TVSS) (mg/L)	21,111.66		2900.00					
Total Solids (TS) (mg/L)	61,402.22	26,220.00						
Carbohydrates (mg/L)			32,900.00					
Protein (mg/L)			1500.00					
Lipids (mg/L)	3811.85							
Total nitrogen (mg/L)	370.20	952.50	900.00	2930.00	630.00	650.50	3570.00	
NH4+ (mg/L)	192.58		39.08					
NO3- (mg/L)	35.00	3732.08	24.30					
CN- (mg/L)			16.40					
Alkalinity (mg^3/L)	536.00							
Total Phosphorus (mg/L)	135.00	2.61	549.00	1625.00	130.00	699.00	160.00	83.00
PO4- (mg/L)	252.38	296.64						
K (mg/L)	2287.68	2336.20	1339.20	13,535.00	1950.00		10,000.00	459.00
Ca (mg/L)	253.44	1116.35	162.60					
Mg (mg/L)	486.85	249.50	243.00					

Beet Vinasse	Sweet Soghum Vinasse	Olive Mill	Poultry slaughterhouse	Abattoir	Dairy farm	Potato chips	Cheese Whey	Beer production
5.10	4.50	5.07	6.85	7.31				8.25
78,300.00	46.00	17,120.00		1617.00		1950.00	11,000.00	2400.00
		58,890.00	6880.00	2004.00	12,000.00	8122.00	42,500.00	4000.00
		40,510.00		1232.00				
		12,590.00	2800.00	1450.00		640.00	2550.00	600.00
					1860.00			
		720,15.00			4540.00		65,000.00	
						19,470.00		
						2880.00	34,000.00	
							6000.00	
		220.00	675.00	550.00	2400.00		7.90	52.50
			335.00[a]	153.00[a]	349.28			
		0.15						
			1315.00					
91.00	1990.00	0.24	8.95	38.00	480.00		143.00	30.00
		673.61						
10,015.00					6200.00			
0.71					1350.00			
1.23					625.00			

(continued)

Table 6.1 (continued)

Components	Palm oil mixed effluents (POME)	Sugarcane Vinasse	Cassava	Spent Wash Sugarcane Molasses	Spent Wash Sugarcane Juice	Spent Wash Corn	Spent Wash Beet Molasses	Wine Vinasse
S (mg/L)	0.14	544.36	50.30	3330.00	1350.00	2990.00	3710.00	40.06
Fe (mg/L)	73.70		7.60					
Zn (mg/L)	2.79	1.66	3.80	1.66				
Mn (mg/L)	2.32		1.20					
Na (mg/L)	390.32	120.60	51.70					
Cl (mg/L)								
Ba (mg/L)		0.41		0.41				
Cd (mg/L)	0.24							0.07
Cr (mg/L)		0.04		0.04				
Cu (mg/L)	1.52	0.35	0.50	0.35				1.73
Co (mg/L)	0.03							
Mo (mg/L)		0.01		0.01				
Ni (mg/L)		0.03		0.03				
Pb (mg/L)								0.95
References	Bala et al. (2015), Tabassum et al. (2015), Ahmad et al. (2016), Ding et al. (2016), and Faria (2016)	Christofoletti et al. (2013), Prado et al. (2013), de Mattos and Bastos (2016), Montalvo et al. (2019)	de Carvalho et al. (2018)	Hoarau et al. (2018)	Wilkie et al. (2000)	Wilkie et al. (2000)	Wilkie et al. (2000)	Christofoletti et al. (2013)

[a]Originally reported as NH_3

Beet Vinasse	Sweet Soghum Vinasse	Olive Mill	Poultry slaughterhouse	Abattoir	Dairy farm	Potato chips	Cheese Whey	Beer production
0.21								
					43.00			
					5.45			
					15.15			
3.79					453.00			
<1								
3.55	37.00				32.35			
					1715.00			
<5								
Christofoletti et al. (2013)	Christofoletti et al. (2013)	Di Caprio et al. (2015), Oz and Uzun (2015)	Debik and Coskun (2009)	Keskes et al. (2012)	Ding et al. (2016)	Mishra et al. (2004)	Carvalho et al. (2013)	Brito et al. (2007)

As biomass is produced, the medium becomes more turbid, impairing light diffusion through the liquid (Sydney et al. 2019). The microalgae growth is, thus, self-limited, unless biomass is periodically removed/recovered. In residual wastewater, the organic matter is mainly in the suspended solids form with complex and often poorly characterized compositions. These solids also scatter and absorb light, and colored dissolved substances contribute to a low light penetration in the waste (de Carvalho et al. 2018; Valev et al. 2020). Initially, microalgae media is filtered or decanted to remove suspended solids (clarification). The organic matter in residual wastewater has complex molecules, such as lipids, proteins, and complex carbohydrates. Although microalgae produce a large variety of extracellular enzymes (Brasil et al. 2017), their low growth rate and considerably lower maximum biomass concentration (maximum of 3–5 g/L) puts microalgae among the worst catalytic-enzyme producers (in terms of concentration in comparison to bacteria, yeast, and fungi). Significant reduction of complex substances is expected to occur only in high retention time cultures, which is not viable when considering the biological treatment of wastes produced in huge volumes.

6.2.3 Normal Destination of Residues

The primary effluents from agro-industry have high polluting potential—the permitted COD for discharge in rivers depends on the local legislation but is generally around 120 mg/ L (Zero Discharge of Hazardous Chemicals Programme—ZDHC 2019), or 17 to 1000 times *lower* than the concentration of raw residues listed in Table 6.1. Treatment is a legal responsibility of the producer but need not be a burden because these effluents can be regarded as diluted energy sources: a diluted but recoverable *value*. The chemical energy contained in the effluents can be reclaimed by an anaerobic treatment step that reduces the organic load and maintains part of the nutrients available for microalgal growth: a biodigester or a biohydrogen reactor, for example (de Carvalho et al. 2018; Martinez-Burgos et al. 2020) is especially interesting in biorefinery integrated processes. The low specific energy content in the residue makes its concentration (e.g., evaporation or tangential filtration) economically unfeasible unless the concentrated streams can be recycled to the process, as is the case in the kraft pulping process. Figure 6.2 illustrates microalgal production integrated into a typical biorefinery process.

6.2.4 Secondary Effluents Composition

Secondary effluents from anaerobic treatment can be adequate for microalgae processing because of their reduced COD and increased transparency—the preliminary anaerobic treatment has high residence times that imply self filtration or settling of the solids inside the anaerobic digester, thus clarifying the effluent. The effluent

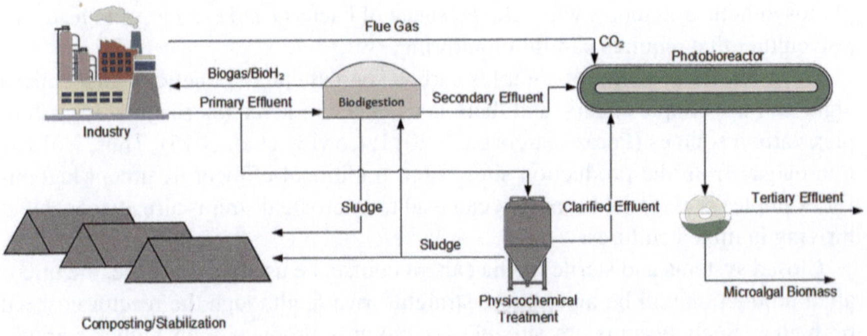

Fig. 6.2 Possible integration of microalgae cultures in agro-industry wastewaters value addition

from these systems may be still rich in minerals; for example, from potato processing, with concentrations of 53 mg/L of PO_4^{3-} and 110 mg/L of NH_4^+ (Moerman et al. 2009). The composition of several anaerobically treated wastewaters was compiled by (Muhmood et al. 2019), with PO_4^{3-} ranging from 50 to 600 mg/L and NH_4^+ ranging from 700 to 6200 mg/L of NH_4^+. Therefore, although the direct use of primary or clarified effluent can be done for algae biomass production, the recovered biogas and stabilized solids from the biodigestion can be products useful for the industry (Fig. 6.1). On the other hand, the secondary effluent still has a very high concentration of nutrients (and, frequently, COD in the form of undigested carbon and volatile fatty acids). The cultivation of microalgae in primary and secondary effluents is discussed in the next session.

6.3 Microalgae Cultivation in Liquid Effluents

6.3.1 Mode of Operation and Contamination

Agroindustrial wastes are usually generated in large volumes, because of the economy allowed by large-scale processing. This fact is on par with posterior microalgal cultivation because microalgal cultures tend to be of relatively low concentration (Kirnev et al. 2018), and large volumes of culture media are required to produce biomass on an industrial scale.

Because of the large volumes required and lower CAPEX, open cultivation systems are more common. Furthermore, these systems are susceptible to climate conditions and contamination naturally carried by air. Since the agro-industry residues tend to have high carbon contents, open systems can lead to bacterial growth instead of a mixotrophic, axenic microalgal growth (Cai et al. 2013; Zhou et al. 2014).

The consequences are that microalgal cultures in residues, processed in open systems, have complex dynamics compared to closed photobioreactors. Light absorption by suspended solids and biosolids may lead to a reduction of photosynthetic efficiency, while the presence of bacteria and even grazers leads to a polyculture that requires careful monitoring.

However, the presence of complex carbon sources can be beneficial: most microalgae can use simple sugars, and there is evidence of some microalgae using complex carbon sources (Perez-Garcia et al. 2011; Lowrey et al. 2015). Thus, isolating microalgae from the production sites (from traditional effluent treatment lagoons, for example) is essential because it can lead to microalgal strains already capable of thriving in mixed cultures.

Closed systems and sterile media can, of course, be used; in that case, the microalgal production will be much more straightforward, although the reactor cost will be higher. Such systems are suitable for biomass products with higher commercial value.

6.3.2 Direct Use × Secondary Effluents in Microalgal Cultures

Each agroindustrial wastewater has a specific composition and that leads to specific microalgae that can be grown on the effluent. Obstacles such as high BOD (may sustain heavy contamination in open ponds) and high suspended solids content (may lead to poor light penetration) can be minimized by using adapted cultures and amended media: pH, salinity, organic load, the concentration of cations, and anions must be evaluated. Table 6.2 presents selected cases of microalgae production in agro-industry effluents.

Productivities shown in Table 6.2 are highly variable but consistent with microalgal production in industrial systems. The data is presented based on information available in the references—the composition of the residue is not always completely known. Cultivating microalgae reduces COD by a small amount (10–15%) but is very efficient in removing nitrogen and phosphate, being consistent with a predominantly autotrophic growth of microalgae in these cultures, and again shows the complementarity of cultivating microalgae in pre-digested effluents which, in a second step (the microalgal culture), will have a reduction of eutrophicating ions, such as HPO_4^{2-}, NO_3^-, and NH_4^+. Table 6.2 also shows that it is common to dilute the residues before cultivation, although selected and adapted microalgae strains can sometimes grow in raw residues (100% concentration).

6.4 Microalgal Products

Production of microalgal biomass from agro-industrial residues is well-referenced in the literature. Table 6.2 presents several examples of classical microalgae produced in effluents. However, the variability of the feedstocks, the need for control of the process, and the relatively high value of nutraceutical biomasses make using chemicals-based, formulated media more common for industrial products. In order of suitability, microalgal biomass from wastes ranges from least stringent to most stringent processes, depending on the use. There is no general legislation for microalgal products, and these usually must comply with existing legislation regarding specific applications. Table 6.3 lists examples:

In terms of processing issues, the work with residues suffers from the natural variability of the input, and for open systems, of climate conditions. This makes microalgae biomass production inherently variable, and that should be taken into account in the economic analysis of the process. On top of that, there are specificities based on the product. Regulation of the microalgae products is in development worldwide (Enzing et al. 2014). For now, there is no general legislation regulating biomass quality, but rather for the algal products, usually by segment. Biofuels, for example, have to comply with energy agencies regulations; feeds, with agricultural ministries and agencies, and food and nutraceutical biomass or fractions, to food and drug agencies regulations.

As can be seen in Table 6.3, there is no concern when the microalgal biomass will be processed into gaseous energy vectors (methane, bioH_2), although heavy metals may accumulate in the solid residue after gasification or biodigestion, and that secondary residue must be destined appropriately. In biomasses that will be used for the production of liquid fuels—biodiesel, third-generation ethanol, or biocrude for further processing, the presence of heavy metals must be monitored, and biomass processing must guarantee that it does not reach the products.

Agricultural-derived products such as biofertilizers require that any plant pathogens are destroyed—either by sterilization or by guaranteeing that processing conditions, for example, anaerobic treatment and microalga cultivation on the centrate. This is essential to avoid any recycling to crops, and an orthogonal application of the fertilizer (i.e., to a production system unrelated to the fertilizer origin) is ideal.

For feed, food, and nutraceutical products of microalgae, production and processing are naturally more stringent. Besides the obvious requirement of the absence of pathogens—which precludes the use of municipal wastewaters for food production, for example—it is necessary that wastewaters have low concentrations of heavy metals, antibiotics, pesticides, hormones, and other xenobiotics. Although processing may limit the presence of these substances in a final product, such as feed biomass or protein-rich meals, some processes, such as the supercritical fluid extraction used in astaxanthin production, can actually selectively extract pesticides, increasing the relative concentration in the product. Quality control of the wastewater and the final product are mandatory (EFSA Panel on Dietetic Products 2014; Liu et al. 2017).

6.5 Integration of Microalgae Processing into Agro-industry Biorefineries

The agro-industry is responsible for transforming raw materials, prolonging their availability, increasing their shelf life, decreasing their seasonality, and adding value to fresh foods. The concerns about the accelerated population growth and the

Table 6.2 Successful cases (primary and secondary effluents)

Condition							Nutrients removal		
Microalga	Wastewater	Reactor system	Inoculum concentration mg.L^{-1}	Target (biomass or Metabolite)	Concentration (% of residue, per liter of culture)	Productivity (mg.L^{-1}. day^{-1})	Terminal COD (mg.L^{-1})	Initial COD (mg.L^{-1})	Initial TN (mg.L^{-1})
Chlamydomonas sp.	POME			Biomass	25%	634	432	512	73
Chlamydomonas sp.	POME			Biomass	16.7%	917	298	341	49
Chlamydomonas sp.	POME			Biomass	12.5%	716	234	256	37
Desmodesmus sp.	Vinasse sugarcane	Stirred batch	1000	Biomass		101			1420
Consortium	Dairy farm	Pound	100	Biomass		213	1.30		
Chlorella sp.	Piggery	Batch	300	Biomass	100%	670			
Chlorella sp.	Piggery	Batch	300	Lipid	25%	155			
Chlorella sp.	Piggery	Semi continuous	300	Biomass	25%	852			
Chlorella sp.	Piggery	Semi continuous	300	Lipid	75%	128			
Nannochloropsis oculata	Pulp and paper	Flat panel photobioreactor		Biomass		950			
Nannochloropsis oculata	Pulp and paper	Flat panel photobioreactor		EPA		42			
Nannochloropsis oculata	Pulp and paper	Flat panel photobioreactor		Total fatty acids		119			
Chlorella pyrenoidosa	Textile		150	Biomass	25%	0.014			
Chlorella pyrenoidosa	Textile		150	Biomass	50%	0.011			
Chlorella pyrenoidosa	Textile		150	Biomass	75%	0.010			
Chlorella pyrenoidosa	Textile		150	Biomass	100%	0.008			
Acutodesmus dimorphus	Dairy farm						256.60		
Scenedesmus sp.	Olive mill	Roux bottle		Biomass		60	28[a]		
Scenedesmus sp.	Olive mill	Roux bottle		Lipid		90			
Scenedesmus sp.	Olive mill	Roux bottle		Carbohydrate		100			
Chlorella pyrenoidosa	Starch	Outdoor	150–200	Biomass	Summer	293			310.80
Chlorella pyrenoidosa	Starch	Outdoor	150–200	Biomass	Autumn	180			313.50
Chlorella pyrenoidosa	Starch	Outdoor	150–200	Biomass	Winter	56			301.70

Terminal TN ($mg.L^{-1}$)	TN removal %	Initial NH_4^+-N ($mg.L^{-1}$)	Terminal NH_4^+- N ($mg.L^{-1}$)	NH_4^+- N removal %	Initial NH_3- N ($mg.L^{-1}$)	Terminal NH_3- N ($mg.L^{-1}$)	NH3 removal %	Initial TP ($mg.L^{-1}$)	Terminal TP ($mg.L^{-1}$)	TP removal %	Initial PO_4-P ($mg.L^{-1}$)	Terminal PO_4-P ($mg.L^{-1}$)	PO_4-P removal %	References
26	35.62	51	5.60	10.98				33.80	15	44.38				Ding et al. (2016)
15	30.61	34	0	0				22.50	10.50	46.67				Ding et al. (2016)
10	27.03	25.50	0	0				17	6.20	36.47				Ding et al. (2016)
	52.10							2.60						de Mattos and Bastos (2016)
		0.90			3.30	0.02	0.64				4.60	0.06	1.24	Hena et al. (2015)
														Kuo et al. (2015)
														Kuo et al. (2015)
														Kuo et al. (2015)
														Kuo et al. (2015)
														Polishchuk et al. (2015)
														Polishchuk et al. (2015)
														Polishchuk et al. (2015)
														Pathak et al. (2015)
														Pathak et al. (2015)
														Pathak et al. (2015)
														Pathak et al. (2015)
		277.40			0.95			5.90						Chokshi et al. (2016)
														Di Caprio et al. (2015)
														Di Caprio et al. (2015)
														Di Caprio et al. (2015)
56.10	18.05							42	1.20	2.86				Chu et al. (2015)
107.10	34.16							35.40	1.80	5.08				Chu et al. (2015)
259.30	85.95							33.80	15.70	46.45				Chu et al. (2015)

(continued)

Table 6.2 (continued)

Condition							Nutrients removal		
Microalga	Wastewater	Reactor system	Inoculum concentration mg.L^{-1}	Target (biomass or Metabolite)	Concentration (% of residue, per liter of culture)	Productivity (mg.L^{-1}. day^{-1})	Terminal COD (mg.L^{-1})	Initial COD (mg.L^{-1})	Initial TN (mg.L^{-1})
Chlorella pyrenoidosa	Starch	Outdoor	150–200	Biomass	Spring	137			301.10
Chlorella sorokiniana	IC reactor[b]	Pond simulator ePBR			6%				
Chlorella sorokiniana	IC reactor[b]	Pond simulator ePBR			25%				
Chlorella sorokiniana	IC reactor[b]	Pond simulator ePBR			50%				
Chlorella sorokiniana	IC reactor[b]	Pond simulator ePBR			100%				
Scenedesmus sp.	Paper Mill	Open circular ponds		Protein		16[a]	2250		
Scenedesmus sp.	Paper Mill	Open circular ponds		Carbohydrate		18.90[a]			
Scenedesmus sp.	Paper Mill	Open circular ponds		Lipid		15.80[a]			
Chlamydomonas incerta	POME			Lipid		0.197			
Chlorella sp.	Meat plant				KILL[c]		3560		327.60
Chlorella sp.	Meat plant				CUT[c]		1019		64.80
Chlorella sp.	Meat plant				MPGP[c]		2035		91.30
Chlorella sp.	Meat plant				REFINERY[c]		1016		117.50
Chlorella sp.	Meat plant			Lipid	KILL[c]	15.80[a]			
Chlorella sp.	Meat plant				CUT[c]	21[a]			
Chlorella sp.	Meat plant				MPGP[c]	25.60[a]			
Chlorella sp.	Meat plant				REFINERY	23.90[a]			
Chlorella sp.	Meat plant			Protein	KILL[c]	63.30[a]			
Chlorella sp.	Meat plant				CUT[c]	51.58[a]			
Chlorella sp.	Meat plant				MPGP[c]	55			
Chlorella sp.	Meat plant				REFINERY[c]	60.33[a]			
Chlorella minutissima	Cassava	Tubular photobioreactor	200	Biomass		2790			

[a]Quantities obtained by means of a calculation basis
[b]Anaerobic treatment reactor Novozymes
[c]Five types of wastewater obtained from different processing steps in a meat processing plant

Terminal TN $(mg.L^{-1})$	TN removal %	Initial NH_4^+-N $(mg.L^{-1})$	Terminal NH_4^+-N $(mg.L^{-1})$	NH_4^+-N removal %	Initial NH_3-N $(mg.L^{-1})$	Terminal NH_3-N $(mg.L^{-1})$	NH3 removal %	Initial TP $(mg.L^{-1})$	Terminal TP $(mg.L^{-1})$	TP removal %	Initial PO_4-P $(mg.L^{-1})$	Terminal PO_4-P $(mg.L^{-1})$	PO_4-P removal %	References
116	38.53							37.80	46	121.69				Chu et al. (2015)
				91.88									71.67	Van Wagenen et al. (2015)
				97.16									82.87	Van Wagenen et al. (2015)
				99.98									98.99	Van Wagenen et al. (2015)
				99.99									99.44	Van Wagenen et al. (2015)
													71.29	Usha et al. (2016)
														Usha et al. (2016)
														Usha et al. (2016)
														Kamyab et al. (2016)
109	33.27				193	88.01	45.60	46.80	30.50	65.17				Lu et al. (2016)
	0				8.20	8.20	100	45.90		0				Lu et al. (2016)
30.30	33.19				2.20	2.20	100	32.90	10.10	30.70				Lu et al. (2016)
38.40	32.68				101.70	61.50	60.47	5.60	5.60	100				Lu et al. (2016)
														Lu et al. (2016)
														Lu et al. (2016)
														Lu et al. (2016)
														Lu et al. (2016)
														Lu et al. (2016)
														Lu et al. (2016)
														Lu et al. (2016)
														Lu et al. (2016)
														Borguetthi (2009)

Table 6.3 Examples of microalgal bioproducts from agricultural wastewaters

	Safety requirements of the feedstock (wastewater) and process drawbacks		Examples		
Biomass use	Microbiological	Chemical	Residue source	Product	References
Bioenergy (biodigestion for biogas or biohydrogen)	None (product is volatile)	Inhibitors, high color The C/N ratio and mineral content must be balanced	Beer production	Biogas from algal biomass digestion	Assemany et al. (2020)
			Piggery wastewater	Biogas from algal biomass digestion	Carminati et al. (2018)
Bioenergy (biodiesel or biocrude)	None (processing destroy biomass)		Biodigested Chicken manure	Biocrude from hydrothermal liquefaction	Li et al. (2018)
			Biodigested "molasses" wastewater	Biomass for biodiesel and heavy metal removal	Yang et al. (2017)
Fractionated for biofertilizers	Processing must ensure the destruction of plant pathogens	Inhibitors, heavy metals	Meat processing, mixed wastewater	Biomass for Micronutrient amendment	de Siqueira et al. (2020)
			Parboiled rice mill effluent	Phosphorus recovery and application using *Chlorella* and cyanobacteria	Mukherjee et al. (2015)
Feed (raw biomass, lipid or pigment amendment)	No animal pathogens. Amendment with sugars enhance lipids but also contamination in open systems	Inhibitors, heavy metals	Starch processing plant, biodigested and settled	*C. pyrenoidosa* biomass, lipid-enriched	Tan et al. (2018)
			Slaughterhouse	*Phormidium autumnale* carotenoids	Rodrigues et al. (2014)
Feed (protein source)	No animal pathogens, limited heterotroph contamination (reduces productivity)	Inhibitors, heavy metals, antibiotics, hormones	Pork, poultry, red meat, dairy, and sugar	Microalgal biomass	Hülsen et al. (2018)
			Potato wastewater	Microalgae-bacteria consortia (data-based simulation)	Spiller et al. (2020)
Food and nutraceuticals, specialty ingredients	Unialgal culture preferable	Inhibitors, heavy metals, xenobiotics	Sugarcane vinasse	*Arthrospira* sp. bioactive peptides	Montalvo et al. (2019)
			Selenium-rich, irrigation drainage water	Selenium-rich brine shrimp fed with mixed microalgae	Schmidt et al. (2013)

difficulties related to producing enough food and its equal distribution have been calling attention to maximizing the use of the raw materials produced by agriculture and livestock. Increasing process yields and using waste generated during processing are crucial strategies. Photosynthetic organisms have been considered a key element in biorefineries and circular economy: they can use natural light and fix carbon dioxide. Microalgae can do even more. Their ability to grow in liquid media, especially in wastewaters, producing a wide range of value-added biomolecules, turns them into a powerful technology for the environment and the economy.

Microalgae have been worldwide commercially explored for the production of food supplements, feed, cosmetics, and pigments (Bhalamurugan et al. 2018). These are high added-value ingredients, mostly produced in synthetic media, usually in open ponds, although those directed for cosmetics are preferentially produced in closed systems (Chew et al. 2017). The high rate of contamination and the presence of anti-nutritional or toxic components previously discussed are the major bottlenecks in using wastewater, especially for food and feed purposes at a commercial scale.

Therefore, much research has been directed for the use of microalgae in the production of biofuels—a valid use in the context of the circular economy, because every industry demands for fuel and energy. Microalgae can accumulate intracellular lipids, sugars, and proteins. The fact that these metabolites are accumulated inside microalgae cells demands methods for breaking or destabilizing the cell wall or membrane, and mostly this is the technical limiting step for biofuel production from microalgae. Economic challenges include biomass drying (Hosseinizand et al. 2017) and medium costs (Neto et al. 2019), the last significantly reduced when wastewaters are used as a nutrient source.

Those microalgae that accumulate lipids are called "oleaginous" and have been used as feedstock for biodiesel production. The total intracellular lipids are extracted generally from the dried biomass using hot solvents and submitted to transesterification. Strategies to effectively extract lipids from wet biomass is widely explored (Ghasemi Naghdi et al. 2016; Neto et al. 2020). When the carbohydrate fraction is desired, the cells are dried, disrupted, and the starch hydrolyzed prior to use as feedstock for ethanol fermentation (Behera et al. 2015). Pyrolysis and gasification are alternative technologies that use the whole biomass for the production of bio-oil or synthesis gas, respectively. They can be used for the production of heat, electricity, combined heat, and power or be transformed through the Fisher-Tropsch process into biochemicals and biofuels (dos Santos and Alencar 2019).

Another bottleneck in waste-integrated commercial scale is the reuse of CO_2-rich emissions. Most industrial gas emissions contain SOx and NOx; the first strongly inhibits microalgae growth, while the second, especially NO, decreases the growth rate at concentrations above 300 ppm (Klinthong et al. 2015). Prior to use, industrial emissions should be submitted to catalytic reduction and desulfurization to remove NOx and SOx, respectively. CO_2 fixation by microalgae systems depends on the species used, bioreactor configuration, CO_2 concentration at the feed, and operating conditions (pH, medium composition, and temperature) (Klinthong et al. 2015; Mistry et al. 2019). In order to optimize CO_2 fixation, tolerant species must be chosen, high contact time between gas and liquid should be allowed, and higher pH should be preferred. The medium composition, when using liquid wastes, is challenging to manipulate. Concerning CO_2 solubilization, the optimal temperature range for most microalgae (25–30 °C) should not influence the solubility in more than 15% (between 1.25 and 1.5 g CO_2/kg water at 25 and 30 °C, respectively) (Duan and Sun 2003); moreover, the addition of CO_2 causes medium acidification and must be controlled to avoid culture collapse.

6.6 Process Development and Perspectives

Even if agro-industrial wastewaters and microalgal production can be a good match, the development of a microalgal production system based on residue use is complex and starts with a few design decisions, followed by a conceptual material balance and then the development of the microalgal culture. The following questions must be addressed:

- What is the primary objective: to use all the residue, or produce a specific product? If the process is residue-centric, then the use of microalgae isolated from the production site, or even polycultures, is advisable. The biomass produced can be returned to a biodigester after pretreatment (de Carvalho et al. 2020), inserting a solar energy fraction in the production of biogas, or processed into biocrude or other products. Preferentially, the liquid residue should be used concentrated or minimally diluted; a physicochemical pretreatment is frequently applied.
- If the process is product-centered, then the microalgal species must be selected and adapted to the culture medium, which may require amendment of the composition. Also, the bioreactor system may have to be closed in order to reduce contamination unless a somewhat extremophile alga is used. The need to control contamination and, consequently, bioreactor design also depends on the destination of the final product. Direct human consumption of microalgae products will demand much more controlled production than biofuels, biofertilizers, or feed, following specific country regulations.
- What volume is to be processed? Can a closed system be used? If relatively low volumes are to be processed—1–10 m^3 of culture media per day—the production system can be closed. For larger volumes, the process will likely use open bioreactors. Because culturing time—and therefore, the residence time in the reactor—can be long, the volume of the bioreactors will be larger than the volume of residue to be processed. If, for example, 10 m^3 of liquid are to be processed daily, a microalgal process with a relatively short cultivation time of seven days will have a volume of 70 m^3. That will require 2200 m of tubes with a 10 cm radius or a 350 m^2 tank with a depth of 20 cm. For high-added value products, even large, closed systems may prove feasible.

After preliminary calculation of volumes, investments and feasibility are made, the process development is done on a small scale, focused on culture development.

Generally, at this scale, sterilized medium is used, medium composition and culture conditions are optimized, and growth kinetics is measured. The final product of a small scale, pilot facility, is a series of reference protocols: what works best and what can be expected.

One of the main hindrances in microalgal production in agroindustrial residues is the competition of heterotrophs for organic carbon. Even cultures thought to be axenic may be contaminated when cultivated with organic carbon sources: bacterial populations feed on organic carbon and can reduce microalgal productivity. One possibility to reduce the competition by heterotrophs is using a preliminary treatment step, such as sedimentation, filtration, or anaerobic digestion. In these processes, suspended solids (and COD) levels are reduced, which increases the transparency and, therefore, the odds in favor of microalgae.

The direct use of residues with high BOD is also possible. However, isolating robust microalgal cultures or using extremophile conditions and strains is essential. These aspects are important for residues with anti-nutritional or toxic components, such as cyanide (cassava processing wastewater) or phenols (olive mill waste), or high salt concentration (as in sugarcane vinasse)—as depicted in Table 6.1. Several researchers successfully adapted microalgae to residues by cultivation in increasing concentrations to determine maximal suitable concentrations. The *direct* addition of a seed culture to a highly concentrated residue can lead to culture death (Sydney et al. 2019).

The development of new processes can start with selected, commercially applied strains such as *Dunaliella salina*, *Chlorella pyrenoidosa*, or *Arthrospira platensis*, which must be adapted to the effluent, with eventual dilution; the process is not dissimilar from the traditional development of new media.

An alternative to classical strains is the isolation of new microalgal strains. The sampling of soil, wastewater in facultative or stabilization ponds, or other biotopes that share the same climate and irradiation where the microalgae processing plant will be established tends to give robust strains that can compete with contaminant microorganisms. Samples can be isolated after cultivation in the residue, guaranteeing its ability to grow in the residue or isolated in rich media and the screened for its growth rate, composition, and other engineering criteria (Sydney et al. 2011; Gonzalez et al. 2020).

After a preliminary evaluation of classical strains or new isolates growth on the residue, the process can be further improved by adding specific nutrients or pretreating the effluent. As can be seen in Tables 6.1 and 6.2, residues are often rich in micronutrients, but they may be imbalanced in terms of nitrogen, phosphate, or vitamins. Specific residues such as vinasse (low pH), cassava processing wastewater (high cyanide concentration), or wastewaters intensely colored or with a high concentration of suspended solids may require specific amendments. These pretreatments must take into account the costs for treatment. Pasteurization or filtration, for example, may require significant energy inputs, while flocculation, pH, and composition amendment requires the addition of fertilizer-grade chemicals.

After developing successful small-scale cultivation, it is time to proceed to scale-up. This step involves maintaining composition but increasing volume while also verifying the development of the culture in in-situ conditions that differ from the ideal laboratory setting: light irradiation and temperatures oscillate in greenhouses and open systems, reagents may not have the same purity, agitation can be irreplicable. The best strategy is to develop preliminary tests and evaluate the culture stability early on, in parallel with the small laboratory cultures. Ideally, process development is also done with a parallel techno-economic projection of costs, from wastewater pretreatment to the final product, to verify the bottlenecks and most critical costs.

As previously stated, microalgae production using agro-industry wastewaters brings together the much necessary effluent treatment and the very welcome large volumes of cheap culture media required for microalgae biomass production. It is a good match, and despite concerns about the biomagnification of contaminants, the biomass produced from agro-industry residues tends to be of high microbiological and chemical quality compared to that produced from other wastewaters. It is a technology destined to thrive in the next decades.

References

Ahmad A, Buang A, Bhat AH (2016) Renewable and sustainable bioenergy production from microalgal co-cultivation with palm oil mill effluent (POME): a review. Renew Sust Energ Rev 65:214–234. https://doi.org/10.1016/j.rser.2016.06.084

Assemany P, de Paula Marques I, Calijuri ML, Reis A (2020) Complementarity of substrates in anaerobic digestion of wastewater grown algal biomass. Waste Biomass Valorization 11:1–12

Bala JD, Lalung J, Ismail N (2015) Studies on the reduction of organic load from palm oil mill effluent (POME) by bacterial strains. Int J Recycl Org Waste Agric 4:1–10. https://doi.org/10.1007/s40093-014-0079-6

Behera S, Singh R, Arora R et al (2015) Scope of algae as third generation biofuels. Front Bioeng Biotechnol 2:1–13. https://doi.org/10.3389/fbioe.2014.00090

Bhalamurugan GL, Valerie O, Mark L (2018) Valuable bioproducts obtained from microalgal biomass and their commercial applications: a review. Environ Eng Res 23:229–241

Borguetthi I (2009) Avaliação do crescimento da microalga Chlorella minutissima em meio de cultura com diferentes concentrações de ivo alberto borghetti avaliação do crescimento da microalga Chlorella minutissima. Universidade Federal Do Paraná

Brasil B d SAF, de Siqueira FG, Salum TFC et al (2017) Microalgae and cyanobacteria as enzyme biofactories. Algal Res 25:76–89. https://doi.org/10.1016/j.algal.2017.04.035

Brito AG, Peixoto J, Oliveira JM et al (2007) Brewery and winery wastewater treatment: some focal points of design and operation. In: Oreopoulou V, Russ W (eds) Utilization of by-products and treatment of waste in the food industry. Springer, New York, pp 109–131

Cai T, Park SY, Li Y (2013) Nutrient recovery from wastewater streams by microalgae: status and prospects. Renew Sust Energ Rev 19:360–369

Carminati P, Gusmini D, Pizzera A et al (2018) Biogas from mono-and co-digestion of microalgal biomass grown on piggery wastewater. Water Sci Technol 78:103–113

Carvalho F, Prazeres AR, Rivas J (2013) Cheese whey wastewater: characterization and treatment. Sci Total Environ 445–446:385–396. https://doi.org/10.1016/j.scitotenv.2012.12.038

Chew KW, Yap JY, Show PL et al (2017) Microalgae biorefinery: high value products perspectives. Bioresour Technol 229:53–62. https://doi.org/10.1016/j.biortech.2017.01.006

Chokshi K, Pancha I, Ghosh A, Mishra S (2016) Microalgal biomass generation by phycoremediation of dairy industry wastewater: an integrated approach towards sustainable biofuel production. Bioresour Technol 221:455–460

Christofoletti CA, Escher JP, Correia JE et al (2013) Sugarcane vinasse: environmental implications of its use. Waste Manag 33:2752–2761. https://doi.org/10.1016/j.wasman.2013.09.005

Chu HQ, Tan XB, Zhang YL, Yang L Bin, Zhao FC, Guo J (2015) Continuous cultivation of Chlorella pyrenoidosa using anaerobic digested starch processing wastewater in the outdoors. Bioresour Technol 185:40–48. https://doi.org/10.1016/j.biortech.2015.02.030

de Carvalho JC, Borghetti IA, Cartas LC et al (2018) Biorefinery integration of microalgae production into cassava processing industry: potential and perspectives. Bioresour Technol 247:1165–1172. https://doi.org/10.1016/j.biortech.2017.09.213

de Carvalho JC, Magalhães AI, de Melo Pereira GV et al (2020) Microalgal biomass pretreatment for integrated processing into biofuels, food, and feed. Bioresour Technol:122719. https://doi.org/10.1016/j.biortech.2019.122719

de Mattos LFA, Bastos RG (2016) COD and nitrogen removal from sugarcane vinasse by heterotrophic green algae Desmodesmus sp. Desalin Water Treat 57:9465–9473. https://doi.org/10.1080/19443994.2015.1028454

de Siqueira CJ, Calijuri ML, Mattiello EM et al (2020) Algal biomass from wastewater: soil phosphorus bioavailability and plants productivity. Sci Total Environ 711:135088

Debik E, Coskun T (2009) Bioresource technology use of the static granular bed reactor (SGBR) with anaerobic sludge to treat poultry slaughterhouse wastewater and kinetic modeling. Bioresour Technol 100:2777–2782. https://doi.org/10.1016/j.biortech.2008.12.058

Di Caprio F, Altimari P, Pagnanelli F (2015) Integrated biomass production and biodegradation of olive mill wastewater by cultivation of Scenedesmus sp. Algal Res 9:306–311. https://doi.org/10.1016/j.algal.2015.04.007

Ding GT, Yaakob Z, Takriff MS et al (2016) Biomass production and nutrients removal by a newly-isolated microalgal strain Chlamydomonas sp in palm oil mill effluent (POME). Int J Hydrog Energy 41:4888–4895. https://doi.org/10.1016/j.ijhydene.2015.12.010

dos Santos RG, Alencar AC (2019) Biomass-derived syngas production via gasification process and its catalytic conversion into fuels by Fischer Tropsch synthesis: a review. Int J Hydrog Energy. https://doi.org/10.1016/j.ijhydene.2019.07.133

Duan Z, Sun R (2003) An improved model calculating CO2 solubility in pure water and aqueous NaCl solutions from 273 to 533 K and from 0 to 2000 bar. Chem Geol 193:257–271

EFSA Panel on Dietetic Products N and A (NDA) (2014) Scientific Opinion on the safety of astaxanthin-rich ingredients (AstaREAL A1010 and AstaREAL L10) as novel food ingredients. EFSA J 12:3757

Enzing C, Ploeg M, Barbosa M, Sijtsma L, authors Vigani M, Parisi C, Rodriguez Cerezo E, editors (2014) Current markets, products and future developments for micro-algae. In: Microalgae-based products for the food and feed sector: an outlook for Europe. EUR 26255. Luxembourg (Luxembourg): Publications Office of the European Union. JRC85709. https://dx.doi.org/10.2791/3339

FAO (2019) Faostat crop production database. In: Faostat Crop Prod. Database. http://www.fao.org/faostat/en/#data/QC. Accessed 14 May 2019

Faria RAP (2016) Potencial de produção de biogás e degradação anaeróbia do POME (Palm Oil Mill Effluent) resultante do processamento da palma em óleo e biodiesel. Federal University of Paraná

Ghasemi Naghdi F, González González LM, Chan W, Schenk PM (2016) Progress on lipid extraction from wet algal biomass for biodiesel production. Microb Biotechnol 9:718–726. https://doi.org/10.1111/1751-7915.12360

Gonzalez EG, de Carvalho JC, Aulestia DTM et al (2020) Bioprospection of green microalgae native to Paraná, Brazil using a multi-criteria analysis: potential for the production of lipids, proteins, and carotenoids. Bioresour Technol Rep 10:100398

Hena S, Fatimah S, Tabassum S (2015) Cultivation of algae consortium in a dairy farm wastewater for biodiesel production. Water Res Ind 10:1–14. https://doi.org/10.1016/j.wri.2015.02.002
Hoarau J, Caro Y, Grondin I, Petit T (2018) Sugarcane vinasse processing: toward a status shift from waste to valuable resource. A review. J Water Process Eng 24:11–25. https://doi.org/10.1016/j.jwpe.2018.05.003
Hosseinizand H, Lim CJ, Webb E, Sokhansanj S (2017) Economic analysis of drying microalgae Chlorella in a conveyor belt dryer with recycled heat from a power plant. Appl Therm Eng 124:525–532. https://doi.org/10.1016/j.applthermaleng.2017.06.047
Hülsen T, Hsieh K, Lu Y et al (2018) Simultaneous treatment and single cell protein production from agri-industrial wastewaters using purple phototrophic bacteria or microalgae–a comparison. Bioresour Technol 254:214–223
Kamyab H, Md Din MF, Ponraj M, Keyvanfar A, Rezania S, Taib SM, Abd Majid MZ (2016) Isolation and screening of microalgae from agro-industrial wastewater (POME) for biomass and biodiesel sources. Desalin Water Treat 57:29118–29125. https://doi.org/10.1080/19443994.2016.1139101
Keskes S, Hmaied F, Gannoun H et al (2012) Performance of a submerged membrane bioreactor for the aerobic treatment of abattoir wastewater. Bioresour Technol 103:28–34. https://doi.org/10.1016/j.biortech.2011.09.063
Kirnev PCS, de Carvalho JC, Miyaoka JT et al (2018) Harvesting Neochloris oleoabundans using commercial organic flocculants. J Appl Phycol. https://doi.org/10.1007/s10811-018-1429-y
Klinthong W, Yang Y-H, Huang C-H, Tan C-S (2015) A review: microalgae and their applications in CO2 capture and renewable energy. Aerosol Air Qual Res 15:712–742. https://doi.org/10.4209/aaqr.2014.11.0299
Kuo CM, Chen TY, Lin TH, Kao CY, Lai JT, Chang JS, Lin CS (2015) Cultivation of Chlorella sp. GD using piggery wastewater for biomass and lipid production. Bioresour Technol 194:326–333. https://doi.org/10.1016/j.biortech.2015.07.026
Li H, Wang M, Wang X et al (2018) Biogas liquid digestate grown Chlorella sp. for biocrude oil production via hydrothermal liquefaction. Sci Total Environ 635:70–77
Liu J, van der Meer JP, Zhang L, Zhang Y (2017) Cultivation of Haematococcus pluvialis for astaxanthin production. In: Microalgal production for biomass and high-value products. CRC Press, Boca Raton, pp 267–293
Lowrey J, Brooks MS, McGinn PJ (2015) Heterotrophic and mixotrophic cultivation of microalgae for biodiesel production in agricultural wastewaters and associated challenges—a critical review. J Appl Phycol 27:1485–1498. https://doi.org/10.1007/s10811-014-0459-3
Lu Q, Zhou W, Min M, Ma X, Ma Y, Chen P, Zheng H, Doan YTT, Liu H, Chen C, Urriola PE, Shurson GC, Ruan R (2016) Mitigating ammonia nitrogen deficiency in dairy wastewaters for algae cultivation. Bioresour Technol 201:33–40. https://doi.org/10.1016/j.biortech.2015.11.029
Martinez-Burgos WJ, Sydney EB, de Paula DR et al (2020) Biohydrogen production in cassava processing wastewater using microbial consortia: process optimization and kinetic analysis of the microbial community. Bioresour Technol 309:123331
Mishra BK, Arora A, Lata (2004) Optimization of a biological process for treating potato chips industry wastewater using a mixed culture of Aspergillus foetidus and Aspergillus niger. Bioresour Technol 94:9–12. https://doi.org/10.1016/j.biortech.2003.11.020
Mistry AN, Ganta U, Chakrabarty J, Dutta S (2019) A review on biological systems for CO 2 sequestration: organisms and their pathways. Environ Prog Sustain Energy 38:127–136. https://doi.org/10.1002/ep.12946
Mitra D, van Leeuwen JH, Lamsal B (2012) Heterotrophic/mixotrophic cultivation of oleaginous Chlorella vulgaris on industrial co-products. Algal Res 1:40–48
Moerman W, Carballa M, Vandekerckhove A et al (2009) Phosphate removal in agro-industry: pilot- and full-scale operational considerations of struvite crystallization. Water Res 43:1887–1892
Molina D, de Carvalho JC, Júnior AIM et al (2019) Biological contamination and its chemical control in microalgal mass cultures. Appl Microbiol Biotechnol 103:9345–9358

Montalvo GEB, Thomaz-Soccol V, Vandenberghe LPS et al (2019) Arthrospira maxima OF15 biomass cultivation at laboratory and pilot scale from sugarcane vinasse for potential biological new peptides production. Bioresour Technol 273:103–113. https://doi.org/10.1016/j.biortech.2018.10.081

Muhmood A, Lu J, Dong R, Wu S (2019) Formation of struvite from agricultural wastewaters and its reuse on farmlands: status and hindrances to closing the nutrient loop. J Environ Manag 230:1–13

Mukherjee C, Chowdhury R, Ray K (2015) Phosphorus recycling from an unexplored source by polyphosphate accumulating microalgae and cyanobacteria—a step to phosphorus security in agriculture. Front Microbiol 6:1421

Neto CJD, Letti LAJ, Karp SG, Vítola FMD, & Soccol CR (2019) Production of biofuels from algae biomass by fast pyrolysis. In Biofuels from algae. Elsevier, pp 461–473. https://doi.org/10.1016/B978-0-444-64192-2.00018-4

Neto CJD, Sydney EB, Candeo ES et al (2020) New method for the extraction of single-cell oils from wet oleaginous microbial biomass: efficiency, oil characterisation and energy assessment. Waste Biomass Valorization 11. https://doi.org/10.1007/s12649-019-00705-x

Oz NA, Uzun AC (2015) Ultrasound pretreatment for enhanced biogas production from olive mill wastewater. Ultrason Sonochem 22:565–572. https://doi.org/10.1016/j.ultsonch.2014.04.018

Pathak VV, Kothari R, Chopra AK, Singh DP (2015) Experimental and kinetic studies for phycoremediation and dye removal by Chlorella pyrenoidosa from textile wastewater. J Environ Manage 163:270–277. https://doi.org/10.1016/j.jenvman.2015.08.041

Pelizer LH, de Carvalho JCM, de Oliveira Moraes I (2015) Protein production by Arthrospira (Spirulina) platensis in solid state cultivation using sugarcane bagasse as support. Biotechnol Rep 5:70–76

Perez-Garcia O, Escalante FME, de-Bashan LE, Bashan Y (2011) Heterotrophic cultures of microalgae: metabolism and potential products. Water Res 45:11–36. https://doi.org/10.1016/j.watres.2010.08.037

Pittman JK, Dean AP, Osundeko O (2011) The potential of sustainable algal biofuel production using wastewater resources. Bioresour Technol 102:17–25

Polishchuk A, Valev D, Tarvainen M, Mishra S, Kinnunen V, Antal T, Yang B, Rintala J, Tyystjärvi E (2015) Cultivation of Nannochloropsis for eicosapentaenoic acid production in wastewaters of pulp and paper industry. Bioresour Technol 193:469–476. https://doi.org/10.1016/j.biortech.2015.06.135

Prado RDM, Caione G, Campos CNS (2013) Filter cake and vinasse as fertilizers contributing to conservation agriculture. Appl Environ Soil Sci 2013:1–8. https://doi.org/10.1155/2013/581984

Rodrigues DB, Flores ÉMM, Barin JS et al (2014) Production of carotenoids from microalgae cultivated using agroindustrial wastes. Food Res Int 65:144–148

Schmidt R, Tantoyotai P, Fakra SC et al (2013) Selenium biotransformations in an engineered aquatic ecosystem for bioremediation of agricultural wastewater via brine shrimp production. Environ Sci Technol 47:5057–5065

Spiller M, Muys M, Papini G et al (2020) Environmental impact of microbial protein from potato wastewater as feed ingredient: comparative consequential life cycle assessment of three production systems and soybean meal. Water Res 171:115406

Sydney EB, da Silva TE, Tokarski A et al (2011) Screening of microalgae with potential for biodiesel production and nutrient removal from treated domestic sewage. Appl Energy 88:3291–3294. https://doi.org/10.1016/j.apenergy.2010.11.024

Sydney EB, Neto CJD, de Carvalho JC et al (2019) Microalgal biorefineries: integrated use of liquid and gaseous effluents from bioethanol industry for efficient biomass production. Bioresour Technol 292:121955. https://doi.org/10.1016/j.biortech.2019.121955

Tabassum S, Zhang Y, Zhang Z (2015) An integrated method for palm oil mill effluent (POME) treatment for achieving zero liquid discharge – a pilot study. J Clean Prod 95:148–155. https://doi.org/10.1016/j.jclepro.2015.02.056

Tan X-B, Zhao X-C, Yang L-B et al (2018) Enhanced biomass and lipid production for cultivating Chlorella pyrenoidosa in anaerobically digested starch wastewater using various carbon sources and up-scaling culture outdoors. Biochem Eng J 135:105–114

Usha MT, Sarat Chandra T, Sarada R, Chauhan VS (2016) Removal of nutrients and organic pollution load from pulp and paper mill effluent by microalgae in outdoor open pond. Bioresour Technol 214:856–860. https://doi.org/10.1016/j.biortech.2016.04.060

Valev D, Santos HS, Tyystjärvi E (2020) Stable wastewater treatment with Neochloris oleoabundans in a tubular photobioreactor. J Appl Phycol 32. https://doi.org/10.1007/s10811-019-01890-x

Van Wagenen J, Pape ML, Angelidaki I (2015) Characterization of nutrient removal and microalgal biomass production on an industrial waste-stream by application of the deceleration-stat technique. Water Res 75:301–311. https://doi.org/10.1016/j.watres.2015.02.022

Wang J, Yang H, Wang F (2014) Mixotrophic cultivation of microalgae for biodiesel production: status and prospects. Appl Biochem Biotechnol 172:3307–3329

Wilkie AC, Riedesel KJ, Owens JM (2000) Stillage characterization and anaerobic treatment of ethanol stillage from conventional and cellulosic feedstocks. Biomass Bioenergy 19:63–102

Woiciechowski AL, Carvalho JC, Spier MR et al (2013) Emprego de resíduos agroindustriais em bioprocessos alimentares. In: Biotecnologia de alimentos, coleção Ciência, tecnologia, engenharia de alimentos e nutrição, editora Atheneu, São Paulo, Rio de Janeiro e Belo Horizonte, pp 143–172

Yang S, Xu J, Wang Z-M et al (2017) Cultivation of oleaginous microalgae for removal of nutrients and heavy metals from biogas digestates. J Clean Prod 164:793–803

Zero Discharge of Hazardous Chemicals Programme (ZDHC) (2019) Wastewater Guidelines Version 1.1. Road Map to Zero, Amsterdam

Zhou W, Chen P, Min M et al (2014) Environment-enhancing algal biofuel production using wastewaters. Renew Sust Energ Rev 36:256–269. https://doi.org/10.1016/j.rser.2014.04.073

Chapter 7
The Neuroprotective and Antioxidant Activities of *Spirulina*

Telma Elita Bertolin, Cíntia Guarienti, Elionio Galvão Frota, and Tiago Fleming Outeiro

Abstract Science has been showing the great impact that diet represents on man's health. The diversified consumption of food is deficient throughout the world. The academy and the food industry need, in line with the consumer, to introduce new foods and new substances and seek prevention and therapy for different chronic noncommunicable diseases (NCDs). Neurodegenerative diseases, especially Alzheimer's and Parkinson's disease, are age-related disorders and are closely associated with stochastic factors that denote the choices of each individual for himself and the environment. These behaviors potentiate the oxidative stress and the inflammatory processes that culminate in the appearance of these pathologies. *Spirulina platensis*, due to its content of proteins, pigments, and vitamins, plays an important role because these elements are considered valuable targets for therapeutic interventions in neurodegenerative diseases.

Keywords Diet · Oxidative stress · Functional food · Neuroprotection

7.1 Introduction

The growth of intensive agriculture has increased the availability of food in the last decades; however, in addition to the strong environmental impacts, the current system of food production is not meeting the appeal of food and nutritional security.

T. E. Bertolin (✉) · E. G. Frota
Postgraduate Program in Food Science and Technology, University of Passo Fundo, Passo Fundo, RS, Brazil
e-mail: telma@upf.br

C. Guarienti
Federal Institute of Education, Science and Technology Farroupilha, Santo Augusto, RS, Brazil
e-mail: cintia.guarienti@iffarroupilha.edu.br

T. F. Outeiro
Department of Experimental Neurodegeneration, Center for Biostructural Imaging of Neurodegeneration, University Medical Center Göttingen, Göttingen, Germany

J. A. V. Costa et al. (eds.), *Microalgal Bioengineering*,
https://doi.org/10.1007/978-3-031-61253-4_7

Today, about one billion people suffer chronically from poor diet, either due to malnutrition or due to obesity. The diversified consumption of food is deficient throughout the world. We are ingesting many processed foods, salt, and sugary drinks (Tilman and Clarck 2014).

These behaviors are also the result of consumer choice, whether due to their purchasing power or the nature of this type of food. This scenario increases the incidence of chronic noncommunicable diseases (NCD) such as cancer, diabetes, cardiovascular diseases, and neurodegenerative diseases, among others (Sieri et al. 2017).

The need to renew and introduce actions for food and nutrition education within the productive and consumer sectors emerges today as a world priority. The 2030 agenda for sustainable development aims to end hunger, achieve food security, improve nutrition, and promote sustainable agriculture. Global policies will need to feed 9 billion people by 2050, and these foods will need to be safe and nutritious. Microalgae, with an emphasis on *Spirulina platensis*, emerge at present as a major superfood for the next generations. *S. platensis* is a cyanobacterium with functional components such as essential amino acids, essential fatty acids, antioxidant pigments, carotenoids, β-carotene, and its main functional component, the phycocyanin pigment (Singh et al. 2023).

These attributes of the *Spirulina* cyanobacterium, according to different scientific investigations, promote the possibility of nutritional benefits; anticancer action; antitoxicological, antioxidant, and neuroprotective actions; and actions in the metabolism of cholesterol and glucose (Priyanka et al. 2023).

Because of the multifaceted properties of *Spirulina*, there is a range of studies on the beneficial effects of its bioactive compounds on health. In this chapter, the emphasis will be on the antioxidant activity of *Spirulina* in the context of neuroprotection.

7.2 Functional Compounds of *Spirulina platensis* and Its Antioxidant Activity

In the late 1970s, edible cyanobacteria, including *S. platensis*, were introduced to the health food market and gained considerable popularity in this sector (Jensen et al. 2016). Since the year 2000, the production of microalgae destined for human and animal feeding (feed) presented a fivefold increase (Moons et al. 2018).

Spirulina or *Arthrospira* is a microscopic, photosynthetic, and filamentous cyanobacterium, also called microalgae (Fig. 7.1), with nutritional importance (Sotiroudis and Sotiroudis 2013) due to the variety of macro and micronutrients it contains (Phang et al. 2000), and has been recommended by the National Aeronautics and Space Administration (NASA) and the European Space Agency (ESA) as one of the main foods for consumption during long-term space missions (Asghari et al. 2016). Cyanobacteria have a high content of proteins, polysaccharides, lipids,

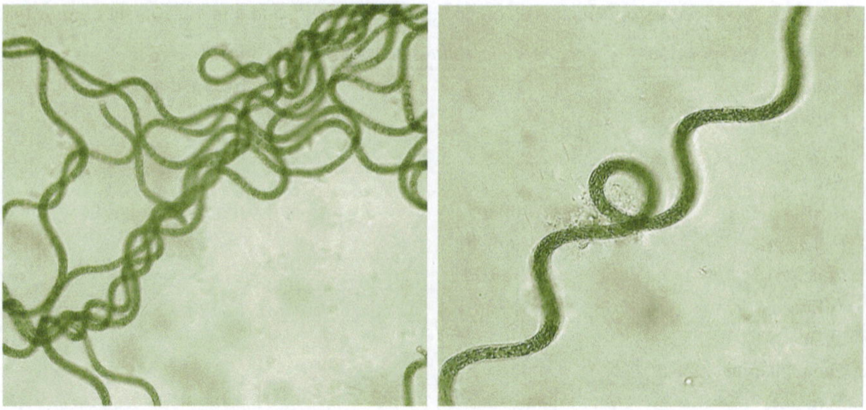

Fig. 7.1 *Spirulina* microalgae. (Guarienti 2018)

amino acids and essential fatty acids, minerals, and vitamins (Hoseini et al. 2013). Table 7.1 shows the averages of the centesimal composition, content of amino acids, minerals, and vitamins present in the *Spirulina* microalgae.

Among the constituents of *Spirulina*, pigments are highlighted. The main groups of pigments present in *Spirulina* are carotenoids, chlorophylls, and phycobiliproteins (Barra et al. 2014; Buchweitz 2016). Chlorophyll and carotenoids are found in small amounts in *Spirulina* (1.1% and 0.37%, respectively) compared to the phycocyanin (20%), with chlorophyll being mainly chlorophyll-a and carotenoids, which are liposoluble photopigments and may present as β-carotene, xanthophylls, zeaxanthin, and lutein (Henrikson 1994). Phycocyanin is a water-soluble pigment composed of a protein part and phycocyanobilin, which is the prosthetic group, an open-chain tetrapyrrole component that gives rise to its intense blue color (Fig. 7.2) (Fernández-Rojas et al. 2014).

In addition to the nutritional contribution, *Spirulina* has been studied for its therapeutic effects, being the antioxidant effect as its most striking feature (De La Jara et al. 2018). The antioxidant effect of *Spirulina* in vivo and in vitro systems is mediated by compounds such as organic acids (caffeic, chlorogenic, salicylic, synaptic, and trans-cinnamic acids), β-carotene, phycocyanin, and other minerals and vitamins, which may act individually or synergistically (Abdel-daim et al. 2013; Cherdkiatikul and Suwanwong 2014; Estrada et al. 2001; Miranda et al. 1998).

Several studies have focused on using natural pigments derived from microalgae because they have health-promoting properties and a wide range of potential industrial applications (Christaki et al. 2015). Microalgae pigments, due to their chemical structure, can be very useful for improving health due to various properties like antioxidant, anticancer, antiobesity, neuroprotective, antiaging, and so on (Chakdar and Pabbi 2017).

For instance, β-carotene, astaxanthin, and, to a lesser extent, other microalgal pigments have a high antioxidant activity, which aims to mediate the harmful effects

Table 7.1 Mean values of centesimal composition, amino acid, and mineral content of *Spirulina*

Chemical composition	% in *Spirulina* on dry basis	Sources
Protein	73.8	Oliveira et al. (2009)
Carbohydrates	10.3	
Lipids	8.6	
Ashes	7.3	
Amino acids	**g 100 g^{-1}**	Volkmann et al. (2008)
Isoleucine	6.49–6.70	
Leucine	9.26–10.17	
Valine	6.45–7.10	
Lysine	4.42–4.99	
Phenylalanine	4.45–5.30	
Tyrosine	5.26–5.69	
Methionine	2.05–2.50	
Cysteine	0.59–0.90	
Tryptophan	0.08–0.30	
Threonine	4.65–6.20	
Histidine	2.20–10.41	
Alanine	8.51–9.50	
Arginine	7.09–8.00	
Asparagine	3.90–11.80	
Glutamine	9.47–13.40	
Glycine	2.98–5.70	
Proline	3.33–4.20	
Serine	4.59–5.10	
Minerals	**mg kg^{-1}**	Falquet (2006)
Calcium	1300–14,000	
Phosphor	6700–9000	
Iron	580–1800	
Magnesium	2000–2900	
Zinc	21–40	
Chrome	2.8	
Sodium	4500	
Potassium	6400–15,400	
Manganese	25–37	
Copper	08–10.0	
Vitamins	**mg kg^{-1}**	Falquet (2006)
B1	34–50	
B2	30–46	
B6	5–8	
B12	1.5–2	
Niacin	130	
Folate	0.5	
Pantothenate	4.6–2.5	
Biotin	0.05	
C	Traces	

Source: Guarienti (2018)

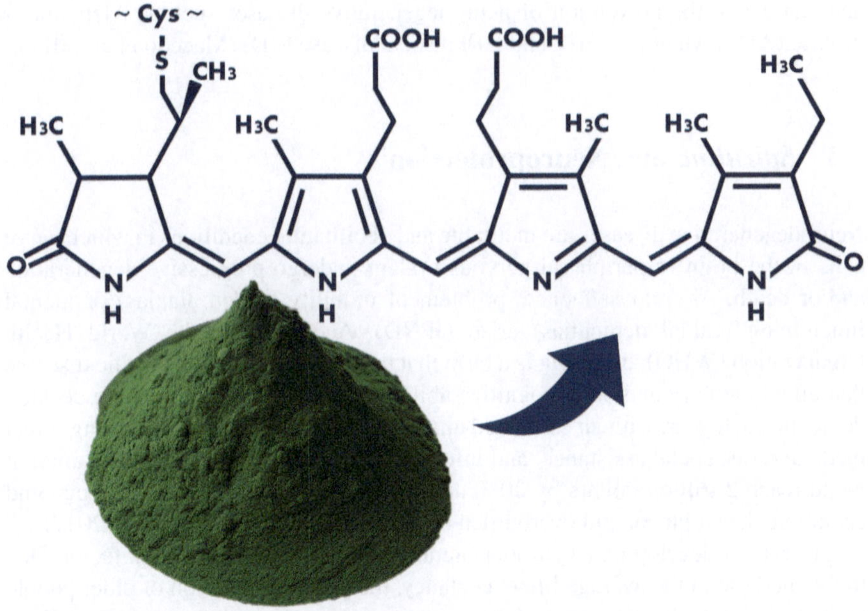

Fig. 7.2 Chemical structure of phycocyanin present in *Spirulina platensis*. (Adapted from Izadi and Fazilati 2018; Hsieh-Lo et al. 2019)

of free radicals, protecting the lipophilic part of lipid peroxidation or capturing reactive oxygen species (ROS) in photooxidative processes (Christaki et al. 2015; Lordan et al. 2011). Phycobiliproteins, such as phycocyanin-C and allophycocyanin, are present in significant amounts within the protein fraction of *Spirulina.* Phycocyanin accounts for 15–25% of the dry weight of the bacteria and has known antioxidant, anti-inflammatory, and neuroprotective effects (Hosseini et al. 2013; Macedo et al. 2017; Reboleira et al. 2019).

The components β-carotene, vitamins C and E, enzyme superoxide dismutase, selenium, and phycocyanin are also indicated as responsible for the metalloprotective action of *Spirulina* (Hoseini et al. 2013). *Spirulina* has played proprietary iron chelating, which is a metal known to catalyze many processes that lead to the onset of free radicals. Therefore, the antioxidant properties of *Spirulina* can arise both from the elimination of free radicals and the chelation of metals, since several iron chelators have been shown to have neuroprotective activity in animal models (Bermejo-Bescós et al. 2008).

Based on scientific data, it is inferred that the ingestion of *Spirulina* may be useful for the treatment of diseases that are aggravated by oxygen reactive species (ROS), as it exerts a significant antioxidant effect that prevents dyslipidemia associated with nontransmissible diseases (Hernández-Lepe et al. 2015); reduces total cholesterol, LDL, and triglycerides; and increases HDL (Serban et al. 2016), preventing cardiovascular diseases (Deng and Chow 2010). It also has a positive impact on patients with chronic obstructive pulmonary disease (COPD) (Ismail et al. 2015) and can act in the prevention of neurodegenerative diseases such as Alzheimer's disease (AD) (Wu et al. 2016) and Parkinson's disease (PD) (Macedo et al. 2017).

7.3 *Spirulina* and Neuroprotection

Neurodegenerative diseases are incurable and debilitating conditions in which nerve cells in the brain or peripheral nervous system undergo progressive degeneration and/or death. As a consequence, problem of mobility (called ataxias) or mental functioning (called dementias) arise (JPND). According to the World Health Organization (WHO), dementia is a term that encompasses several specific diseases that affect memory and other cognitive abilities. Nearly 10 million people develop dementia each year, with an estimated annual cost of $ 818 billion, including direct medical costs, social assistance, and informal care. The forecast is that this amount could reach 2 trillion dollars by 2030 and, as a result, could undermine social and economic development and overburden social and health services (WHO 2017).

The risk of developing a neurodegenerative disease has age as a risk factor. Due to the increase in the average life expectancy, the global population of older people tends to increase. Thus, a critical demand arises for a better understanding of the causes of these diseases and the development of alternative forms for prevention and treatment (Broom et al. 2019).

According to the National Institute of Environmental Health Science (NIEHS), the scientific community recognizes that specific genes increase susceptibility to a particular neurodegenerative disease. However, the combination with the conditions of the environment to which it is exposed during life directly affects this predisposition (USA 2018).

Evidence is growing about the relationship between dementia and lifestyle-related risk factors (lack of physical activity, tobacco use, harmful alcohol use, diabetes mellitus, middle-aged hypertension, obesity, and imbalance of diet). It is necessary to reduce exposure to modifiable factors from childhood and strengthen choices that promote health (WHO 2017).

The incidence and occurrence of neurodegenerative disorders are directly related to eating habits and the practice of physical exercises, which reflect the population's lifestyle (Broom et al. 2019; Brown et al. 2019; Lourenço et al. 2019; Williams and Spencer 2012). Diet directly compromises cognition, as it can affect many brain processes through the regulation of neurotransmitter pathways, synaptic transmission, changes in membrane permeability, and signal transduction pathways (Gómez-Pinilla 2008). Nutritional deficiency can lead to inflammatory processes. An inappropriate diet with a lack of physical exercise aggravates the risks of cognitive decline and AD (Blumenthal et al. 2019; Duffy et al. 2019; Graham et al. 2016). When feeding rats with a high-fat, high-sugar diet for 1 year, Niu et al. (2015) verified that the animals became obese and presented metabolic syndrome (MS) and histopathological characteristics of AD.

Among neurodegenerative diseases, AD is the most prevalent dementia in older individuals (60–80% of cases) and is increasing as the world's population ages, with more than 35 million people affected worldwide (Pugazhenthi et al. 2017). PD is the second late-onset neurodegenerative disease (Wang et al. 2018).

In these pathologies some proteins develop incorrectly and induce other proteins of the same type to fold and cling to each other, forming clusters that are the triggers of the pathological processes that lead to neurodegeneration (Tenreiro et al. 2014; Wang et al. 2018). In AD there is a progressive accumulation of insoluble aggregates of amyloid beta protein (Aβ) between neurons (Huynh and Holtzman 2018) and in PD α-synuclein, which is mainly located in the nerve terminals when under normal conditions; it forms intraneuronal aggregates called Lewy bodies (Tenreiro et al. 2014; Villar-Piqué et al. 2016; Walker 2018).

In AD, two main alterations affect the brain: the formation of Aβ plaques between neurons, which interferes with synaptic communication, and the entangled hyperphosphorylated TAU within neurons, which makes it difficult to transport nutrients and other essential molecules (Alzheimer's Association 2018). The disease causes loss of memory and other cognitive functions, behavior change, mood, and personality, affecting people's daily life, family structuring, and the health system itself (Terraciano and Sutin 2019). Oxidative stress is linked to AD, evidenced by oxidative damage in characteristic lipid and neuronal proteins (Cheignon et al. 2018).

PD is a debilitating condition that affects millions of people around the world and is characterized by the selective loss of dopaminergic neurons in substantia nigra pars compacta (SNpc) (Kalia and Lang 2015; Braak et al. 2004). The loss of dopaminergic neurons leads to motor impairment with the onset of symptoms such as bradykinesia, rest tremor, rigidity, and postural instability, as well as nonmotor symptoms such as sleep disorders, depression, and cognitive deficits (Rodriguez-Oroz et al. 2009). Among the causes of dopaminergic neuronal susceptibility in PD, mitochondrial dysfunction, neuroinflammation, dopamine metabolism, and environmental factors are increasingly accepted as key determinants (Macedo et al. 2017; Ryan et al. 2015).

Chronic neurodegenerative pathologies, in general, are characterized by mitochondrial dysfunction and neuroinflammation. These processes may excessively release reactive oxygen species (ROS) and nitrogen (RNS) and, together with the reduction of endogenous antioxidant defenses, may cause oxidative stress (Lima et al. 2017; Tan et al. 2019). In this context, diets enriched with antioxidant and anti-inflammatory foods can effectively aid in the modulation of neurodegeneration (Gómez-Pinilla 2008; Krzysztoforska et al. 2019; Mohammad-Beigi et al. 2019; Macedo et al. 2018; Pabon et al. 2012).

Venkatesan et al. (2015) reported an accumulation of evidence indicating that using phytochemicals in the diet can prevent or reverse neurodegenerative diseases because it improves mental and physical performance and increases the survival of neuronal cells and boosts the antioxidant system. The authors point out that the main responsible for this action is the activity of phytochemicals as an antioxidant, removing free radicals; its role in enhancing the phagocytic activity of immunological cells, which aids in the clearance of Aβ plaques in AD and Lewy bodies in PD;

and its ability to inhibit the cleavage enzymes of the neurotransmitters, leading to the increase of their concentration and the regulation of neutrophils, which are important for the survival, maintenance, and regeneration of specific neuronal populations in the brain.

Other works support the hypothesis that foods rich in antioxidant compounds may delay the onset and/or delay progression of neurodegenerative disorders (Mendes et al. 2018; Williams and Spencer 2012). Phenolic compounds are also components of the diet with potential neuroprotective action involving various effects on the brain, including its potential ability to suppress neuroinflammation, and promote memory, learning, and cognitive function (Aguilera et al. 2015).

The outstanding positive performance of the bioactive compounds of the diet in front of neurodegenerative disorders directs the search for foods or ingredients that present these characteristics. In this sense, because they have anti-inflammatory, antioxidant, and metal-chelating properties, *Spirulina* have shown promising results with its neuroprotective effect, which has been reported in several in vitro and in vivo studies and mechanisms of action (Blesa et al. 2015; Khalil et al. 2018).

The potential modulator that *Spirulina* and its derivatives play on the central nervous system could be considered a part of therapeutic regimens for neurodegenerative diseases such as AD and PD and remyelination associated with multiple sclerosis and cerebral ischemia (Pabon et al. 2012; Pentón-Rol et al. 2018; Thaakur and Sravanthi 2010). In addition, it has been shown that the effects of neurotoxicity on the development of neurological disorders are related to the radical generation to protect dopaminergic neurotransmission (Tobón-Velasco et al. 2013).

A study carried out by Zhang et al. (2015) demonstrates that the polysaccharide extract obtained from *Spirulina platensis* plays a neuroprotective role in mice with induced Parkinson's disease. In this study, pretreatment with *S. platensis* polysaccharide extract attenuated the deleterious effects of 1-methyl-4-phenyl-1,2,3,6-tetrahydropyridine (MPTP) (decrease in levels and increase in the rates of dopamine metabolism, reduced activity of the enzymes superoxide dismutase (SOD) and glutathione peroxidase (GSH-Px), decrease in immunoreactivity and expression of mRNA of tyrosine hydroxylase (TH) and dopamine transporter (DAT), and the antioxidant properties of these polysaccharides are likely to underlie their neuroprotective effect).

Bermejo-Bescós et al. (2008) demonstrated that the *S. platensis* protein extract exerted antioxidant activity, eliminating ROS due to its chelating properties; protecting the activity of cellular antioxidants GPx total, GPx-Se, and GR; and increasing the reduced glutathione in the cells against oxidative stress induced by iron. Considering that excessive iron levels produce oxidative stress and decline neuronal functions, these results suggest that the *S. platensis* extract is a powerful antioxidant capable of interfering with free radical-mediated cell death. This may be useful in diseases known to be aggravated by ROS and in the development of novel treatments for neurodegenerative disorders, if iron has been implicated in the neuropathology.

Inflammation in the brain, particularly microglial activation, is associated with the pathogenesis of Parkinson's disease and several other neurodegenerative disorders (Pabon et al. 2012). Microglial cells are involved in host defense and tissue repair in the central nervous system, but overactivation of these cells can cause damage to brain tissues (Chen et al. 2012). The study by Pabon et al. (2012) reports that the pretreatment with *Spirulina* can protect the neuronal loss induced by α-synuclein, being the main responsible for this effect the reduction in the microglial activation. However, the authors mention that many active components in *Spirulina* can interact with microglia and suggest the study of phycocyanin in isolation.

The effect of phycocyanin on alpha-synuclein (α-Syn) toxicity in a standard yeast model was evaluated (Macedo et al. 2017). This protein present at the presynaptic terminals of neurons can undergo denaturation and formation of characteristic intracellular filamentous aggregates that are associated with PD (Goedert et al. 2013). The accumulation of α-Syn in the dopaminergic neurons reduces the activity of the mitochondrial complex I of the respiratory chain, increasing ROS production leading to cell death (Martin et al. 2006). Thus, toxicity caused by α-Syn aggregation is strongly associated with increased levels of oxidative stress, which is a central feature in PD (Blesa et al. 2015).

The results showed that phycocyanin, the pigment of *Spirulina platensis*, had neuroprotective action by mitigating α-Syn-induced oxidative stress through the modulation of antioxidant defense cellular pathways, increasing cellular viability (Macedo et al. 2017). The inhibitory effect on lipid peroxidation may be related to the ability of phycocyanin to eliminate the hydroxyl radicals, which are the initiators of lipid peroxidation (Macedo et al. 2017).

In another study evaluating the efficacy of *Spirulina* in modulating neurotoxic changes induced by subacute exposure of rats to lead acetate, treatment with phycocyanin resulted in decreased lead concentration in the body as evidenced by the decline in elevated levels of lead in exposed rats, as well as by the inhibition of lead toxicity. In addition, administration of *Spirulina* reversed the changes induced by lead acetate in antioxidant enzymes, and these findings were similar to those of previous studies (Ponce-Canchihuamán et al. 2010). This effect is due to the improvement of mechanisms of antioxidant capacity of brain tissues due to the greater supply of antioxidants, thus reducing the oxidative damage represented by reduction of malonaldehyde (MDA) and protein carbonyl (PC) formation. This statement implies that *Spirulina* effectively prevented lead-induced neurotoxic damage in rats (Khalil et al. 2018).

The action of phycocyanin on cortical tissues was detected after oral administration in a rat neurotoxicity model (Mitra et al. 2015). The observations that phycocyanin can cross the blood-brain barrier make this compound extremely promising in neurodegenerative disorders (Macedo et al. 2017). Min et al. (2015) determined that noninvasive (intranasal) administration of phycocyanin in rats with middle cerebral artery occlusion (MCAO) can be used to treat oxidative brain damage. This neuroprotective effect is based on the activation of oxidized astrocytes by protecting and repairing the ischemic brain with the combinatorial effects of antioxidative, neurotrophic, and anti-inflammatory mechanisms that strengthened the astrocytes around the sites of brain injury, improving resistance to oxidative stress associated with injury, without promoting cytotoxicity for astrocytes and normal neurons.

Therefore, considering the reports of the antioxidant and neuroprotective activities performed by *Spirulina* and its compounds, mainly phycocyanin, the identification of new intervention strategies for the correction of different types of neuropathologies, such as oxidative stress damage or protein aggregation, is essential.

7.4 Final Considerations

The incidence of neurodegenerative diseases presents a projection of global growth and raises a series of social, economic, and health implications. The search for alternatives to minimize the occurrence and complications resulting from these diseases is eminent.

The lifestyle adopted by the individual directly interferes with the risk of developing neurodegenerative disorders. Among the modifiable factors that influence this probability is the consumption of healthy and balanced diets, not forgetting the practice of physical exercises. Foods or ingredients with bioactive components, mainly antioxidant and anti-inflammatory, have been studied and have shown promising results for prevention, stagnation, and/or attenuation of deleterious effects caused by neurodegenerative diseases.

The food industry needs to commit its efforts to developing and delivering nutritious, practical, healthy, and affordable food to the consumer. Food choices and concerns about public health are expected to fuel a growing demand for health-conscious products.

Spirulina is recognized as a beneficial ingredient in health. Different components are pointed out as responsible for these effects, and they can act in isolation or in synergy. In the neuroprotection it is possible to verify that the majority of the studies indicates the antioxidant activity of *Spirulina* as the main responsible for the beneficial effects, with prominence of performance its main pigment, phycocyanin. Therefore, studies are needed to reinforce the neuroprotective evidence of this cyanobacterium, elucidating the mechanism of action of the bioactive compounds, analyzing possible synergistic or antagonistic interactions, and evaluating its application in matrices and food processes in order to broaden its therapeutic applications.

References

Abdel-Daim MM, Abuzead SM, Halawa SM (2013) Protective role of *Spirulina platensis* against acute deltamethrin-induced toxicity in rats. PLoS One 8(9). https://doi.org/10.1371/journal.pone.0072991

Aguilera Y, Martin-Cabrejas MA, González de Mejia E (2015) Phenolic compounds in fruits and beverages consumed as part of the mediterranean diet: their role in prevention of chronic diseases. Phytochem Rev 15(3):405–423. https://doi.org/10.1007/s11101-015-9443-z

Alzheimer's Association (2018) Alzheimer's disease facts and figures. Alzheimers Dement 14(3):367–429. https://doi.org/10.1016/j.jalz.2018.02.001

Asghari A, Fazilati M, Latifi AM, Salavati H, Choopani A (2016) A review on antioxidant properties of *Spirulina*. J Appl Biotechnol Rep 3(1):345–351

Barra L, Chandrasekaran R, Corato F, Brunet C (2014) The challenge of ecophysiological biodiversity for biotechnological applications of marine microalgae. Mar Drugs 12:1641–1675. https://doi.org/10.3390/md12031641

Bermejo-Bescós P, Piñero-Estrada E, Del Fresno ÁMV (2008) Neuroprotection by *Spirulina platensis* protean extract and phycocyanin against iron-induced toxicity in SH-SY5Y neuroblastoma cells. Toxicol in Vitro 6:1496–1502. https://doi.org/10.1016/j.tiv.2008.05.004

Blesa J, Trigo-Damas I, Quiroga-Varela A, Jackson-Lewis VR (2015) Oxidative stress and Parkinson's disease. Front Neuroanat 9:91. https://doi.org/10.3389/fnana.2015.00091

Blumenthal JA, Smith PJ, Mabe S et al (2019) Lifestyle and neurocognition in older adults with cognitive impairments A randomized trial. Neurology 92(3). https://doi.org/10.1212/WNL.0000000000006784

Braak H, Ghebremedhin E, Rüb U, Bratzke H, Del Tredici K (2004) Stages in the development of Parkinson's disease-related pathology. Cell Tissue Res 318:121–134. https://doi.org/10.1007/s00441-004-0956-9

Broom GM, Shaw IC, Rucklidge JJ (2019) The ketogenic diet as a potential treatment and prevention strategy for Alzheimer's disease. Nutrition 60:118–121. https://doi.org/10.1016/j.nut.2018.10.003

Brown BM, Peiffer J, Rainey-Smith SR (2019) Exploring the relationship between physical activity, beta-amyloid and tau: a narrative review. Ageing Res Rev 50:9–18. https://doi.org/10.1016/j.arr.2019.01.003

Buchweitz M (2016) Natural solutions for blue colors in food. In: Carle R, Schweiggert RM (eds) Handbook on natural pigments in food and beverages, pp 355–384. https://doi.org/10.1016/b978-0-08-100371-8.00017-8

Chakdar H, Pabbi S (2017) Algal pigments for human health and cosmeceuticals. Algal Green Chem:171–188. https://doi.org/10.1016/b978-0-444-63784-0.00009-6

Cheignon C, Tomas M, Bonnefont-Rousselot D et al (2018) Oxidative stress and the amyloid beta peptide in Alzheimer's disease. Redox Biol 14:450–464. https://doi.org/10.1016/j.redox.2017.10.014

Chen JC, Liu KS, Yang TJ, Hwang JH, Chan YC, Lee IT (2012) *Spirulina* and C-phycocyanin reduce cytotoxicity and inflammation-related genes expression of microglial cells. Nutr Neurosci 15(6):252–256. https://doi.org/10.1179/1476830512Y.0000000020

Cherdkiatikul T, Suwanwong Y (2014) Production of the a and b subunits of spirulina allophycocyanin and C-phycocyanin in *Escherichia coli*: a comparative study of their antioxidant activities. J Biomol Screen 19(6):959–965. https://doi.org/10.1177/1087057113520565

Christaki E, Bonos E, Florou-Paneri P (2015) Innovative microalgae pigments as functional ingredients in nutrition. In: Kim S (ed) Handbook of marine microalgae, pp 233–243. https://doi.org/10.1016/b978-0-12-800776-1.00014-5

De La Jara A, Ruano-Rodriguez C, Polifrone M et al (2018) Impact of dietary Arthrospira (*Spirulina*) biomass consumption on human health: main health targets and systematic review. J Appl Phycol 30:2403–2423. https://doi.org/10.1007/s10811-018-1468-4

Deng R, Chow T-J (2010) Hypolipidemic antioxidant and antiinflammatory activities of microalgae spirulina. Cardiovasc Ther 28(4). https://doi.org/10.1111/cdr.2010.28.issue-4, https://doi.org/10.1111/j.1755-5922.2010.00200.x

Duffy CM, Hofmeisterm JJ, Nixon JP, Butterick TA (2019) High fat diet increases cognitive decline and neuroinflammation in a model of orexin loss. Neurobiol Learn Mem 157:41–47. https://doi.org/10.1016/j.nlm.2018.11.008

Estrada JE, Bescós P, Villar Del Fresno AM (2001) Antioxidant activity of different fractions of *Spirulina platensis* protean extract. Il Farmaco 56:497–500. https://doi.org/10.1016/S0014-827X(01)01084-9

Falquet J (2006) The nutritional aspects of *Spirulina*. Antenna Technology

Fernández-Rojas B, Hernández-Juárez J, Pedraza-Chaverri J (2014) Nutraceutical properties of phycocyanin. J Funct Foods 11:375–392. https://doi.org/10.1016/j.jff.2014.10.011

Goedert M, Spillantini MG, Del Tredici K, Braak H (2013) 100 years of Lewy pathology. Nat Rev Neurol 9:13–24. https://doi.org/10.1038/nrneurol.2012.242

Gómez-Pinilla F (2008) Brain foods: the effects of nutrients on brain function. Nat Rev Neurosci 9(7):568–578. https://doi.org/10.1038/nrneurol.2012.242

Graham LC, Harder JM, Soto I et al (2016) Chronic consumption of a western diet induces robust glial activation in aging mice and in a mouse model of Alzheimer's disease. Sci Rep 6:21568. https://doi.org/10.1038/srep21568

Guarienti C (2018) Micropartículas de *Spirulina sp.* aplicadas em mistura para bebida de base láctea. Tese. Centro de Ciências Rurais, Universidade Federal de Santa Maria. Santa Maria—RS, p 131

Henrikson R (1994) Microalga *Spirulina*—superalimento del futuro. Urano, Barcelona

Hernández-lepe MA et al (2015) Spirulina y su efecto hipolipemiante y antioxidante en humanos: una revisión sistemática. Nutr Hosp 32(2):494–500

Hosseini SM, Khosravi-Darani K, Mozafari MR (2013) Nutritional and medical applications of *Spirulina* microalgae. Mini-Rev Med Chem 13(8):1231–1237. https://doi.org/10.2174/1389557511313080009

Hsieh-Lo M, Castillo G, Ochoa-Becerra MA, Mojica L (2019) Phycocyanin and phycoerythrin: Strategies to improve production yield and chemical stability. Algal Res 42:101600. https://doi.org/10.1016/j.algal.2019.101600

Huynh TPV, Holtzman DM (2018) Amyloid-β 'seeds' in old vials of growth hormone. Nature 564:354–355. https://doi.org/10.1038/d41586-018-07604-6

Ismail MD et al (2015) Effect of spirulina intervention on oxidative stress antioxidant status and lipid profile in chronic obstructive pulmonary disease patients. Biomed Res Int 1–7. https://doi.org/10.1155/2015/486120

Izadi M, Fazilati M (2018) Extraction and purification of phycocyanin from spirulina platensis and evaluating its antioxidant and anti-inflammatory activity. Asian J Green Chem 2(4):364–379. https://doi.org/10.22034/AJGC.2018.63597

Jensen GS et al (2016) Clinical Safety of a high dose of phycocyanin-enriched aqueous extract from *Arthrospira (Spirulina) platensis*: results from a randomized, double-blind, placebo-controlled study with a focus on anticoagulant activity and platelet activation. J Med Food 19(7):645–653. https://doi.org/10.1089/jmf.2015.0143

JPND—EU Joint Programe—Neurodegenerative Disease Reseach. What is neurodegenerative disease? http://www.neurodegenerationresearch.eu/about/what/. Accessed January 2019

Kalia LV, Lang AE (2015) Parkinson's disease. Lancet 386(9996):896–912. https://doi.org/10.1016/S0140-6736(14)61393-3

Khalil SR, Khalifa HA, Abdel-Motal SM et al (2018) *Spirulina platensis* attenuates the associated neurobehavioral and inflammatory response impairments in rats exposed to lead acetate. Ecotoxicol Environ Saf 157:255–265. https://doi.org/10.1016/j.ecoenv.2018.03.068

Krzysztoforska K, Mirowska-guzel D, Widy-tyszkiewicz E (2019) Pharmacological effects of protocatechuic acid and its therapeutic potential in neurodegenerative diseases: review on the basis of in vitro and in vivo studies in rodents and humans. Nutr Neurosci 22(2):72–82. https://doi.org/10.1080/1028415X.2017.1354543

Lima FAV, Joventino IP, Joventino FP et al (2017) Neuroprotective activities of *Spirulina platensis* in the 6-OHDA model of Parkinson's disease are related to its anti-inflammatory effects. Neurochem Res 42(12):3390–3400. https://doi.org/10.1007/s11064-017-2379-5

Lordan S, Paul Ross R, Stanton C (2011) Marine bioactives as functional food ingredients: potential to reduce the incidence of chronic diseases. Mar Drugs 9:1056e1100. https://doi.org/10.3390/md9061056

Lourenço MV, Frozza RL, Freitas GB et al (2019) Exercise-linked FNDC5/irisin rescues synaptic plasticity and memory defects in Alzheimer's models. Nat Med 25:165–175. https://doi.org/10.1038/s41591-018-0275-4

Macedo D, Bertolin TE, Oro T et al (2017) Phycocyanin protects against Alpha-Synuclein toxicity in yeast. J Funct Foods 38(A):553–560. https://doi.org/10.1016/j.jff.2017.09.044

Macedo D, Jardim C, Figueira I et al (2018) (Poly)phenol-digested metabolites modulate alpha-synuclein toxicity by regulating proteostasis. Sci Rep 8:6965. https://doi.org/10.1038/s41598-018-25118-z

Martin LJ, Pan Y, Price AC et al (2006) Parkinson's disease alpha-synuclein transgenic mice develop neuronal mitochondrial degeneration and cell death. J Neurosci 26:41–50. https://doi.org/10.1523/JNEUROSCI.4308-05.2006

Mendes D, Oliveira MM, Moreira PI, Coutinho J, Videira RA (2018) Beneficial effects of white wine polyphenols-enriched diet on Alzheimer's disease-like pathology. J Nutr Biochem 55:165–177. https://doi.org/10.1016/j.jnutbio.2018.02.001

Min SK, Park JS, Luo L, Kwon YS, Lee HC, Shim HJ, Kim I-D, Lee J-K, Shin HS (2015) Assessment of C-phycocyanin effect on astrocytes-mediated neuroprotection against oxidative brain injury using 2D and 3D astrocyte tissue model. Sci Rep 5(1). https://doi.org/10.1038/srep14418

Miranda MS et al (1998) Antioxidant activity of the microalga *Spirulina maxima*. Braz J Med Biol Res 31:1075–1079. https://doi.org/10.1590/S0100-879X1998000800007

Mitra S, Siddiqui WA, Khandelwal S (2015) C-Phycocyanin protects against acute tributyltin chloride neurotoxicity by modulating glial cell activity along with its anti-oxidant and anti-inflammatory property: A comparative efficacy evaluation with N-acetyl cysteine in adult rat brain. Chem Biol Interact 238:138–150. https://doi.org/10.1016/j.cbi.2015.06.016

Mohammad-Beigi H, Aliakbari F, Sahin C et al (2019) Oleuropein derivatives from olive fruit extracts reduce α-synuclein fibrillation and oligomer toxicity. J Biol Chem 11:4215–4232. https://doi.org/10.1074/jbc.RA118.005723

Moons I, Barbarossa C, De Pelsmacker P (2018) The determinants of the adoption intention of eco-friendly functional food in different market segments. Ecol Econ 151:151–161. https://doi.org/10.1016/j.ecolecon.2018.05.012

Niu L, Han D, Xu RL, Han B et al (2015) A high-sugar high-fat diet induced metabolic syndrome shows some symptoms of Alzheimer's disease in rats. J Nutr Health Aging 20(5):509–513. https://doi.org/10.1007/s12603-015-0601-1

Oliveira EG, Rosa GS, Moraes MA, Pinto LAA (2009) Characterization of thin layer drying of *Spirulina platensis* utilizing perpendicular air flow. Bioresour Technol 100:1297–1303. https://doi.org/10.1016/j.biortech.2008.05.052

Pabon MM, Jernberg JN, Morganti J et al (2012) *Spirulina*-enhanced diet provides neuroprotection in an α-synuclein model of Parkinson's disease. PLoS One 7(9). https://doi.org/10.1371/journal.pone.0045256

Pentón-Rol G, Marín-Prida J, Falcón-Cama V (2018) C-Phycocyanin and phycocyanobilin as remyelination therapies for enhancing recovery in multiple sclerosis and ischemic stroke: a preclinical perspective. Behav Sci 8:15. https://doi.org/10.3390/bs8010015

Phang SM, Miah MS, Yeoh BG, Hashim MA (2000) *Spirulina* culture in digested sago starch factory waste water. J Appl Phycol 12:395–400. https://doi.org/10.1023/A:1008157731731

Ponce-Canchihuamán JC, Pérez-Méndez O, Hernández-Muñoz R et al (2010) Protective effects of Spirulina maxima on hyperlipidemia and oxidative-stress induced by lead acetate in the liver and kidney. Lipids Health Dis 9(1):35. https://doi.org/10.1186/1476-511X-9-35

Priyanka S, Varsha R, Verma R, Surendra Babu A (2023) Spirulina: a spotlight on its nutraceutical properties and food processing applications. J Microbiol Biotechnol Food Sci e4785. https://doi.org/10.55251/jmbfs.4785

Pugazhenthi S, Qin L, Reddy PH (2017) Common neurodegenerative pathways in obesity, diabetes, and Alzheimer's disease. Biochim Biophys Acta Mol basis Dis 1863(5):1037–1045. https://doi.org/10.1016/j.bbadis.2016.04.017

Reboleira J, Freitas R, Pinteus S, Silva J, Alves C, Pedrosa R, Bernardino S (2019) *Spirulina*. In: Nonvitamin and nonmineral nutritional supplements, pp 409–413. https://doi.org/10.1016/b978-0-12-812491-8.00055-2

Rodriguez-Oroz MC, Jahanshahi M, Krack P et al (2009) Initial clinical manifestations of Parkinson's disease: features and pathophysiological mechanisms. Lancet Neurol 8:1128–1139. https://doi.org/10.1016/S1474-4422(09)70293-5

Ryan BJ, Hoek S, Fon EA, Wade-Martins R (2015) Mitochondrial dysfunction and mitophagy in Parkinson's: from familial to sporadic disease. Trends Biochem Sci 40:200–210. https://doi.org/10.1016/j.tibs.2015.02.003

Serban M-C, Sahebkar A, Dragan S, Stoichescu-Hogea G, Ursoniu S, Andrica F, Banach M (2016) A systematic review and meta-analysis of the impact of Spirulina supplementation on plasma lipid concentrations. Clin Nutr 35(4):842–851. https://doi.org/10.1016/j.clnu.2015.09.007

Sieri S, Agnoli C, Pala V et al (2017) Dietary glycemic index, glycemic load, and cancer risk: results from the EPIC-Italy study. Sci Rep 7:9757. https://doi.org/10.1038/s41598-017-09498-2

Singh SK, Shukla L, Yadav N, Singh PK, Singh SM, Yadav MK, Kaushalendra, Kumar A (2023) Spirulina: From ancient food to innovative super nutrition of the future and its market scenario as a source of nutraceutical. In: Neilan B, Passarini MRZ, Singh PK, Kumar A (eds) Cyanobacterial biotechnology in the 21st century. Springer Nature Singapore, Singapore, pp 51–61.

Sotiroudis TG, Sotiroudis GT (2013) Health aspects of *Spirulina (Arthrospira)* microalga food supplement. J Serb Chem Soc 78(3):395–405. https://doi.org/10.2298/JSC121020152S

Tan SH, Karri V, Tay NWR et al (2019) Emerging pathways to neurodegeneration: dissecting the critical molecular mechanisms in Alzheimer's disease, Parkinson's disease. Biomed Pharmacother 111:765–777. https://doi.org/10.1016/j.biopha.2018.12.101

Tenreiro S, Eckermann K, Outeiro TF (2014) Protein phosphorylation in neurdegeneration: friend or foe? Front Mol Neurosci 7:42. https://doi.org/10.3389/fnmol.2014.00042

Terracciano A, Sutin AR (2019) Personality and Alzheimer's disease: an integrative review. Personal Disord 10(1):4–12. https://doi.org/10.1037/per0000268

Thaakur S, Sravanthi R (2010) Neuroprotective effect of *Spirulina* in cerebral ischemia–reperfusion injury in rats. J Neural Transm 117(9):1083–1091. https://doi.org/10.1007/s00702-010-0440-5

Tilman D, Clark M (2014) Global diets link environmental sustainability and human healt. Nature 515:518–522. https://doi.org/10.1038/nature13959

Tobón-Velasco JC, Palafox-Sánchez V, Mendieta L et al (2013) Antioxidant effect of *Spirulina* (Arthrospira) maxima in a neurotoxic model caused by 6-OHDA in the rat striatum. J Neural Transm 120(8):1179–1189. https://doi.org/10.1007/s00702-013-0976-2

USA. National Institute of Environmental Health Science (NIEHS) (2018) Neurodegenerative disease. https://www.niehs.nih.gov/research/supported/health/neurodegenerative/index.cfm. Accessed January 2019

Venkatesan R, Ji E, Kim SY (2015) Phytochemicals that regulate neurodegenerative disease by targeting neurotrophins: a comprehensive review. Biomed Res Int 1–22. https://doi.org/10.1155/2015/814068

Villar-Piqué A, da Fonseca TL, Outeiro TF (2016) Structure function and toxicity of alpha-synuclein: the Bermuda triangle in synucleinopathies. J Neurochem 139(S1):240–255. https://doi.org/10.1111/jnc.2016.139.issue-S1. https://doi.org/10.1111/jnc.13249

Volkmann H, Imianovsky U, Oliveira JLB, Sant'Anna ES (2008) Cultivation of *arthrospira (spirulina) platensis* in desalinator wastewater and salinated synthetic medium: protein contant and amino-acid profile. Braz J Microbiol 39:98–101. https://doi.org/10.1590/S1517-83822008000100022

Walker LC (2018) Sabotage by the brain's supporting cells helps fuel neurodegeneration. Nature 557:499–500. https://doi.org/10.1038/d41586-018-04988-3

Wang C, Telpoukhovskaia MA, Bahr BA et al (2018) Endo-lysosomal dysfunction: a converging mechanism in neurodegenerative diseases. Curr Opin Neurobiol 48:52–58. https://doi.org/10.1016/j.conb.2017.09.005

WHO. World Health Organization (2017) Global action plan on the public health response to dementia 2017–2025, 44 p. https://www.who.int/mental_health/neurology/dementia/action_plan_2017_2025/en/. Accessed January 2019

Williams RJ, Spencer JPE (2012) Flavonoids, cognition, and dementia: actions, mechanisms, and potential therapeutic utility for Alzheimer disease. Free Radic Biol Med 52(1):35–45. https://doi.org/10.1016/j.freeradbiomed.2011.09.010

Wu Q, Liu L, Miron A, Klímová B, Wan D, Kuča K (2016) The antioxidant immunomodulatory and anti-inflammatory activities of Spirulina: an overview. Arch Toxicol 90(8):1817–1840. https://doi.org/10.1007/s00204-016-1744-5

Zhang F, Lu J, Zhang J, Xie J (2015) Protective effects of a polysaccharide from *Spirulina platensis* on dopaminergic neurons in an MPTP-induced Parkinson's disease model in C57BL/6J mice. Neural Regen Res 10(2). https://doi.org/10.4103/1673-5374.152387

Chapter 8
Plastome Engineering in Microalgae: The Future of the Green Biotechnology

Maria da Glória Esquível, Rute Gonçalves Matos, and Cecília Maria Arraiano

Abstract The chloroplast is a green sophisticated organelle that is often considered as the biosynthetic center of the plant cell. The first successful chloroplast transformation was performed in the microalgae *Chlamydomonas*, and since then the algae chloroplast appears very attractive for modifying biochemical pathways, either by changing expression of existing genes or by adding genes designed to produce new valuable products with medical and industrial applications. We discuss a series of technical achievements developed for the success of algal chloroplast transformation. However, there are limitations of the existing methodology which leads to relatively low expression yields for the desired products in microalgae. Therefore, it is really necessary to increase the knowledge of transcriptional and posttranscriptional control of gene expression in the chloroplast and the role of the algae plastid RNases. *Chlamydomonas* to this day constitutes the only algae routinely used in transplastomic technology for engineering metabolic pathways and molecular farming. We also review recent products such as enzymes and pharmaceutical proteins which have been successfully produced with plastome engineering.

Keywords Chloroplast · Plastome engineering · Ribonucleases · Expression systems · Protein therapeutics

M. d. Esquível (✉)
LEAF – Linking Landscape, Environment, Agriculture and Food, Research Center, Associate Laboratory TERRA, Instituto Superior de Agronomia (ISA), Universidade de Lisboa, Lisbon, Portugal
e-mail: gesquivel@isa.ulisboa.pt

R. G. Matos · C. M. Arraiano
ITQBNOVA – Instituto de Tecnologia Química e Biológica António Xavier, Universidade Nova de Lisboa, Avenida da República, Oeiras, Portugal

J. A. V. Costa et al. (eds.), *Microalgal Bioengineering*,
https://doi.org/10.1007/978-3-031-61253-4_8

8.1 Introduction

The chloroplast has a prokaryotic origin and is one of the most important organelles in the cell of plants and algal species. The plastid transcription and translation systems are very similar to that of eubacteria, as are ribosomal size and the spectrum of antibiotics sensitivities. Chloroplasts are the best-known plastid type and have sophisticated machinery to convert light into chemical energy. For this process, they use the photosynthetic electron transport chain in the extensive system of internal membranes known as thylakoids. The proteins essential to the light reactions of photosynthesis have a unique orientation embedded in the thylakoid membranes and form complexes, the cytochrome b6f and photosystems I and II, and the light-absorbing antenna systems associated with the photosystems. The electron transport chain drives the production of ATP, by the enzyme complex ATP synthase, and NADPH which are used for the production of carbohydrates. The chloroplast stroma contains the enzyme ribulose 1,5-bisphosphate carboxylase/oxygenase (EC4.1.1.39, Rubisco) which is the protein most abundant on Earth. Rubisco catalyzes the initial step of carbon metabolism, the fixation of carbon dioxide, in photosynthetic eukaryotes (Spreitzer and Salvucci 2002). It is responsible for CO2 sequestration. The remaining enzymes of the Calvin cycle for carbon reduction are also present in the stroma as well as the enzymes for the nitrogen and sulfur assimilation. In addition, chloroplasts also synthesize photosynthetic pigments, chlorophylls and carotenoids, starch, fatty acids, lipid soluble vitamins (B1, K1, and E), phytohormones, amino acids, nucleotides and secondary metabolites such as alkaloids, and isoprenoids. The chloroplast with this exclusive and complex metabolism has been a good target of genetic engineering to increase the effectiveness of photosynthetic organisms and for the biotechnological creation of new products, especially therapeutic proteins (antibodies, immunotoxins, antigens, enzymes, hormones, clotting factors, and bioactive peptides). Recombinant proteins can be detected by conventional techniques like immunoblotting and purified easily from liquid cultures.

The first successful chloroplast transformation was performed in the microalgae *Chlamydomonas reinhardtii* (Boynton et al. 1988). Chloroplast transformation has permitted disruption and site-specific mutagenesis of genes on the chloroplast genome, allowing considerable gain in the knowledge of photosynthesis and chloroplast gene expression. Recent works are focusing on biotechnological aspects using as a tool the transformation of chloroplast to obtain new valuable proteins with medical and industrial applications (Taunt and Stoffels Purton 2018).

Algal chloroplast genomes (or plastomes) exhibit great variation conversely to land plants in which they are highly conserved. The plastome of the model algae *Chlamydomonas reinhardtii* has 203.8 kbp (GenBank accession number BK000554) contains only 99 genes, including 5 rRNA genes, 17 ribosomal protein genes, 30 tRNAs specifying all of the amino acids, and 5 genes encoding the catalytic core of a eubacterial-type RNA polymerase, and has a very large number of repeated sequences (>20%) (Maul et al. 2002). Transformation in the algal chloroplast is particularly attractive since this organelle contains a small genome suitable for rapid engineering, and numerous species have only a single chloroplast. Several microalgae with a single chloroplast have a genome present in many copies which effectively amplifies expression of introduced genes compared to the nuclear genome (Harris 2009). For instance, the chloroplast genome of *Chlamydomonas* has approximately 80 copies. However, this advance may also hamper a careful survey of the first generations of the transformants to check if they have a plastid homoplasmic or heteroplasmic (in this case, with wild-type and modified copies of the genome). To decrease this effect, the protocols for *Chlamydomonas* chloroplast transformation include the treatment of the cells with 0.5 mM 5-fluorodeoxyuridine (FdUrd) for approximately six generations prior to transformation to reduce the chloroplast DNA copy number.

High gene expression levels can be achieved in the chloroplast, and it is possible to target transgenes to predictable insertion due to homologous recombination (Goldschmidt-Clermont 1998).

However, due to the prokaryotic origin of the chloroplast, it lacks the machinery to perform complex posttranslational modifications such as glycosylation (the glycosylated proteins come from the endoplasmic reticulum and Golgi apparatus), but, unlike bacteria, the *Chlamydomonas* chloroplast performs phosphorylation and disulfide-thiol exchange of Cys residues (review by Grabsztunowicz and Koskela 2017). Recombinant proteins yields are low (typically 0.5–5% of total soluble protein) compared to established microbial platforms (Taunt and Stoffels Purton 2018). A better understanding of chloroplast gene expression regulation should lead to significant improvements. Recombinant proteins can be detected by conventional techniques like immunoblotting and purified from liquid cultures.

The ability of engineering chloroplasts was possible by a series of technical achievements: (1) improvement of methods that work well for the incorporation of foreign DNA into the chloroplast, (2) selection of an appropriate reporter gene to assure that the chloroplast was transformed, and (3) development of promoter systems to allow expression after the DNA incorporation.

8.2 Methods for Incorporating Foreign DNA in the Chloroplast and Selection of Reporter Genes

The major challenge of chloroplast transformation is the delivery of DNA across the cell wall structure and the three membranes: plasma membrane and the two plastid membranes (Goldschmidt-Clermont 1998).

Chlamydomonas chloroplast was the first to be transformed three decades ago (Boynton et al. 1988). A mutant that lacks the atpB gene (whose product is one of the most important chloroplastic subunits of ATP synthase) was used as the host of transformation. This mutant was deficient in photosynthesis and requires acetate for growth due to deletion of the 2.5 kb locus that contains atpB. Transformed cells were recovered by selecting photosynthetic competent cells on a minimal medium (without acetate) in the light. This method requires sophisticated equipment first designed using a modified shotgun. An apparatus with compressed helium had replaced the gunpower for propulsion of the particles, commercialized by Bio-Rad Laboratories. The biolistic method (microparticle bombardment) has proved to be the most efficient and highly reproducible method for delivering exogenous DNA into microalgal plastids and is a routine practice (Bock 2015). The transformed DNA is released from the particles and integrated into the chloroplast genome through homologous recombination.

The large limitation of this method is the size of the available golden particles and the small size of the plastids from many microalgae species. The biolistic plastid method had a great success with *Chlamydomonas* (Chlorophyta) since it has a quite large chloroplast. The DNA fragments to be delivered were originally coated with tungsten and latter with submicron gold particles (Harris 2009). Numerous Rubisco large subunit mutants of the Spreitzer collection (now available in *Chlamydomonas* Resource Center—www.chlamycollection.org) were achieved by the biolistic method. The microalga *Haematococcus pluvialis*, an exclusive producer of pigment astaxanthin, is another Chlorophyta with a single chloroplast transformed by the biolistic method (Gutiérrez et al. 2012; Galarza et al. 2018) as well as *Platymonas subcordiformis* (Cui et al. 2014). DNA was also introduced into the chloroplast of the red alga *Porphyridium* sp. (Lapidot et al. 2002) and the Euglenoid *Euglena gracilis* (Doetsch et al. 2001) by biolistic.

However, for small-size microalgal species of which the diameters are approximately two microns (e.g., *Nannochloropsis* sp. and *Chlorella* sp.), the plastome cannot be genetic engineering using biolistic tools.

Glass bead methods were also applied to chloroplast transformation but required using wall-less cells (Kindle et al. 1991). The cells are agitated with DNA and the glass beads to allow the entrance of the DNA into the cell. This method is used more often to incorporate DNA into the nucleus. Silicon carbide whiskers are also used instead of the glass beads. Electroporation employs electrical change to temporarily open pores in the cell membrane to introduce DNA into the nucleus. This method was also used with success to transform the chloroplast of small size microalga *Nannochloropsis oceanica* (Gan et al. 2018). Regarding the application of CRISPR/Cas9 editing tools, is rarely reported in chloroplast gene expression, which is likely due to the difficulty to deliver both the Cas9 protein and the guide RNA to the chloroplasts or expressing these two critical components in the chloroplasts simultaneously (Kang et al. 2021).

Improving appropriated reporter gene that allows to certify that the chloroplast was transformed was a high priority in chloroplast transformation methodologies. One of the strategies that had great success in the *Chlamydomonas* chloroplast was to express the bacterial *aadA* reporter gene which product the aminoglycoside adenyl transferase (AAD) that can inactivate spectinomycin and streptomycin, antibiotics that target the 70S ribosome from bacteria and chloroplast (Goldschmidt-Clermont 1991). This selectable marker was often applied in *Chlamydomonas* and for the transformation of other microalgae including *Euglena gracilis* (Doetsch et al. 2001) and *Haematococcus pluvialis* (Gutiérrez et al. 2012; Galarza et al. 2018).

The bacterial *aph46* gene, which encodes an aminoglycoside phosphotransferase that give resistance to kanamycin and amikacin, was used in *Chlamydomonas* as a selectable marker (Bateman and Purton 2000). However, reporter genes coding for antibiotic resistance that are commonly utilized in other organisms for transformation could be not appropriated because the algae are naturally resistant to this kind of compound.

The beta-glucuronidase (GUS) structural bacterial gene (uidA) is a widely used reporter. Higher plants and algae lack intrinsic beta-glucuronidase activity. The enzyme is stable and is highly active. The expression of GUS can be measured accurately using colorimetric and fluorometric assays (Jefferson et al. 1987). GUS has been expressed early as a reporter gene in *Chlamydomonas* chloroplast transformation (Salvador et al. 1993; Ishikura et al. 1999). However, the screening of GUS expression is a destructive method, which is a great disadvantage when working with unicellular living organisms. The small green fluorescent protein (GFP: 26.9 kDa) was first isolated from the jellyfish *Aequorea victoria* and is commonly used as a reporter gene in living cells. The development of a GFP fluorescence reporter gene for *Chlamydomonas reinhardtii* chloroplast was done by Franklin and coworkers (2002). The coral *Renilla reniformis* luciferase (Minko et al. 1999) and firefly luciferase (Mayfield and Schultz 2004) were also adapted to the chloroplast usage.

Deletion gene mutants that lead to defects in photosynthesis were identified in chloroplasts, and they could be cloned, providing a source of selectable markers for the transformation of the corresponding mutant host (Boynton et al. 1988; Goldschmidt-Clermont 1998). These mutants need a source of carbon, acetate, for growing; in minimal medium in the light, they died. The transformants were recovered by selecting for photosynthetic competence on minimal medium in the light. This strategy was used extensively for the study of Rubisco large subunit *rbcL* mutants (Spreitzer 1998; Marin-Navarro and Moreno 2005; Esquível et al. 2006).

Normally, the foreign gene is introduced under the control of the *C. reinhardtii* chloroplast UTRs of highly expressed genes such as *rbcL*, *psbA*, *psbD*, and *atpA* (rubisco large subunit; D1 and D2 proteins of the photosystem II reaction center and α subunit of ATP synthase, respectively) by homologous recombination. Esland et al. (2018) reviewed this topic of selectable markers and reporter genes for engineering the chloroplast of *Chlamydomonas reinhardtii*, which provide additional perspectives that will further advance the microalgae chloroplast as an expression platform.

8.3 Studies on Gene Expression to Improve Chloroplast Engineering

Chloroplasts are specialized organelles where photosynthesis and major anabolic pathways occur within green plant cells. Their structural proteins and enzymes, such as ribonucleases and RNA binding proteins, are encoded by both the nuclear and the chloroplast genomes (Ruwe et al. 2011; Nakamura et al. 2012; Stoppel and Meurer 2012). Studying gene expression and stability of the RNA in the chloroplast and the role of the plastid RNases are important requests for improving chloroplast engineering.

High-throughput techniques have revealed that both strands of the genome of the chloroplast are transcribed (Hotto et al. 2011; Zmienko et al. 2011; Zhelyazkova et al. 2012), which emphasizes the importance of RNA metabolism in these organelles. For example, the *Chlamydomonas* chloroplast genome contains about 100 genes which are transcribed into monocistronic or polycistronic messages. The chloroplasts harbor a prokaryotic-like transcription machinery with eubacterial promoter sequences and a RNA polymerase composed of eubacterial-type subunits. There are two types of RNA polymerases, a plastid encoded and a nucleus-encoded enzyme (abbreviated PEP and NEP, respectively), but in *Chlamydomonas* only PEP seems to be involved in transcription. The core complex of RNA polymerase binds DNA nonspecifically, and promoter selectivity is conferred to the core complex by exchangeable sigma subunits. To date, only few types of promoters have been identified. Transcription is terminated when RNA polymerase pauses and the newly synthesized transcript is released from the ternary RNA polymerase/DNA/RNA complex. Stem-loop structures have defined the 3′ends of transcripts, but these structures are not good terminators. The chloroplast diverges from bacteria in that the 3′end of mRNAs is determined posttranscriptionally by processing. Chloroplast mRNAs do not possess a 5′cap-like cytoplasmic RNAs although they can become polyadenylated at the 3′terminus. In most cases, transcript levels peak early in the light period and are lowest at the beginning of the dark period. These fluctuations are caused by changes in both transcription and RNA degradation (Hwang et al. 1996). The transcription seems to be in most cases under circadian control, whereas RNA degradation seems to be primarily influenced by light conditions (Hwang et al. 1996). In fact, photosynthetic electron transport promotes RNA degradation via a redox-regulated pathway (Salvador and Klein 1999).

During chloroplast biogenesis there are constant adjustments of RNA levels, which are accurately regulated by posttranscriptional mechanisms mediated by several ribonucleases. Contrary to what is verified in bacteria, the stability of the transcripts has a considerably longer stability with half-lives going from 20 min to hours, and the majority of the RNA molecules suffer an extensive maturation process by RNases in order to be functional. Most of the ribonucleases found are homologues from the bacterial counterparts, but their function in chloroplasts is scarcely described. Most plastidial ribonucleases process polycistronic rRNAs, tRNAs, and mRNAs, producing monocistronic mRNAs, similar to what is verified for eukaryotes (Bollenbach et al. 2007; Barkan 2011a, b). The processing and degradation of RNAs in the chloroplast involves the combined action of exoribonucleases which cleave the RNA from one of the termini, and endoribonucleases which cleave the RNA internally. RNA degradation in the chloroplasts was found to be very analogous to that of bacteria (Hayes et al. 1999; Schuster et al. 1999; Monde et al. 2000). It usually starts with an endonucleolytic cleavage of the RNA molecule, followed by the addition of a poly(A) tail (Komine et al. 2000) or a poly(A)-rich tail (Lisitsky et al. 1996). The polyadenylated products are then rapidly degraded by plastidial exonucleases (Lisitsky et al. 1997; Lisitsky and Schuster 1999).

We will present and discuss here what is known regarding RNA processing and degradation within chloroplasts, namely, in *Chlamydomonas*. The exoribonucleases which have known to be involved are PNPase, RNase II-like, and Xrn1. The endonucleases possibly involved, which are predicted in the genome, are RNase J1, YbeY, and CSP41.

PNPase is an exoribonuclease from the PDX family of enzymes. As a member of this family, PNPase is a phosphate-dependent exoribonuclease. PNPase is a multifunctional protein since it can perform distinct activities. PNPase typically degrades RNA in the $3' \rightarrow 5'$ direction in a processive way, releasing nucleoside diphosphates as products of degradation (Grunberg-Manago 1963). However, it has also the ability to synthesize single-stranded heteropolymeric RNA tails from nucleoside diphosphates when the inorganic phosphate concentration is low (Littauer and Soreq 1982). Structural studies revealed that PNPase is a homotrimer that presents a ring-like architecture (Symmons et al. 2000; Shi et al. 2008; Nurmohamed et al. 2009). In eukaryotes, PNPase is found in chloroplasts and mitochondria. In chloroplasts, PNPase seems to act as a tRNA-degrading enzyme (Walter et al. 2002), being also involved in mRNA and 23S rRNA 3′ processing and RNA degradation (Yehudai-Resheff et al. 2001; Walter et al. 2002; Kishine et al. 2004; Germain et al. 2011). Considering that no poly A polymerase (PAP) is found in the chloroplasts, PNPase is responsible for both polyadenylation and degradation (Yehudai-Resheff et al. 2001). However, the green algae *C. reinhardtii* only has homogenous poly (A) tails, which suggests that it may have a Poly A polymerase (PAP) (Komine et al. 2000). The first chloroplast PNPase was identified in *Chlamydomonas* (called PNP1), and the deletion of this gene affected the degradation of poly A transcripts and the ability of the algae to adapt to phosphate starvation (Nishimura et al. 2004; Yehudai-Resheff et al. 2007). Therefore in *C. reinhardtii*, it was shown that PNPase has a role in the phosphate limitation response, since it can be responsible for altering the metabolic flux. In this way, it not only determines the fate of RNA and DNA but also influences photosynthesis and the phosphate economy of the cell as a whole (Yehudai-Resheff et al. 2007).

RNase II-like proteins belong to the RNB family of enzymes, which comprise enzymes that progressively degrade RNA from the 3′ to the 5′ end releasing 5′-nucleotide monophosphates (Cannistraro and Kennell 1994). This family of hydrolytic enzymes is widely distributed and includes bacterial RNase II and RNase R and the eukaryotic Rrp44/Dis3, Dis3L1, and Dis3L2. The main feature of this family is the presence of the very well conserved RNB domain which gives the name to this family. An enzyme of this family is found to locate in *C. reinhardtii* plastid (called RNB2 by Yehudai-Resheff et al. 2007), and after being expressed, it was biochemically characterized and has properties quite similar to the *E. coli* RNase II (Frazão et al. 2006; Matos et al. 2009), since it is sensible to secondary structures and releases a final product of 4 nts (Rute G. Matos personal communication).

The exoribonuclease Xrn1 degrades RNA in the 5′ to 3′ direction and is the major 5′-3′ exoribonuclease involved in mRNA decay (Parker and Song 2004), and one homologue, XRN1, has a predicted organelle targeting sequence.

RNase J proteins were the first described bacterial 5′ to 3′ exoribonucleases (Mathy et al. 2007). Additionally, they were shown to act both as endo- and as exoribonucleases (Even et al. 2005). The activity of these proteins is regulated by the 5′ phosphorylation of the substrate. They play important functions in RNA stability and maturation (Mader et al. 2008).

In *Arabidopsis*, RNase J localizes in the chloroplast and is important for RNA surveillance and 16S and 23S rRNA maturation (Sharwood et al. 2011; Luro et al. 2013). Interestingly, the absence of RNase J leads to the accumulation of antisense RNA (Sharwood et al. 2011). The *Chlamydomonas* genome also encodes homologues of bacterial RNase J1, and alternative spliced variants RNJ1 and RNJ2 may be chloroplast targeted.

YbeY is metalloendoribonuclease from the UPF0054 family. It was first identified in *E. coli* as a heat shock protein involved in translation (Rasouly et al. 2009). Later on, it was shown that YbeY was in fact involved in rRNA maturation and ribosome biogenesis (Davies et al. 2010; Rasouly et al. 2010). In *E. coli*, deletion of YbeY results in a defective processing of 16S, 23S, and 5S rRNAs (Davies et al. 2010). In *Sinorhizobium meliloti*, YbeY homologue regulates the accumulation of small RNAs and their target mRNAs (Pandey et al. 2011) and more recently was described as an endoribonuclease with unprecedented catalytic features that can also act as a gene-silencing enzyme (Saramago et al. 2017). In *Arabidopsis*, YbeY localizes in the chloroplast (Liu et al. 2015). Its absence reduces chlorophyll content, impairs chloroplast development and thylakoid formation, and is lethal at the seedling stage under standard photoautotrophic growth conditions (Liu et al. 2015). Some biochemical data obtained with YbeY from *Arabidopsis* suggest that YbeY may also participate in mRNA and tRNA decay or metabolism in the chloroplast (Liu et al. 2015). This RNase YbeY is also present in the *Chlamydomonas* genome. However, no studies regarding its role in RNA metabolism have yet been performed.

In chloroplasts, several RNA-binding proteins (RNPs) have been described. Although most of them may bind to RNA in a nonspecific manner, others were shown to be involved in pre-mRNA maturation (Schuster and Gruissem 1991; Hayes et al. 1996). The use of gel mobility shift (GMS) assays allowed to find three proteins that associate with stem loop structures with 55, 41, and 29-kD. These proteins were named Chloroplast Stem-Loop Binding Proteins (CSP) 55, CSP41 and CSP29 (Chen and Shyu 1995; Chen et al. 1995).

CSP41 was shown to be an endoribonuclease that accumulates in photosynthetic tissues and is induced by light (Yang et al. 1996). Its RNA binding activity is sequence specific, contrary to its endonuclease activity, which was shown to be nonspecific (Yang et al. 1996). It exists in two forms, CSP41a and CSP41b, which interact with each other and associate with ribosomes (Beligni and Mayfield 2008; Bollenbach et al. 2009). *Chlamydomonas* encode proteins similar to CSP41a and CSP41b of land plants. These proteins co-purified with chloroplast ribosomes and were named RAP41 and RAP38, respectively (Yamaguchi et al. 2003).

The mechanism that determines the distinct half-lives of different chloroplast mRNAs and the specific role of each of the ribonucleases remain to be elucidated. However, it is certainly an important goal for the future since these principles will certainly affect the levels of mRNAs and consequently the levels of their respective proteins. In chloroplasts from eukaryotic algae, codon bias has been shown to be one of the most critical elements for protein expression (Surzycki et al. 2009). Therefore, the choice of the appropriate codon usage can be important for the production of the recombinant proteins.

In conclusion, much is still to be learned regarding transcriptional and posttranscriptional control of gene expression in *Chlamydomonas* chloroplasts. However, some of the bottlenecks are being overcome for the successful engineering of recombinant proteins in chloroplasts.

8.4 Recent Product Examples of Vaccines, Enzymes, and Other Pharmaceutical Proteins Successfully Accomplished with Plastome Engineering

The genetic tools for *C. reinhardtii* are relatively advanced. This organism is easy to grow, nontoxic, and classified as "generally regarded as safe" (GRAS). Therefore, some important products, including vaccines, antibodies, and immunotoxins, have started to be produced with plastome engineering. The first demonstration of mammalian protein produced in the *C. reinhardtii* chloroplast was a human monoclonal antibody directed against herpes simplex virus (HSV) (Mayfield et al. 2003). The factors affecting the expression of vaccines in microalgae chloroplast were studied by Surzycki et al. (2009) with special attention to the VP28 protein to make a vaccine against white spot syndrome virus (WSSV). They have concluded that codon optimization and protease activity were the main factors affecting the expression of the recombinant protein into the chloroplast. The Pfs25 and Pfs28 which elicit antibodies that inhibit malaria transmission were already expressed and characterized in *C. reinhardtii* chloroplasts, and they can contribute to the eradication of malaria (Gregory et al. 2012). The same laboratory has also produced the Ara h1 protein core domain that confers protection from peanut induced anaphylaxis (Gregory et al. 2016). The attenuated form of the E7 oncoprotein of the human papillomavirus (HPV) was also expressed in *C. reinhardtii*, and it functions as a vaccine that induces specific tumor protection (Demurtas et al. 2013). An orally immunogenic protein targeting the p210 epitope implicated in atherosclerosis immunotherapies was also successfully expressed in the algae chloroplast (Beltrán-López et al. 2016). Wannathong et al. (2016) demonstrated that functional hGH, the pituitary human growth hormone (hGH), can be also produced in the *C. reinhardtii* chloroplast. Industrial enzymes like xylanase, alfa-galactosidases, and phytases were well expressed in the chloroplast of *C. reinhardtii* and the marina green algae *Dunaliella tertiolecta* (Georgianna et al. 2013). Recently, *C. reinhardtii* strains with specific recombinant proteins in the chloroplast can be toxic to mosquito larvae. They have been recently proposed for insect pest control (Kang et al. 2018) or malaria antigen diagnosis tests (Shamriz and Ofoghi 2019). Many other examples could be cited. The examples of active recombinant proteins produced in algae chloroplasts are growing exponentially, and they certainly promise an exciting future for green biotechnology.

References

Barkan A (2011a) Expression of plastid genes: organelle-specific elaborations on a prokaryotic scaffold. Plant Physiol 155:1520–1532

Barkan A (2011b) Studying the structure and processing of chloroplast transcripts. Methods Mol Biol 774:183–197

Bateman JM, Purton S (2000) Tools for chloroplast transformation in *Chlamydomonas*: expression vectors and a new dominant selectable marker. Mol Gen Genet 263:404–410

Beligni MV, Mayfield SP (2008) *Arabidopsis thaliana* mutants reveal a role for CSP41a and CSP41b, two ribosome-associated endonucleases, in chloroplast ribosomal RNA metabolism. Plant Mol Biol 67:389–401

Beltrán-López JI, Romero-Maldonado A, Monreal-Escalante E, Bañuelos-Hernández B, Paz-Maldonado LM, Rosales-Mendoza S (2016) *Chlamydomonas reinhardtii* chloroplasts express an orally immunogenic protein targeting the p210 epitope implicated in atherosclerosis immunotherapies. Plant Cell Rep 35:1133–1141

Bock R (2015) Engineering plastid genomes: methods, tools, and applications in basic research and biotechnology. Ann Rev Plant Biol 66:211–241

Bollenbach TJ, Vetsigian K, Kishony R (2007) Evolution and multilevel optimization of the genetic code. Genome Res 17:401–404

Bollenbach TJ, Sharwood RE, Gutierrez R, Lerbs-Mache S, Stern DB (2009) The RNA- binding proteins CSP41a and CSP41b may regulate transcription and translation of chloroplast-encoded RNAs in *Arabidopsis*. Plant Mol Biol 69:541–552

Boynton JE, Gillham NW, Harris EH, Hosler JP, Johnson AM et al (1988) Chloroplast transformation in *Chlamydomonas* with high velocity microprojectiles. Science 240:1534–1538

Cannistraro VJ, Kennell D (1994) The processive reaction mechanism of ribonuclease II. J Mol Biol 243:930–943

Chen CY, Shyu AB (1995) AU-rich elements: characterization and importance in mRNA degradation. Trends Biochem Sci 20:465–470

Chen Q, Adams CC, Usack L, Yang J, Monde RA, Stern DB (1995) An AU-rich element in the 3′ untranslated region of the spinach chloroplast petD gene participates in sequence- specific RNA-protein complex formation. Mol Cell Biol 15:2010–2018

Cui Y, Qin S, Jiang P (2014) Chloroplast transformation of *Platymonas (Tetraselmis) subcordiformis* with the bar gene as selectable marker. PLoS One 9:e98607. https://doi.org/10.1371/journal.pone.0098607

Davies BW, Kohrer C, Jacob AI, Simmons LA, Zhu J et al (2010) Role of *Escherichia coli* YbeY, a highly conserved protein, in rRNA processing. Mol Microbiol 78:506–518

Demurtas OC, Massa S, Ferrante P, Venuti A, Franconi R, Giuliano G (2013) A *Chlamydomonas*-derived human papillomavirus 16 E7 vaccine induces specific tumor protection. PLoS One 8:e61473. https://doi.org/10.1371/journal.pone.0061473

Doetsch NA, Favreau MR, Kuscuoglu N, Thompson MD, Hallick RB (2001) Chloroplast transformation in *Euglena gracilis*: splicing of a group III twintron transcribed from a transgenic psbK operon. Curr Genet 39:49–60

Esland L, Larrea-Alvarez M, Purton S (2018) Selectable markers and reporter genes for engineering the chloroplast of *Chlamydomonas reinhardtii*. Biol 7:46. https://doi.org/10.3390/biology7040046

Esquível MG, Pinto TS, Marín-Navarro J, Moreno J (2006) Substitution of tyrosine residues at the aromatic cluster around the betaA/betaB loop of rubisco small subunit affects the structural stability of the enzyme and the in vivo degradation under stress conditions. Biochemistry (USA) 45:5745–5753. https://doi.org/10.1021/bi052588y

Even S, Pellegrini O, Zig L, Labas V, Vinh J et al (2005) Ribonucleases J1 and J2: two novel endoribonucleases in B. subtilis with functional homology to *E.coli* RNase E. Nucleic Acids Res 33:2141–2152

Franklin S, Ngo B, Efuet E, Mayfield SP (2002) Development of a GFP reporter gene for *Chlamydomonas reinhardtii* chloroplast. Plant J 30:733–744

Frazão C, McVey CE, Amblar M, Barbas A, Vonrhein C, Arraiano CM, Carrondo MA (2006) Unravelling the dynamics of RNA degradation by ribonuclease II and its RNA- bound complex. Nature 443:110–114

Galarza JI, Gimpel JA, Rojas V, Bertha O, Arredondo-Veja BO, Henríquez V (2018) Overaccumulation of astaxanthin in *Haematococcus pluvialis* through chloroplast genetic engineering. Algal Res 31:291–297

Gan Q, Jiang J, Han X, Wang S, Lu Y (2018) Engineering the chloroplast genome of oleaginous marine microalga *Nannochloropsis oceanica*. Front Plant Sci 9:439. https://doi.org/10.3389/fpls.2018.00439

Georgianna DR, Hannon MJ, Marcuschi M, Wu S, Botsch K, Lewis AJ et al (2013) Production of recombinant enzymes in the marine alga *Dunaliella tertiolecta*. Algal Res 2:2–9

Germain A, Herlich S, Larom S, Kim SH, Schuster G, Stern DB (2011) Mutational analysis of *Arabidopsis* chloroplast polynucleotide phosphorylase reveals roles for both RNase PH core domains in polyadenylation, RNA 3′-end maturation and intron degradation. Plant J 67:381–394

Goldschmidt-Clermont M (1991) Transgenic expression of aminoglycoside adenine transferase in the chloroplast: a selectable marker of site-directed transformation of *Chlamydomonas*. Nucleic Acids Res 19:4083–4089

Goldschmidt-Clermont M (1998) Chloroplast transformation and reverse genetics. In: Rochaix J-D, Goldschmidt-Clermont M, Merchant S (eds) The molecular biology of chloroplast and mitochondria in *Chlamydomonas*. Kluwer Academic, Dordrecht, pp p139–p149

Grabsztunowicz M, Koskela MM, Mulo P (2017) Post-translational modifications in regulation of chloroplast function: recent advances. Front Plant Sci 23(8):240. https://doi.org/10.3389/fpls.2017.00240. eCollection 2017

Gregory JA, Li F, Tomosada LM, Cox CJ, Topol AB, Vinetz JM, Mayfield S (2012) Algae-produced pfs25 elicits antibodies that inhibit malaria transmission. PLoS One 7(5):e37179. https://doi.org/10.1371/journal.pone.0037179

Gregory JA, Shepley-McTaggart A, Umpierrez M, Hurlburt BK, Maleki SJ, Sampson HA, Mayfield SP, Berin MC (2016) Immunotherapy using algal-produced Ara h1 core domain suppresses peanut allergy in mice. Plant Biotechnol J:141541–141550

Grunberg-Manago M (1963) Polynucleotide phosphorylase. Prog Nucl Acid Res Mol Biol 1:93–133

Gutiérrez CL, Gimpel J, Escobar C, Marshall SH, Henríquez V (2012) Chloroplast genetic tool for the green microalgae *Haematococcus pluvialis* (chlorophyceae, volvocales). J Phycol 48:976–983. https://doi.org/10.1111/j.1529-8817.2012.01178.x

Harris EH (2009) The *Chlamydomonas* sourcebook, vol 1, 2nd edn. Elsevier Science and Technology, New York

Hayes R, Kudla J, Schuster G, Gabay L, Maliga P, Gruissem W (1996) Chloroplast mRNA 3′-end processing by a high molecular weight protein complex is regulated by nuclear encoded RNA binding proteins. EMBO J 15:1132–1141

Hayes R, Kudla J, Gruissem W (1999) Degrading chloroplast mRNA: the role of polyadenylation. Trends Biochem Sci 24:199–202

Hotto AM, Schmitz RJ, Fei Z, Ecker JR, Stern DB (2011) Unexpected diversity of chloroplast noncoding RNAs as revealed by deep sequencing of the Arabidopsis transcriptome. G3 (Bethesda) 1:559–570

Hwang S, Kawazoe R, Herrin DL (1996) Transcription of tufA and other chloroplast- encoded genes is controlled by a circadian clock in *Chlamydomonas*. Proc Natl Acad Sci USA 93:996–1000

Ishikura K, Takaoka Y, Kato K, Sekine M, Yoshida K, Shinmyo A (1999) Expression of a foreign gene in *Chlamydomonas reinhardtii* chloroplast. J Biosci Bioeng 87:307–314

Jefferson RA, Kavanagh TA, Bevan MW (1987) GUS fusions: beta-glucuronidase as a sensitive and versatile gene fusion marker in higher plants. EMBO J 6:3901–3907

Kang BC, Bae SJ, Lee S, Lee JS, Kim A, Lee H, Baek G, Seo H, Kim J, Kim JS (2021) Chloroplast and mitochondrial DNA editing in plants. Nat Plants 7(7):899-905. https://doi.org/10.1038/s41477-021-00943-9

Kang S, Odom OW, Malone CL, Thangamani S, Herrin DL (2018) Expression of a synthetic gene for the major cytotoxin (Cyt1Aa) of *Bacillus thuringiensis* subsp. *israelensis* in the chloroplast of wild-type *Chlamydomonas*. Biology 7:29. pii: E29. https://doi.org/10.3390/biology7020029

Kindle KL, Richards KL, Stern DB (1991) Engineering the chloroplast genome: techniques and capabilities for chloroplast transformation in *Chlamydomonas reinhardtii*. Proc Natl Acad Sci USA 88:1721–1725

Kishine M, Takabayashi A, Munekage Y, Shikanai T, Endo T, Sato F (2004) Ribosomal RNA processing and an RNase R family member in chloroplasts of Arabidopsis. Plant Mol Biol 55:595–606

Komine Y, Kwong L, Anguera MC, Schuster G, Stern DB (2000) Polyadenylation of three classes of chloroplast RNA in *Chlamydomonas reinhadtii*. RNA (New York) 6:598–607

Lapidot M, Raveh D, Sivan A, Arad SM, Shapira M (2002) Stable chloroplast transformation of the unicellular red alga *Porphyridium* species. Plant Physiol 129:7–12

Lisitsky I, Schuster G (1999) Preferential degradation of polyadenylated and polyuridinylated RNAs by the bacterial exoribonuclease polynucleotide phosphorylase. Eur J Biochem 261:468–474

Lisitsky I, Klaff P, Schuster G (1996) Addition of destabilizing poly (A)-rich sequences to endonuclease cleavage sites during the degradation of chloroplast mRNA. Proc Natl Acad Sci USA 93:13398–13403

Lisitsky I, Kotler A, Schuster G (1997) The mechanism of preferential degradation of polyadenylated RNA in the chloroplast. The exoribonuclease 100RNP/polynucleotide phosphorylase displays high binding affinity for poly(A) sequence. J Biol Chem 272:17648–17653

Littauer YZ, Soreq H (1982) Polynucleotide phosphorylase. Academic Press, New York

Liu J, Zhou LG, Yang C, Sun Y, Wu W et al (2015) The conserved endoribonuclease YbeY is required for chloroplast ribosomal RNA processing in Arabidopsis. Plant Physiol 168:205–221

Luro S, Germain A, Sharwood RE, Stern DB (2013) RNase J participates in a pentatricopeptide repeat protein-mediated 5′ end maturation of chloroplast mRNAs. Nucleic Acids Res 41:9141–9151

Mader U, Zig L, Kretschmer J, Homuth G, Putzer H (2008) mRNA processing by RNases J1 and J2 affects *Bacillus subtilis* gene expression on a global scale. Mol Microbiol 70:183–196

Marin-Navarro J, Moreno J (2005) Cysteines 449 and 459 modulate the reduction- oxidation conformational changes of ribulose 1,5-bisphosphate carboxylase/oxygenase and the translocation of the enzyme to membranes during stress. Plant Cell Environ 29:898–908

Mathy N, Benard L, Pellegrini O, Daou R, Wen T, Condon C (2007) 5′-to-3′ exoribonuclease activity in bacteria: role of RNase J1 in rRNA maturation and 5′ stability of mRNA. Cell 129:681–692

Matos RG, Barbas A, Arraiano CM (2009) RNase R mutants elucidate the catalysis of structured RNA: RNA-binding domains select the RNAs targeted for degradation. Biochem J 423:291–301

Maul JE, Lilly JW, Cui L, dePamphilis CW, Miller W, Harris EH, Stern DB (2002) The *Chlamydomonas reinhardtii* plastid chromosome: islands of genes in a sea of repeats. Plant Cell 14:2659–2679

Mayfield SP, Schultz J (2004) Development of a luciferase reporter gene, LuxCt, for *Chlamydomonas reinhardtii* chloroplast. Plant J 37:449–458

Mayfield SP, Franklin SE, Lerner RA (2003) Expression and assembly of a fully active antibody in algae. Proc Natl Acad Sci USA 100:438–442

Minko I, Holloway SP, Nikaido S, Carter M, Odom OW, Johnson CH, Herrin DL (1999) *Renilla* luciferase as a vital reporter for chloroplast gene expression in *Chlamydomonas*. Mol Gen Genet 262:421–425

Monde RA, Greene JC, Stern DB (2000) The sequence and secondary structure of the 3′- UTR affect 3′-end maturation, RNA accumulation, and translation in tobacco chloroplasts. Plant Mol Biol 44:529–542

Nakamura T, Yagi Y, Kobayashi K (2012) Mechanistic insight into pentatricopeptide repeat proteins as sequence-specific RNA-binding proteins for organellar RNAs in plants. Plant Cell Physiol 53:1171–1179

Nishimura Y, Kikis EA, Zimmer SL, Komine Y, Stern DB (2004) Antisense transcript and RNA processing alterations suppress instability of polyadenylated mRNA in *Chlamydomonas* chloroplasts. Plant Cell 16:2849–2869

Nurmohamed S, Vaidialingam B, Callaghan AJ, Luisi BF (2009) Crystal structure of *Escherichia coli* polynucleotide phosphorylase core bound to RNase E, RNA and manganese: implications for catalytic mechanism and RNA degradosome assembly. J Mol Biol 389:17–33

Pandey SP, Minesinger BK, Kumar J, Walker GC (2011) A highly conserved protein of unknown function in *Sinorhizobium meliloti* affects sRNA regulation similar to Hfq. Nucleic Acids Res 39:4691–4708

Parker R, Song H (2004) The enzymes and control of eukaryotic mRNA turnover. Nat Struct Mol Biol 11:121–127

Rasouly A, Schonbrun M, Shenhar Y, Ron EZ (2009) YbeY, a heat shock protein involved in translation in *Escherichia coli*. J Bacteriol 191:2649–2655

Rasouly A, Davidovich C, Ron EZ (2010) The heat shock protein YbeY is required for optimal activity of the 30S ribosomal subunit. J Bacteriol 192:4592–4596

Ruwe H, Kupsch C, Teubner M, Schmitz-Linneweber C (2011) The RNA-recognition motif in chloroplasts. J Plant Physiol 168:1361–1371

Salvador ML, Klein U (1999) The redox state regulates RNA degradation in the chloroplast of *Chlamydomonas reinhardtii*. Plant Physiol 121:1367–1374

Salvador ML, Klein U, Bogorad L (1993) 5′ sequences are important positive and negative determinants of the longevity of *Chlamydomonas* chloroplast gene transcripts. Proc Natl Acad Sci USA 90:1556–1560

Saramago M, Peregrina A, Robledo M, Matos RG, Hilker R, Serrania J, Becker A, Arraiano CM, Jimenez-Zurdo JI (2017) *Sinorhizobium meliloti* YbeY is an endoribonuclease with unprecedented catalytic features, acting as silencing enzyme in riboregulation. Nucleic Acids Res 45:1371–1391

Schuster G, Gruissem W (1991) Chloroplast mRNA 3′ end processing requires a nuclear- encoded RNA-binding protein. EMBO J 10:1493–1502

Schuster G, Lisitsky I, Klaff P (1999) Polyadenylation and degradation of mRNA in the chloroplast. Plant Physiol 120:937–944

Shamriz S, Ofoghi H (2019) Expression of recombinant PfCelTOS antigen in the chloroplast of *Chlamydomonas reinhardtii* and its potential use in detection of malaria. Mol Biotechnol 61:102–110

Sharwood RE, Halpert M, Luro S, Schuster G, Stern DB (2011) Chloroplast RNase J compensates for inefficient transcription termination by removal of antisense RNA. RNA (New York) 17:2165–2176

Shi Z, Yang WZ, Lin-Chao S, Chak KF, Yuan HS (2008) Crystal structure of *Escherichia coli* PNPase: central channel residues are involved in processive RNA degradation. RNA (New York) 14:2361–2371

Spreitzer RJ (1998) Genetic engineering of rubisco. In: Rochaix J-D, Goldschmidt- Clermont M, Merchant S (eds) The molecular biology of chloroplast and mitochondria in *Chlamydomonas*. Kluwer Academic, Dordrecht, pp 515–527

Spreitzer RJ, Salvucci ME (2002) Rubisco: structure, regulatory interactions, and possibilities for a better enzyme. Annu Rev Plant Biol 53:449–475

Stoppel R, Meurer J (2012) The cutting crew - ribonucleases are key players in the control of plastid gene expression. J Exp Bot 63:1663–1673

Surzycki R, Greenham K, Kitayama K, Dibal F, Wagner R, Rochaix JD et al (2009) Factors effecting expression of vaccines in microalgae. Biologicals 37:133–138

Symmons MF, Jones GH, Luisi BF (2000) A duplicated fold is the structural basis for polynucleotide phosphorylase catalytic activity, processivity, and regulation. Structure 8:1215–1226

Taunt HN, Stoffels Purton S (2018) Green biologics: the algal chloroplast as a platform for making biopharmaceuticals. Bioengineered 9:48–54

Walter M, Kilian J, Kudla J (2002) PNPase activity determines the efficiency of mRNA 3′-end processing, the degradation of tRNA and the extent of polyadenylation in chloroplasts. EMBO J 21:6905–6914

Wannathong T, Waterhouse JC, Young RE, Economou CK, Purton S (2016) New tools for chloroplast genetic engineering allow the synthesis of human growth hormone in the green alga *Chlamydomonas reinhardtii*. Appl Microbiol Biotechnol 100:5467–5477

Yamaguchi K, Beligni MV, Prieto S, Haynes PA, McDonald WH, Yates JR 3rd, Mayfield SP (2003) Proteomic characterization of the *Chlamydomonas reinhardtii* chloroplast ribosome. Identification of proteins unique to th e70 S ribosome. J Biol Chem 278:33774–33785

Yang J, Schuster G, Stern DB (1996) CSP41, a sequence-specific chloroplast mRNA binding protein, is an endoribonuclease. Plant Cell 8:1409–1420

Yehudai-Resheff S, Hirsh M, Schuster G (2001) Polynucleotide phosphorylase functions as both an exonuclease and a poly(A) polymerase in spinach chloroplasts. Mol Cell Biol 21:5408–5416

Yehudai-Resheff S, Zimmer SL, Komine Y, Stern DB (2007) Integration of chloroplast nucleic acid metabolism into the phosphate deprivation response in *Chlamydomonas reinhardtii*. Plant Cell 19:1023–1038

Zhelyazkova P, Sharma CM, Forstner KU, Liere K, Vogel J, Borner T (2012) The primary transcriptome of barley chloroplasts: numerous noncoding RNAs and the dominating role of the plastid-encoded RNA polymerase. Plant Cell 24:123–136

Zmienko A, Guzowska-Nowowiejska M, Urbaniak R, Plader W, Formanowicz P, Figlerowicz M (2011) A tiling microarray for global analysis of chloroplast genome expression in cucumber and other plants. Plant Methods 7:29

Chapter 9
Microalgae and Cyanobacteria EPS: Producing Strains, Accumulation and Extraction Strategies, and Valorizations

Céline Laroche

Abstract Microalgae and cyanobacteria are photosynthetic organisms that can produce/accumulate biomolecules of industrial interest. Among these molecules, EPS are macromolecular polysaccharide compounds with biological activities and physicochemical properties that make them suitable for a variety of commercial markets, including cosmetics, therapeutics, nutraceuticals, and hydrocolloids. The number of strains of microalgae and cyanobacteria described to produce such EPS has increased in recent years, and a diversity of monosaccharide compositions has been discovered, underlining the versatility of these organisms. While certain production strategies can be applied to increase EPS production yields, it appears that case-by-case studies are needed to promote EPS synthesis by a strain, as numerous responses exist. This chapter reviews the diversity of EPS-producing strains of microalgae and cyanobacteria and the variability of their compositions. Polymer production and extraction strategies are also discussed. Finally, an overview of biological activities and physicochemical properties enables us to envisage their use in several commercial markets.

Keywords Microalgae · Exopolysaccharides · Composition · Biological activities · Physicochemical properties · Valorization

C. Laroche (✉)
Institut Pascal, Université Clermont Auvergne, CNRS, Clermont Auvergne INP, Clermont-Ferrand, France
e-mail: celine.laroche@uca.fr

J. A. V. Costa et al. (eds.), *Microalgal Bioengineering*,
https://doi.org/10.1007/978-3-031-61253-4_9

9.1 Introduction

In recent decades, microalgae and cyanobacteria have been studied mainly for their potential use as feedstocks for biofuel production (Gonzalez Bautista and Laroche 2021; Sathya et al. 2023). However, in addition to lipids or starch for biofuels, polyunsaturated fatty acids (PUFAs), proteins, pigments, and polysaccharides can also be extracted.

Polysaccharides (PS) are one of the most promising compounds that can be extracted from microalgae. PS are high molecular weight carbohydrate macromolecules present in all living organisms with a wide diversity of structures and biochemical functions. They can be composed of a single repeated monosaccharide (homopolysaccharides) or composed of two or more distinct sugars (heteropolysaccharides). They can be linear or branched and can carry different substituents on their backbone, such as methyl or sulfate groups (Laroche 2022; Dimitriu 2004). On the basis of their location and function, the PS of photosynthetic microorganisms can be divided into three families: (i) storage PS which are intracellular such as starch, (ii) structural PS of cell walls such as hemicellulose, and (iii) extracellular PS. Many microalgae produce and excrete a polysaccharide mucilage presumably to protect them against various biotic and abiotic stresses, such as desiccation, high light fluxes, and so on, to facilitate the fixation of essential elements and micronutrients present in the environment, to contribute to the attachment of the microorganism to a solid substrate, and to play also a role in the formation of colonies for cyanobacteria (Pereira et al. 2009; De Philippis and Vincenzini 1998; Palmer et al. 2007; Tamaru et al. 2005; Shnyukova and Zolotareva 2017; Delattre et al. 2016; Arad and Levy-Ontman 2010; Ramus 1981). EPS from microalgae and notably from *Cyanobacteria* also contributes to stabilization of soils by the formation of biological soil crusts (Belnap and Lange 2013; Chen et al. 2014; Colica et al. 2014; Mager and Thomas 2011).

These extracellular polysaccharides can be totally excreted into the surrounding medium or remained more or less tightly bound to the cells. These later are often called capsular polysaccharides (CPS), bound polysaccharides (BPS), but also slime or sheath (Pereira et al. 2009; Underwood and Paterson 2003), whereas the PS that are released in the medium can be found under the denomination of exopolysaccharide (EPS), mucilage, or released exopolysaccharide (RPS). In many cases, microalgae strains produce both RPS and BPS, as shown for red microalgae for which around 50–70% of polysaccharide could remain bounded to the cell (Singh et al. 2000), but this amount could vary depending on the culture conditions. For the specific case of the cyanobacteria, it has been generally admitted that the exocellular PS are mainly constituted of CPS, with only a small amount found as RPS (Delattre et al. 2016; Pereira et al. 2009). Nevertheless, this aspect is controversial, as for some species RPS amounts exceeded those of CPS, and some others does not produce CPS at all (Mugnai et al. 2018). However, there is no clear evidence for compositional differences between RPS and BPS produced by a strain, but the RPS have a greater molecular weight (Bernaerts et al. 2018). It has then been proposed that these PS are produced following the same metabolic pathway, first as bound polymer, and then progressively released as function of some operating parameters. This assumption is correlated with the fact that the RPS/BPS ratio increase with the age of the culture.

Many studies have been conducted, both in vitro and in vivo, highlighting some biological activities of microalgal EPS, such as antibacterial, antioxidant, anti-inflammatory, antiparasitic, immunomodulatory, antitumor, or anticoagulant properties (Laroche 2022). Some original rheological properties of the microalgal EPS have also been highlighted allowing to consider them as new gelling and thickening agents. To date, the market is mainly occupied by PS from seaweeds and bacteria, but the atypical properties associated with the decrease of the production costs in the near future may open them new applications. All these interesting features (huge microalgae and cyanobacteria diversity, original structures, physicochemical properties, and/or biological activities) make these organisms attractive for valorization of such compounds in various industrial sectors (Borowitzka 2013; Gaignard et al. 2019a).

These EPS, depending on species, could be produced along with growth or under specific conditions that could be associated with nutrient depletion, high salinity, high light fluxes, and so on (Delattre et al. 2016; Villay et al. 2013; Soanen et al. 2016; Gaignard et al. 2019a; Pereira et al. 2009), opening the possibility to develop processes to enhance their production. The development of industrial cultivation systems, the improvement of downstream processes, and the implementation of biorefinery strategies, combined with a better understanding of the factors affecting production, could help to develop the marketing chain in the near future (Molina Grima et al. 2003; Norsker et al. 2011).

Several review papers have been published on this subject (Delattre et al. 2016; Laroche 2022). The objective of this chapter is then to provide an up-to-date overview of EPS from microalgae and *Cyanobacteria*, including producing strains and composition, on the extraction and purification methods, on their biological activities or physicochemical properties, and finally on the potent valorization markets.

9.2 Diversity of Producing Strains and Structures

The first publication highlighting the production of EPS by a microalga was published by Tischer and Moore (1964) and revealed the monosaccharide composition of the polymer produced from *Palmella mucosa*. Since then, many strains identified as producers of EPS have been described. These strains are spread into several phyla such as Chlorophyta, Charophyta, Rhodophyta, Haptophyta, Ochrophyta, and *Cyanobacteria* (Delattre et al. 2016; Rossi and De Philippis 2016; De Philippis et al. 1998; Kiemle et al. 2007; Pereira et al. 2009; Hu et al. 2003; Gugi et al. 2015; Senni et al. 2011). Data available in literature regarding EPS compositions have been gathered in the Tables 9.1a, 9.1b, 9.1c, 9.1d, 9.1e and 9.1f including molar % of each monosaccharide. For a more complete list of producing microalgae,

Table 9.1a Monosaccharide composition of EPS produced by Charophyta (molar %) (Part 1)

Genera	Species	Rha	Gal	Ara	Glc	GlcA	GalA	Man	Fuc	Xyl	Rib	SO_4^{2-}	Proteins	Other informations	References
Closterium	sp.	1	4	8	5	32		1	28	21					Domozych et al. 1993
Cosmarium	sp.	4	15	14	11			10	33	11	2			Uronic acids	Kiemle et al. 2007
Cosmarium	sp.	6	17	11	24			9	17	15	1				Kiemle et al. 2007
Klebsormidium	*flaccidum* (749B)	21	7	8	5			38	12	9				Uronic acids	Barberousse et al. 2006
	flaccidum (748A)	13	2	17	1			28	24	15				Uronic acids	
	flaccidum (446C)	49	3	2	2			31	4	8				Uronic acids	
Netrium	*digitus*	9	8	10	7			3	45	18				Uronic acids	Kiemle et al. 2007
	interruptum	4	11	20	4			2	21	33	5			Uronic acids	
	oblongum	1	48	8	6			3	7	27				Uronic acids	
Penium	*cylindrus*	4	6	12	3			2	40	33				Uronic acids	Kiemle et al. 2007
	spirostriolatum	2	18	43	5			2	10	19	1			Uronic acids	
	margaritaceum		4	5	3	12		1	18	54				Methyl groups	Domozych et al. 2005
Pleurotaenium	*trabecula*	7	15	2	12			5	24	28	6			Uronic acids	Kiemle et al. 2007
Spondylosium	*panduriforme*		30	3	5	24			30	8					Paulsen and Vieira 1994
Staurastrum	*iversenii*	10	40						43	10					Freire-Nordi et al. 2006
	orbiculare	4	11	2	4	17	9	1	42	9				GlcN	Giroldo et al. 2005
Stauroneis	sp.		10		1	55		5	6	22				GlcN	Jin et al. 2018
Tetmemorus	*brebissonii*	5	21	6	18			6	38	3	1			Uronic acids	Kiemle et al. 2007

Main monosaccharide · *Not detected* · *2nd monosaccharide* · *Not determined* · *Other monosaccharide* · *Presence* · *Trace*

including some strains for which no compositional analysis has already been performed, see Laroche (2022). It is to notice that the compositions reported by authors could strongly depend on culture conditions, downstream processes, but also on the analytical methodologies (Cheng et al. 2015; Magaletti et al. 2004; Phelippé et al. 2019). This fact has been highlighted, for instance, on *Arthrospira platensis* EPS by Phelippé et al. (2019).

Despite a few exceptions, EPS from microalgae and cyanobacteria are very complex heteropolymers, with very high molecular mass, often found >10^6 Da, often containing between 3 and 8 monosaccharides. These monosaccharides can be neutral (xylose, glucose, galactose, mannose, rhamnose, arabinose, fucose), or acidic (glucuronic or galacturonic acids) conferring a negative charge to the polymers. They also often wear sulfate, methyl, or acetyl groups in their structure. However, acidic monosaccharides and substituents are not always quantified leading in non-complete published composition.

When analyzed, the uronic acids represent sometimes more than 30% (mol), such as in *Corynoplastis japonica*, *Neorhodella cyanea*, *Exanthemacrysis* sp., or *Phormidium corium* (Borjas Esqueda et al. 2022; Gaignard et al. 2019b; Barberousse et al. 2006). In some cases, one of these uronic acids is even the main or second

Table 9.1b Monosaccharide composition of EPS produced by Chlorophyta (molar %) (Part 2)

Genera	Species	Rha	Gal	Ara	Glc	GlcA	GalA	Man	Fuc	Xyl	Rib	SO_4^{2-}	Proteins	Other informations	References
Ankistrodesmus	*densus*	8	3		5			8	44					Methyl groups	Paulsen et al. 1998
Botryococcus	*braunii*	2	36	9					32	4					Allard and Casadevall 1990
Bracteacoccus	sp.	18	19	5	20			15	9	15				Uronic acids	Barberousse et al. 2006
Chlamydomonas	*agustae*		13	15	42	24		3		3					Allard and Tazi 1992
	humicola	4	5	16	16	15				44					Flaibani et al. 1989
	peterfii	3	55	13	3	23				4					
	sajao	2	59	11	2	23				4					Flaibani et al. 1989
	mirabilis (678F)	16	28	19	5			11	7	15				Uronic acids	Barberousse et al. 2006
	mirabilis (7410G)	5	49	3	15			6	18	4				Uronic acids	
	ellipsoidea	2	54	7	12			14	7	4				Uronic acids	
Chlorococcum	sp.		8		44			14		33					Harun and Danquah 2011
Desmococcus	*olivaceus*	7	29	13	28			6	2	12				Methyl groups	Hu et al. 2003
Dunaliela	*tertiolecta*	5	1	1	85			4		1					Brown 1991
	sp.	21	28	5	1	4	11	13		17					Gaignard et al. 2019b
Geminella	*terricola*	16	7	11	6			36	10	14				Uronic acids	Barberousse et al. 2006
Prasinococcus	sp.	4	74	3	7	7			2	9					Gaignard et al. 2019b
Stichococcus	*bacillaris* (772B, 747C)	38	17	7	6			12	1	19				Uronic acids	Barberousse et al. 2006
	bacillaris (774E, 677A)	27	22	10	11			16	5	7				Uronic acids	
Tetraselmis	*globosa*	61	20			2				4					Gaignard et al. 2019b
	rubens	6	89				2								
	sp.	3	35	19	8	4	2	7		14					

Main monosaccharide *2nd monosaccharide* *Other monosaccharide* *Trace*
Not detected *Not determined* *Presence*

Table 9.1c Monosaccharide composition of EPS produced by Haptophyta (molar %) (Part 3)

Genera	Species	Rha	Gal	Ara	Glc	GlcA	GalA	Man	Fuc	Xyl	Rib	SO_4^{2-}	Proteins	Other informations	References
Chrysotila	*dentata*		36	33		4		8		16					Gaignard et al. 2019b
	sp.		33	11	45	2				8					
Diacronema	sp.	37	23	8	16	4				10					
Exanthemachrysis	sp.	3	14	14	7	32	5	13		12					
Isochrysis	*braarudii*	4	23	47	3	2				19					
Pavlova	*gyrans*	4	21	18	7	11	7		6	25					
	sp.	36	11	23	10	2	2		2	14					
	sp.	7	40	4	35	3	2	4	7	2					
	sp.	30	29	14	17	4				6					

Main monosaccharide *2nd monosaccharide* *Other monosaccharide* *Trace*
Not detected *Not determined* *Presence*

Table 9.1d Monosaccharide composition of EPS produced by Ochrophyta (molar %) (Part 4)

Genera	Species	Rha	Gal	Ara	Glc	GlcA	GalA	Man	Fuc	Xyl	Rib	SO_4^{2-}	Proteins	Other informations	References
Achnanthes	*longipes*	3	25	3	11	10		16	23	10					Wustman et al. 1997
Amphora	*coffeaeformis*	9	13		40			13	9	9					
	holsatica	41		10	45				5					Uronic acids	Leandro et al. 2003
	salina	2	3		48	6		41		1				GlcNAc	Tesson and Hildebrand 2013
Aulacoseira	*granulata*	18	10		15	9		12	16	17					Gouvêa et al. 2005
Chaetoceros	*decipiens*	28	20		4			16	28	4					Urbani et al. 2012
Coscinodiscus	*nobilis*				16	10		19	49	5		Presence			Percival et al. 1980
	radiatus	2			5	12		80		2					Tesson and Hildebrand 2013
Cylindrotheca	*closterium*	15	12		23			4		46			Presence	Uronic acids	Staats et al. 1999
	fusiformis	1	11		80					6	6			Uronic acids	Chiovitti et al. 2005
Cymbella	*cistula*	2	69	1	4			3		18					Wustman et al. 1997
	mexicana	6	57	1	5	1		12	2	16					
Cyclotella	*nana*	33	14		11			10		7	8			Uronic acids	Allan et al. 1972
Glossomastix	sp.	20	2			10	11		57						Gaignard et al. 2019b
	sp.	19				8	13		57						
	sp.	27	3			7	15		47						
Melosira	*nummuloides*	40	2		39					4				Uronic acids	Leandro et al. 2003
Navicula	*directa*	54	14		14					8				Uronic acids	Leandro et al. 2003
	curvilineata	2	3		46	2		40		46					Tesson and Hildebrand 2013
	incerta	33	8					10	20	9				Uronic acids	Allan et al. 1972
	salinarum	6	19		42			14		20		Presence	Presence	Uronic acids	Staats et al. 1999
Nitzschia	*angularis*	20	17	8				7	16	7				Uronic acids	Allan et al. 1972
	frustulum	24	8					34						Uronic acids	Allan et al. 1972
Phaeodactylum	*tricornutum*	7	24		22			6	1	7	31			methyl groups	Chiovitti et al. 2005
Triceratium	*dubium*				6	2	6	67		6			Presence	GlcNAc	Tesson and Hildebrand 2013

Main monosaccharide · *2nd monosaccharide* · *Other monosaccharide* · *Trace*
Not detected · *Not determined* · *Presence*

monosaccharide in EPS compositions. In addition, some other uronic acids have been occasionally described, such as mannuronic acid, detected only in the EPS of *Heterosigma akashiwo* (Lopes et al. 2012), but other nonidentified uronic acids have been suggested in the EPS of *Oscillatoria* sp. and *Anabaena cylindrica* (Richert et al. 2005; Lama et al. 1996).

Some osamines (GlcN and GalN) and their acetylated derivatives (GlcNAc and GalNAc) have also been detected in a few compositions, and mainly for Cyanobacteria, due to similarities with the peptidoglycan layer of non-photosynthetic bacteria. For example, GlcN has been described for *Rhabdoderma rubrum* (Richert et al. 2005), *Synechocystis aquatilis* (Di Pippo et al. 2013), *Stauroneis* sp. (Jin et al. 2018), *Amphora salina* (Tesson and Hildebrand 2013), *Amphora* sp. (Jin et al. 2018), *Aulacoseira granulata* (Gouvêa et al. 2005), *Thalassiosira* sp. (Giroldo et al.

Table 9.1e Monosaccharide composition of EPS produced by Rhodophyta (molar %) (Part 5)

Genera	Species	Rha	Gal	Ara	Glc	GlcA	GalA	Man	Fuc	Xyl	Rib	SO_4^{2-}	Proteins	Other informations	References
Flintiella	*sanguinaria*	10	21	2	6	14				47				Methyl and acetyl groups	Gaignard et al. 2018
	marinum		21		18	6			1	47					Soanen et al. 2016
	purpureum		42		25	2			1	30					Roussel et al. 2015
	sordidum		33		23	5				39				Methyl groups	[107]
	sordidum	1	21		13	18	1			46					Borjas et al., 2022
	sp.	1	31	1	17			1		43				Uronic acids, methyl groups	Geresh et al. 1992
Rhodella	*grisea*	15	7	1	4	17				29				Methyl groups	Capek et al. 2008
	maculata	5	45	2	1	5				42					Roussel et al. 2015
	violacea	2	52	1	7	3				34					Villay et al. 2013
Timspurckia	*oligopyrenoides*		28	1	2	18			1	51					Borjas et al., 2022
Erythrolobus	*coxiae*	1	12	2	1	21	1		4	58					Borjas et al., 2022
	madagascarensis	1	22	3	2	21	1		2	49					Borjas et al., 2022
Corynoplastis	*japonica*	14	9		1	26				50					Borjas et al., 2022
Neorhodella	*cyanea*	1	11		2	25				62					Borjas et al., 2022
Chroodactylon	*ornatum*	5	19	2	30	6	6		10	20					Borjas et al., 2022
Chroothece	*richteriana*	1	58		6	6	2		1	28					Borjas et al., 2022
Bangiopsis	*subsimplex*	1	78	1	6	9	1		1	6					Borjas et al., 2022
Rhodaphanes	*brevistipitata*	1	87	1	3	2				7					Borjas et al., 2022
Rhodospora	*sordida*	1	31	1	6	2	1		1	59					Borjas et al., 2022

Main monosaccharide *2nd monosaccharide* *Other monosaccharide* *Trace*
Not detected *Not determined* *Presence*

2003), *Triceratium dubium* (Tesson and Hildebrand 2013), various *Anabaena* species (Di Pippo et al. 2013; Gouvêa et al. 2005; Nicolaus et al. 1999), *Calothrix* sp. (Di Pippo et al. 2013), *Chroococcus minutus* and *submarinus* (Fischer et al. 1997; Richert et al. 2005), *Gloeocapsa kuetzingigiana* (Rossi et al. 2012), *Gloeocapsa crepidinum* (Rossi et al. 2012), *Leptolyngbya* sp. (Rossi et al. 2012), *Microcoleus vaginatus* (Hu et al. 2003), *Microcystis aeruginosa* (Gouvêa et al. 2005), *Phormidium tenue* (Hu et al. 2003; Hokputsa et al. 2003), *Plectonema* sp. (Rossi et al. 2012), *Synechocystis* sp. (Panoff et al. 1988), *Tolypothrix tenuis* (Nicolaus et al. 1999), or *Trichormus variabilis* (Di Pippo et al. 2013).

The non-sugar groups such as sulfates are not systematically analyzed in available studies. For EPS in which sulfate groups have been quantified, it appears that this feature is widespread among diversity with, depending of strains, some amounts >10% as for cyanobacteria *Arthrospira platensis* (Majdoub et al. 2009), *Nostoc* sp. (Ge et al. 2014a), *Microcoleus vaginatus* (Ge et al. 2014b), *Gloeothece* sp. (Tease et al. 1991) or *Johannesbaptistia pellucida, Chroococcus submarinus, Rhabdoderma rubrum, Aphanocapsa halophytia* and *Pseudophormidium battersii* (Richert et al. 2005), or microalgae *Amphora* sp. (Zhang et al. 2008), *Gyrodinium impudicum*

Table 9.1f Monosaccharide composition of EPS produced by Cyanobacteria (molar %) (Part 6)

Genera	Species	Rha	Gal	Ara	Glc	GlcA	GalA	Man	Fuc	Xyl	Rib	SO_4^{2-}	Proteins	Other informations	References
Anabaena	*augstmalis*		12		48		28	6	3					GalN	Di Pippo et al. 2013
	sp.		23		23			5		50				Uronic acids, pyruvate	Moreno et al. 2000
	spiroides	16	4		14	16		14	10	11					Gouvêa et al. 2005
Aphanocapsa	*halophytia*	2	3		25			15	53	3					Sudo et al. 1995
	platensis	50	6		4	15	17	1		6					Majdoub et al. 2009
	platensis	2	16		16	25	25		5	11				Methyl groups	Filali-Mouhim et al. 1993
Calothrix	*pulvinata*	15	17	7	28			10	8	16				Uronic acids	Barberousse et al. 2006
	sp.		13	9	56	1	1	6	2	6				GlcN	Di Pippo et al. 2013
Chlorogloeopsis	sp.	4	20	1	37			30	7						Nicolaus et al. 1999
Chroococcus	*minutus*	14	13	7	26	4	1	14	6	15				GlcN, methyl groups	Fischer et al. 1997
	submarinus	11	11	5	16	5	5	5	16	26				GlcN	Richert et al. 2005
Cyanospira	*capsulata*			17	17		33	16	16						Vincenzini et al. 1993
	sp.				76					17	6				Parikh and Madamwar 2006
Fischerella	*musicola*		21		57			7	7	7					Nicolaus et al. 1999
Geitlerinema	sp.				99										Richert et al. 2005
Gloeocapsa	*gelatinosa*														Raungsomboon et al. 2006
	kuetzingigiana	9	7	21	21	2	8		5	21	5			GalN, traces GlcN	Rossi et al. 2012
Gloeocapsosis	*crepidinum*	3	7	1	17	13	4		49	3	2			GlcN, GalN	Rossi et al. 2012
	sp.	11	29		34			19		7				Uronic acids,	Tease et al. 1991
Johannesbaptistia	*pellucida*	4	20	20	20	4	4	20	4	4					Richert et al. 2005
Leptolyngbya	*foveolarum*	22	19		24			20		16				Uronic acids	Barberousse et al. 2006
	tenuis	19	17		34			18	6	6				Uronic acids	
	sp.	7	1	2	39	16			13	14	9			GalN, traces GlcN	Rossi et al. 2012
	vaginatus	6	18	9	20	3	4	21	4	8				GlcNAc, methyl groups	Hu et al. 2003
Microcystis	*aeruginosa*	18	16		13		9	7	10	20				GlcNAc, GalNAc	Gouvêa et al. 2005

(continued)

Table 9.1f (continued)

Genera	Species	Rha	Gal	Ara	Glc	GlcA	GalA	Man	Fuc	Xyl	Rib	SO_4^{2-}	Proteins	Other informations	References
	aeruginosa	67			14			19							
	aeruginosa flos-aquae				94			6							Forni et al. 1997
	viridis	11			79			5	5						
	calcicola	3	9	3	22	7	4	7	3	8					Flaibani et al. 1989
	carneum							67		33					Parikh and Madamwar 2006
	commune		19	10	12	13			6	40					Huang et al. 1998
Nostoc	*flagelliforme*		21	3	46			7		23					
	insulare	1		19	44	21	3	2	9						Fischer et al. 1997
	sp.	3	22		44			2		21				Methyl groups	Hu et al. 2003
	sp.	2	10	8	27	5	9	7	5	25					Di Pippo et al. 2013
Oscillatoria	sp.				74					19	7				Parikh and Madamwar 2006
	autumnale	66			3		3	3			22				Di Pippo et al. 2013
	ambiguum	24	18	7	19			6	10	16				Uronic acids	
	corium (442D, 746B)	22	12	4	29			6	13	15				Uronic acids	
Phormidium	*corium* (444A)	10	28	4	24			13	6	14				Uronic acids	Barberousse et al. 2006
	corium (674A)	24	12		13			24		27				Uronic acids	
	corium (743D)	17	16		28			22		17				Uronic acids	
	tenue	10	1	44	32			3	2	5				GlcNAc	Hu et al. 2003
	battersii		18	6	24	6	6	18	12	12					
Plectonema	*golenkinianum*	13	13		26	7		7	8	26					Richert et al. 2005
Rhabdoderma	*rubrum*		8		24			8	16	40				GlcN	
Scytonema	*javanicum*	7	23	10	25			23		6				Methyl groups	Hu et al. 2003
Synechococcus	sp.		27		49					20					Gaignard et al. 2009b
Synechocystis	*aquatilis*	8	3		26	8	8	3	16	9				GlcN	Di Pippo et al. 2003
Tolypothrix	*tenuis*	1	6	2	13			9	5					GlcN	Nicolaus et al. 1999
Trichormus	*variabilis* (VRUC162)		9	1	55			5	8	11	7			GalN	Di Pippo et al. 2003
	variabilis (VRUC168)	1	13	4	36	4	9	7	4	21				GlcN	

Main monosaccharide *2nd monosaccharide* *Other monosaccharide* *Trace*
Not detected *Not determined* *Presence*

(Yim et al. 2003), *Cylindrotheca fusiformis* (Magaletti et al. 2004), *Porphyridium marinum* (Soanen et al. 2016), *Porphyridium cruentum* and *Porphyridium aerugineum* (Percival and Foyle 1979), *Tetmemorus brebissonii* and *Pleurotaenium trabecula* (Kiemle et al. 2007) or *Timspurckia oligopyrenoides, Neorhodella cyanea, Chroodactylon ornatum, Chroothece richteriana, Bangiopsis subsimplex, Rhodospora sordida,* and *Rhodaphanes brevistipitata* with for some of these latter strains' contents even greater than 20% (Borjas Esqueda et al. 2022). Until really recently, only a small number of EPS from red microalgae had been studied (mostly *Porphyridium* and *Rhodella* strains) regarding their diversity, and it has been admitted for years that rhodophyta produced sulfated polysaccharides. Interestingly, the

EPS from *Flintiella sanguinaria* was found to be a 5.1% (w/w) methylated and 3.2% (w/w) acetylated polymer (Gaignard et al. 2018), with no sulfate groups (<0.6% w/w). This is also the case for *Corynoplastis japonica, Erythrolobus coxiae,* and *Erythrolobus madagascarensis* EPS, which have a sulfation degree lower than 2%. In Borjas Esqueda et al.'s (2022) study, and even if no quantification was performed on *O*-methyl and *O*-acetyl groups, their presence is suspected in the 11 EPS from porphyridiophyceae and rhodellophyceae strains studied, i.e., *Erythrolobus coxiae, Erythrolobus madagascarensis, Timspurckia oligopyrenoides, Porphyridium sordidum, Neorhodella cyanea,* and *Corynoplastis japonica.* The presence of methyl groups was already described for other microalgae (Tables 9.1a, 9.1b, 9.1c, 9.1d, 9.1e and 9.1f), whereas acetyl groups were only identified in *Nostoc insulare* (Volk et al. 2006). However, in most studies, these methyl or acetyl groups are not quantified. Identification of the position of these methyl groups is even more rare, but it can be cited, i.e., 3-*O*-Me-Xylose, 4-*O*-Me-Xylose, 2,3-di-*O*-Me-Rhamnose, and 2,3-di-*O*-Me-Fucose for *Rhodella grisea* (Capek et al. 2008), while di-*O*-Me-hexose, 4-*O*-Me-galactose (Geresh et al. 1992), and 6-*O*-Me-mannose (Levy-Ontman et al. 2011) have been detected in EPS of *Porphyridium* sp. Additionally, 3-*O*-Me-Rhamnose have been identified in EPS from *Arthrospira platensis* (Lee et al. 1998). Earlier work from Percival and Foyle (1979) on EPS from *Porphyridium cruentum* has also shown that methyl groups can be found on uronic acids, as 2-*O*-Me-glucuronic acid was highlighted.

Gaignard et al. (2019b) have performed a meta-analysis of 81 EPS compositions from chlorophyta, charophyta, rhodophyta, ochrophyta, and haptophyta, intending to highlight some specificities related to taxonomic affiliation of EPS producers. Interestingly, this study has allowed to observe significant differences in compositions between phyla. As an example, glucose is the main monosaccharide in EPS from cyanobacteria and xylose for rhodophyta (Gaignard et al. 2019b).

If some EPS compositions have been published, complete structure characterization are still lacking, probably linked to the high complexity of these polymers. Up to date, only partial information has been published and on a very limited number of EPS produced by rhodophyta (*Porphyridium* sp., *Porphyridium cruentum, Porphyridium aerugineum, Rhodella reticulata*) or cyanobacteria (*Cyanothece* sp., *Cyanospira capsulata, Mastigocladus laminosus, Nostoc commune*, and *Nostoc insulare*). These incomplete structures have been reported in the review of Laroche (2022). This lack of complete structures prevents establishing clear relationship between structures and biological activities and/or physicochemical properties, which could partially explain the low industrial valorization of EPS.

9.3 Operating Parameters for EPS Production

Actually, the optimization of production implies a systematic study of each strain, in order to define case by case the most appropriate operating parameters. Indeed, if environmental parameters such as nutrient deprivations are known to enhance EPS production for some strains, many comportments can be observed depending on species. It is thus often difficult to determine the most accurate parameter. A better understanding of the metabolic pathways involved in the future could thus bring valuable information for an easier optimization of operating parameters. Currently, if some piecemeal information is available for cyanobacteria (Pereira et al. 2009, 2013), or red microalgae (Arad and Levy-Ontman 2010), these data remain uncomplete. As an example, floridoside has been identified as a precursor of both EPS and starch for *Porphyridium*, but the mechanism by which carbon flux is directed toward one or the other PS is still unknown (Li et al. 2001, 2002).

Studying the EPS production by a microalgae or cyanobacteria strain requires first an efficient method to detect and quantify the polymer release. The easiest method for detection could be by assaying sugars in the culture supernatant, using classical sugar quantifications such as Dubois or Monsigny methods (Dubois et al. 1956; Monsigny et al. 1988). However, attention must be paid to possible interferences of culture medium compounds with the assay reagents. For EPS that exhibit viscosity in solution, this behavior can be an indirect way to evaluate the release of polymers, as shown by Mancuso Nichols et al. (2009) who compared the viscosity of nearly 800 microalgae and cyanobacteria culture supernatants with those of culture media alone. In the same way, a tool has been developed and successfully applied to detection of viscous algal polysaccharides (Badel et al. 2011), even with really low variations in viscosity. However, attention has to be paid that EPS could be produced in some specific operating conditions, and no detection is not necessarily sufficient to say that the strain is an nonproducing one. Moreover, these methods are not suitable for BPS which need an extraction before quantification.

Strategies to enhance EPS production could thus be assimilated to stresses for the cells such as availability of macro and micronutrients (and any condition of deficiency associated), salinity, temperature, or irradiance. Positive or negative effect of the main stresses (N or P deprivation, high salt or high light) on some microalgae or cyanobacteria strains are reported on Fig. 9.1.

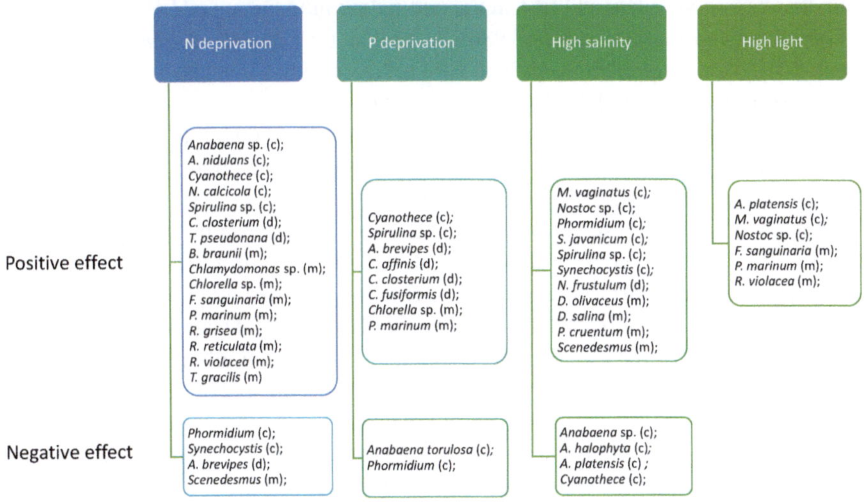

Fig. 9.1 Effect of nutrient deprivations, high light, or high salt on EPS synthesis. (c) ***Cyanobacteria***; (d) diatom; (m) microalgae

9.3.1 Nutrient Limitations

The nutrient limitation is known to induce the accumulation of compounds such as the pigments PUFAs and PS.

The production of EPS can be induced by the deficiency and/or the limitation of available nitrogen in the culture media, as shown in many studies conducted on different microorganisms such as microalgae like *Botryococcus braunii* (Lupi et al. 1994), *Chlamydomonas* (Kroen and Rayburn 1984), *Chlorella* sp. (Brányiková et al. 2011), *Flintiella sanguinaria* (Gaignard et al. 2018), *Porphyridium* (Ramus 1972; Soanen et al. 2016), *Rhodella* (Arad et al. 1992; Villay et al. 2013; Dubinsky et al. 1988), or *Tetraselmis gracilis* (Lourenco et al. 1998) but also for Cyanobacteria *Anabaena* sp. (Moreno et al. 1998), *Anacystis nidulans* (Sangar and Dugan 1972), *Cyanothece* (De Philippis et al. 1993, 1998; Reddy et al. 1996), *Nostoc calcicola* (Flaibani et al. 1989), or *Spirulina* sp. (Nicolaus et al. 1999), and diatoms *Thalassiosira pseudonana* (Ai et al. 2015) or *Cylindrotheca closterium* (Underwood et al. 2004). Nevertheless, this same limitation in nitrogen has no effect on the production of EPS in *Phormidium* (Fattom and Shilo 1984, 1985) or *Synechocystis* (Panoff et al. 1988) or *Scenedesmus* (Angelaalincy et al. 2017). For strains producing both soluble and bound polymers, the nitrogen starvation can impact the repartition between both fractions, increasing the RPS amount, as shown for *Porphyridium* sp. (Arad 1988) or *Rhodella reticulata* (Arad et al. 1992).

If the mechanism by which nitrogen amount promote the EPS synthesis remains unclear, it has been proposed that a high C/N ratio may impact the carbon metabolism (Kumar et al. 2007; Otero and Vincenzini 2004). This hypothesis is supported by the fact that many studies show that the supplementation in CO_2 during the culture affects the growth and the production of various metabolites by the microalgae, including the EPS (Nicolaus et al. 1999; Iqbal and Zafar 1993). Moreover, stimulating carbon metabolism by increasing CO_2 fixation rate and inhibiting photorespiration thanks to glyoxylate addition in the culture medium enhanced EPS synthesis by *Anabaena cylindrical* (Bergman 1986), as well as for *Chlorella pyrenoidosa* (Yang et al. 2010) or *C. capsulata* for which EPS production has increased by 43% after a short exposure to glyoxylate (De Philippis et al. 1996).

However, nitrogen deprivation can also be a suitable strategy to accumulate lipids for some microalgae strains, and only elucidation of metabolic pathways would help understand the underlying mechanisms.

Some studies have also shown that the EPS production can be influenced by the nature of the nitrogen source, such as for *Anabaena cylindrica* (Lama et al. 1996), or for *Botryococcus braunii* (Lupi et al. 1994), while it had no effect on *Anabaena flos-aquae* cultivated using KNO_3 and NH_4Cl (Tischer and Davis 1971). A significant increase in EPS production while supplementing medium with nitrogen in the N_2 form was observed for the cyanobacterium *Anabaena* sp. (Moreno et al. 1998), for *Synechococcus* (Roux 1996), and several *Anabaena* strains (Lama et al. 1996; Nicolaus et al. 1999), while it was not the case for *Nostoc* sp. (Otero and Vincenzini 2004) or *Phormidium laminosum* (Fresnedo and Serra 1992).

Phosphorus and sulfur depletions can also promote EPS production. For the diatom *Achnanthes brevipes*, EPS synthesis is increased by phosphate starvation, whereas nitrogen depletion had no effect. However, combined nitrogen and phosphorus deprivations were shown to have a negative effect, as it induced intracellular carbohydrates accumulation (Guerrini et al. 2000). Positive impact of phosphorus deprivation has been shown for the cyanobacteria *Spirulina* sp. (Nicolaus et al. 1999) and *Cyanothece* (De Philippis et al. 1993), microalgae *Chlorella* sp. (Boonchai et al. 2014) and *Porphyridium marinum* (Soanen et al. 2016), and diatoms *Cylindrotheca closterium* (Underwood et al. 2004; Urbani et al. 2005), *Thalassiosira pseudonana* (Urbani et al. 2005), *Chaetoceros affinis* (Myklestad et al. 1972), and *Cylindrotheca fusiformis* (Magaletti et al. 2004). Conversely, phosphorus limitation had no effect on the production of EPS by *Phormidium* (Nicolaus et al. 1999; Fattom and Shilo 1984) and *Anabaena torulosa* (Nicolaus et al. 1999), and it may even be reduced for *Anabaena cylindrica* (Lama et al. 1996).

For the sulfur starvation, no modification of production yields was noticed for *Rhodella reticulata*, but it has been demonstrated that the degree of sulfation of polymers was reduced by half (Arad 1988; Arad et al. 1992). Alike, the degree of sulfation of EPS produced by *Porphyridium cruentum* was improved with a 104 mM concentration in $MgSO_4$ instead of 21 mM (De Jesus Raposo et al. 2014).

The micronutrients may also have a role in the production of EPS. Indeed, this last is improved in deficiency of calcium cation for *Phormidium* (Fattom and Shilo 1984). For *Cyanospira capsulata,* a 17% increase in the production of EPS was observed in condition of magnesium or calcium deficiency (De Philippis et al. 1991). Conversely, the productivity was found lower for *Cyanothece* in calcium (379 $mg.L^{-1}$) or magnesium (238 $mg.L^{-1}$) deficiency conditions, as compared with the control (500 $mg.L^{-1}$) (De Philippis et al. 1993).

Such nutrient deficiency, if it allows enhancing EPS accumulation, may affect the composition of monosaccharides. Glucose is the main monosaccharide of EPS of *Cylindrotheca closterium, Thalassiosira pseudonana*, and *Skeletonema costatum* during the exponential growth phase in a replete medium (Urbani et al. 2005). In case of phosphorus depletion, its content decreased for *Cylindrotheca closterium,* while for *Thalassiosira pseudonana,* the fucose amount was significantly reduced (from 30% molar ratio in the EPS to 5%) (Urbani et al. 2005). When grown in a replete medium, *Cylindrotheca closterium, Navicula perminta*, and *Amphora exigua* produced a complex EPS containing Rha, Fuc, Xyl, Man, Gal, Glc, and uronic acids (Shnyukova and Zolotareva 2015). An analysis of the monosaccharide composition of the EPS produced by *Cylindrotheca fusiformis* has shown that a nitrogen deficiency led to a specific decrease of the content in Gal, Man, and Xyl (Magaletti et al. 2004) and an increase in the level of Fuc and Rha. At the opposite, in P deficiency condition, the Rha and Gal amounts increased, while the Glc and Man decreased. Additionally, a significant increase (50%) of the uronic acid level in the polysaccharide has been observed in P starvation condition, whereas the sulfation degree decreased by half in N limitation condition (Magaletti et al. 2004).

For the strains for which EPS accumulation is enhanced by nutrient deficiency, the main production is thus observed during the stationary phase. However, depending on the strain, the EPS production can also be observed throughout the growth phase, and entry into stationary phase did not lead to an increase in EPS release. As an example, the EPS production by *Chaetoceros affinis* is even more important during the exponential growth phase, comparing to its production during the stationary phase (Myklestad et al. 1989). This is also the case for *Arthrospira platensis* (Filali-Mouhim et al. 1993), or *Nostoc* strains (Gantar et al. 1995; Mehta and Vaidya 1978). For these strains, other modulations of operating conditions could be preferred for stimulation of EPS production.

9.3.2 Other Stresses

Other environmental parameters such as salinity, irradiance, and temperature of the culture medium can also be used to induce EPS production depending on the strains.

The high salinity strategy even if limited to halophilic or at least halotolerant microalgae and cyanobacteria could be a suitable strategy. As an example, for a 5 M NaCl concentration, the EPS production by *Dunaliella salina* is 17 times more important than for 0.5 M (944 mg.L^{-1} and 56 mg.L^{-1}, respectively) (Mishra and Jha 2009). A similar phenomenon has been observed for *Porphyridium cruentum* (Decamp et al. 2023), *Microcoleus vaginatus* (Chen et al. 2006; Hu et al. 2003), *Scytonema javanicum, Phormidium, Nostoc* sp., *Desmococcus olivaceus* (Hu et al. 2003), *Spirulina* sp. (Nicolaus et al. 1999), and *Scenedesmus* (Angelaalincy et al. 2017). On the contrary, the EPS production by *Aphanocapsa halophyta* is strongly reduced beyond 4% NaCl and is avoided at 10% (Sudo et al. 1995). For *Anabaena* sp. the production was found to decrease from 13.5 to 9.2 and 3.9 g.L^{-1} for NaCl concentrations of 0, 0.85, and 510 mM, respectively (Moreno et al. 1998). For *Synechocystis*, results are contradictory as Ozturk and Aslim (2010) described a positive effect of salinity, whereas for Panoff et al. (1988), an increase in NaCl concentration up to 0.5 M failed to improve the EPS production for *Cyanothece* sp. for NaCl concentrations up to 2 M (De Philippis et al. 1993). For *Nitzschia frustulum,* strong variations in EPS levels were observed with NaCl concentration, without impact on cell growth (Allan et al. 1972). However, the EPS composition was affected when the strain was grown with 20.8 g.L^{-1} of NaCl instead of 1.3 g.L^{-1}, with levels of Rha and Xyl increased by a factor of 2 and 7, while Man, Gal, and Fuc were reduced by a factor of 3, 4, and 6.

The release of EPS can also be improved by a continuous supply of light (Lupi et al. 1994) and high irradiances (Moreno et al. 1998) that can in some cases be assimilated to a stress for the cells. The effect of irradiance has been studied in *Rhodella violacea* (Villay et al. 2013), *Porphyridium marinum* (Soanen et al. 2016), or *Flintiella sanguinaria* (Gaignard et al. 2018), showing that the best EPS productivities were reached for the irradiance of saturation, just before the one inducing the photoinhibition. EPS yield produced by the Cyanobacteria *Nostoc* sp. or *Microcoleus vaginatus* increased with light intensity, but no differences in monosaccharide composition were noticed (Ge et al. 2014a, b). For *Arthrospira platensis,* EPS up to 25–30% of dry weight were observed for cultivation under 800 μmol photons.m^{-2}.s^{-1}, against about 10–15% DW at 100 μmol photons.m^{-2}.s^{-1} (Phelippé et al. 2019). The EPS composition was also modified by this high light condition, with amount of Glc increased up to 77% (% molar ratio in the EPS), detrimentally to the other neutral sugars and uronic acid amounts.

Modification of the light spectrum can also affect the EPS production. Indeed, You and Barnett (2004) have compared the EPS production by *Porphyridium cruentum* while using red light (600–700 nm), blue light (400–500 nm), or white light. Red light was found to stimulate polysaccharide production (both capsular and soluble fractions) by the Cyanobacteria *Nostoc flagelliforme* (Han et al. 2015). On the contrary, long-term (5 months) continuous cultivation of *Arthrospira platensis* under red light (135 μmol photons.m^2.s^{-1}) has shown no differences in growth rate and EPS production as compared to white light (Farges et al. 2009).

Finally, temperature is also a parameter that was shown to promote EPS synthesis for some strains. *Nostoc* sp. produces more of EPS when it is grown at 30 °C rather than 35 °C in the presence of NO_3- ions (Otero and Vincenzini 2004). This phenomenon is not observed when N_2 is the sole source of nitrogen. An increase in the EPS production by a factor of 5 was also observed for *Anabaena* sp. when the culture temperature is shifted from 30–35 °C to 40–45 °C (Moreno et al. 1998). For *Botryococcus braunii* the best EPS production could be observed at 28 °C, with a decrease for lower (20–23 °C) or greater temperature (32 °C) and is zero at 33 °C (Lupi et al. 1991). Productivity in biomass being similar at 20, 23, and 28 °C, it is possible to conclude of an impact of temperature on the production of EPS for this strain.

Finally, in a recent study, Deamici et al. 2021 have studied the effect of static magnetic field on *Arthrospira platensis* EPS production and composition. If static magnetic fields have an impact on growth rate, pigment content, and intracellular polysaccharide accumulation, it seems that the EPS production yield is not increased. However, the EPS composition was found modified, with a greater content in in glucuronic acid, fucose, arabinose, and rhamnose and a decrease in glucose and xylose.

9.3.3 Culture Strategies

For strains for which EPS production is induced (or at least improved) by a condition of nutrient depletion, the main drawback is that a low rate of growth and a possible decline of biomass will be encountered in these conditions (Gonzalez-Fernandez and Ballestros 2012). Usually, the production of compounds of interest is then carried out in batch mode. After a growth phase in a replete medium, the production of EPS will start as soon as the conditions of depletion are fulfilled (entry into stationary phase). In some cases, a semicontinuous strategy can also be used, in order to improve the process productivity. This strategy has been used recently by Soanen et al. (2016) on the red microalga *Porphyridium marinum* leading to a process productivity of 0.031 gEPS.h^{-1} instead of 0.020 gEPS.h^{-1} in batch mode.

In order to control the nature of nutrient deficiency in batch mode culture, the initial N/P ratio of the medium can be adjusted. Ratios between 35 and 50, corresponding to a limitation in phosphate, have improved the production of *Porphyridium cruentum* biomass, while a ratio N/P < 4.9, corresponding to a nitrogen limitation has improved by 2, with the production of PS (Razaghi et al. 2014). Accordingly,

N/P ratios around 4 had allowed to significantly improve EPS production for *Rhodella violacea* (Villay et al. 2013), *Flintiella sanguinaria* (Gaignard et al. 2018), and *Porphyridium marinum* (Soanen et al. 2016). In this latter case, the impact of the P and N deprivations on EPS synthesis was studied, showing a greater specific productivity (0.0622 mg.10^{-6} cells) for N depletion than for P (0.0503 mg.10^{-6} cells). However, it could not be always the case, depending on species studied.

For some strains, a nitrogen-limiting condition (and not complete deprivation) could be sufficient to improve EPS production, as shown for several *Cyanothece* strains (De Philippis et al. 1993, 1998), and then allowing continuous mode strategy.

For other induction methods, a two-step culture mode could be needed, as the parameter allowing induction of production could be inconsistent with the optimal growth parameters. As an example, high incident light fluxes can induce photoinhibition and lead to cellular death. After a growth stage, when biomass concentration has increased, the same incident light flux could be bearable for cells due to shading of cultures. The light stress can then be applied. Another example is the high salinity that may inhibit growth, without leading to cellular mortality. In that case, the stress should be applied after the growth phase as a second step.

9.4 Extraction and Purification Processes

After EPS production, some methods need to be implemented to recover the molecules, with different methods required to recover RPS or BPS.

9.4.1 EPS Released to the Culture Medium

The soluble EPS in the culture medium needs first to be separated from biomass by centrifugation, or membrane filtration (Zhang and Santschi 2009; Li et al. 2011a; Ye et al. 2005; Zhang et al. 2008). Depending on the use of the EPS, a purification step should be necessary. This could be achieved by dialysis, tangential ultrafiltration, or selective alcoholic precipitation (Li et al. 2011a; Marcati et al. 2014; Moore and Tischer 1964; Patel et al. 2013). Indeed, the soluble fraction may be precipitated by alcohols of varying polarities such as methanol, ethanol, or isopropanol (De Brouwer and Stal 2004; Gloaguen et al. 2004; Moreno et al. 1998; Lupi et al. 1991; Liu et al. 2015; Moore and Tischer 1964; Patel et al. 2013). The precipitation will also be enhanced by cold temperatures. Alcoholic precipitation is widely used at industrial level to purify PS from bacteria or higher plants, and the possibility of alcohol recycling makes this process ecofriendly. The main drawback regarding its use for EPS from microalgae and cyanobacteria is the low purity obtained by this method, due to the coprecipitation of salts from the culture medium (Patel et al. 2013). The tangential ultrafiltration and dialysis could then be used as alternatives to the alcoholic precipitation. A comparative study between the techniques of alcoholic precipitation, dialysis, and tangential ultrafiltration was conducted by Patel et al. (2013). The tangential ultrafiltration with a 300 kDa molecular weight cutoff (MWCO) polyethersulfone membrane was the most effective method, leading to a higher purity, but with the drawback that a small fraction of the EPS was lost in the filtrate. Zhang and Santschi (2009) compared three techniques of concentrations/ precipitations (ethanol precipitation, diafiltration in cell Amicon, and tangential ultrafiltration) in order to recover the EPS produced by the diatom *Amphora* sp. The results show that the filtration techniques are the most effective to isolate the EPS from a culture medium rich in salts. Other authors have used the tangential filtration (with a membrane of 10 kDa MWCO) to concentrate the EPS from *Nostoc commune*, before a step of alcoholic precipitation (Helm et al. 2000). Additionally, Li et al. (2011a) have developed a pilot-scale extraction and purification process for EPS from *Chaetoceros muelleri*, *Chlorella pyrenoidosa*, *Spirulina platensis*, *Haematococcus pluvialis*, *Nostoc commune*, and *Nostoc sphaeroides*, with a microfiltration process (polypropylene membrane) to remove biomass, followed by an ultrafiltration with a 5 kDa polyethersulfone membrane for purification and concentration.

Several parameters need to be taken into account for efficiency of membrane filtration for EPS purification: viscosity and concentration of EPS, pore size distribution (MWCO), flow rate, and transmembrane pressure. A high MWCO will present the risk of losing part of the EPS, but the purity will be great (as almost all proteins will be eliminated in filtrate) while maintaining a reasonable flow rate, while for a low MWCO, the flow rate will be significantly lower, allowing to recover almost all sample but probably with contaminating proteins. Some drawbacks related to the use of tangential ultrafiltration exist. Among them, the phenomenon of clogging of the membranes related to the viscosity of the EPS is the most frequent (Rickman et al. 2012; Ye et al. 2005; Zhang et al. 2008). The fouling of membranes resulted therefore in reducing all the flow performances, increasing the operating prices (Ye et al. 2005; Zhang et al. 2008). In addition, a large quantity of water is necessary to isolate, purify, and concentrate the EPS by ultrafiltration (Li et al. 2011a). However, Jhaveri and Murthy (2016) have published a review on the development of nanocomposites membranes (SiO_2, TiO_2, etc.) to prevent clogging. Moreover, the use of inorganic and organic membranes such as AZT (aluminum/ zirconium/titanium oxide) and polyacrylonitrile could help in controlling clogging phenomenon while harvesting *Arthrospira platensis* (Rossi et al. 2004, 2005).

Even if the cheapest industrial extraction remains the alcoholic precipitation, the development of membranes processes is a good alternative for the desalting and concentration of the EPS solutions, even if it constitutes a very limiting step because of the high viscosities of solutions and the high cost of the membranes.

9.4.2 Cell-Bound PS

For the specific case of PS that remain more or less tightly bound to the cells, specific methods have to be used to selectively recover this fraction. Consequently, it does not exist a universal method to release these PS, and different methods can be used depending on the strain, such as hot water treatments, ethylenediaminetetraacetic acid (EDTA), sodium hydroxide (NaOH), sonication, and so on (Brown and Lester 1980; Comte et al. 2006; Frølund et al. 1996; Underwood and Paterson 2003).

For hot water treatment, the biomass concentration, the pH, the temperature (30–100 °C), and the duration of treatment (generally between 1 and 5 h) need to be adjusted to control release of PS. However, if cellular lysis has occurred, EPS could be contaminated by proteins, pigments, but also other high molecular weight polysaccharide such as storage carbohydrate (starch or glycogen). These classical methods will then fail in separating the two polysaccharides. Even if the extraction method is conducted carefully, these contaminations are frequently encountered, as discussed by De Brouwer and Stal (2004) and described for *Thalassiosira pseudonana, Phaeodactylum tricornutum, Cylindrotheca fusiformis, Craspedostauros australis, Pinnularia viridis* (Chiovitti et al. 2004), *Navicula phyllepta, Nitzschia epithemioides* (Bellinger et al. 2005), but also *Navicula jeffreyi* (Takahashi et al. 2009), or *Cylindrotheca closterium* (De Brouwer et al. 2002; Bellinger et al. 2005). As most of EPS from microalgae and cyanobacteria are anionic polymers, due to the presence of uronic acids and/or sulfate groups, the method developed by Ramus (1972) for extraction of the cell-bound polysaccharide from *Porphyridium* can be efficiently used. Briefly, after the hot water treatment, the polysaccharide is selectively precipitated using the cationic compound cetylpyridinium chloride, converted to its calcium salt, and reprecipitated with ethanol. Filali-Mouhim et al. (1993) have proposed a method to extract BPS from *Arthrospira platensis*. First, during the hot treatment, biomass is suspended in a TAPS buffer (0.05 M), supplemented with EDTA (0.025 M) and NaCl (0.025 M), to limit the cellular lysis. After centrifugation, the supernatant containing the polymer is precipitated by addition of 10 volumes of 3% cetyl trimethyl ammonium bromide (CTAB), which was assumed to allow specific precipitation of acidic polysaccharides. The limiting step of this method is that the PS should then be separated from the CTAB. The pellet is then dissolved in 1 M KCl before precipitation with ethanol. These latter solubilization/precipitation steps are repeated two times with decreasing KCl concentration solutions (0.75 and 0.3 M) in order to destabilize complex between CTAB and polysaccharide. The same protocol was successfully applied to *Porphyridium* by Gloaguen et al. (2004). In addition, an ultrafiltration step can be added at the end of the protocol to remove remaining salts, allowing to increase purity of samples (Borjas Esqueda et al. 2022). However, even if this method allows efficient extraction and selective purification of the anionic cell-bound PS, the high number of successive steps makes the method unsuitable at industrial level.

Some authors have also described the use of fixative agents such as formaldehyde or glutaraldehyde that can chemically react with hydroxyl, sulfhydryl, carbonyl, or amino groups of the outer membrane, to prevent cell lysis during extraction (Cloete et al. 1998; McDonnell and Russell 1999). The advantage is that the PS could be further recovered by classical ethanolic precipitation. However, some chemical modifications of the EPS have been noticed (Azeredo et al. 2003; Comte et al. 2006; Liu and Fang 2002; Takahashi et al. 2009).

A few studies have also dealt with treatments using cationic resins. These resins have been shown to disturb interaction of BPS with the cellular membrane without cellular lysis, allowing their recovery (Liu and Fang 2002; Takahashi et al. 2009). If this method seems promising regarding the purity and the extraction yields that can be reached, its use is actually not possible at the industrial level due to the high cost of resins and the low amount of sample that can be treated, limiting its application to structural analysis purpose.

Finally, in a biorefinery context aiming to covalorization of several compounds, the study of Marcati et al. (2014) allowed the sequentially purification of BPS and pigments (B-phycoerythrin, B-PE) from *Porphyridium cruentum*. In this work, authors used high pressure treatment to achieve cellular lysis and release B-PE, with, as a consequence of the mechanical constraints, the simultaneous release of BPS. They then developed a two-step ultrafiltration process, using successively 300 kDa and 10 kDa MWCO polyethersulfone membranes, to further purify both fractions.

9.5 Valorization

9.5.1 Biological Activities

The EPS produced by microalgae and cyanobacteria have been the subject of numerous publications regarding their biological activities. This includes, among other, antioxidant, antiviral, antifungal, antibacterial, anti-aging, anticancer, and immunomodulatory agents. This part has therefore no claim to be exhaustive regarding the huge number of available studies.

Antiviral activities are probably the most studied biological activities for EPS from microalgae and cyanobacteria, even if the mechanism is still not well understood. This is the case of the EPS isolated from *Cochlodinium polykrikoides* which has antiviral properties in vitro against the respiratory viruses (RSV-A and RSV-B) and influenza virus types A and B (Hasui et al. 1995). In addition, the EPS inhibits the replication of HIV-1 virus at a concentration IC50 equal to 1.7 $\mu g.mL^{-1}$ (Hasui et al. 1995). The sulfated EPS from *Porphyridium* sp. also has an antiviral effect, against the herpesvirus (HSV-1 and HSV-2), with optimal concentration about 100 $\mu g.mL^{-1}$ (Huheihel et al. 2001, 2002), whereas the sulfated polysaccharide p-KG03 isolated from *Gyrodinium impudicum* has a strong activity against the encephalomyocarditis virus (EMCV) (Yim et al. 2003). Finally, EPS from *Arthrospira platensis* has shown strong inhibiting activity against human cytomegalovirus, herpes simplex virus types 1 and 6 (HSV-1, HSV-6), and human immunodeficiency virus type 1 (HIV-1), while only weak or no inhibition was noted for Epstein-Barr virus and influenza A virus (Hayashi et al. 1996a, b; Rechter et al. 2006).

Many EPS from microalgae and cyanobacteria have been shown to exhibit antioxidant activities (scavenging abilities on superoxide radicals, hydroxyl radicals, and hydroxyl peroxide) including *Spirulina platensis* (Herrero et al. 2005; Li et al.

2007), *Porphyridium* sp., *P. cruentum*, and *Rhodella reticulata* (Sun et al. 2009; Tannin-Spitz et al. 2005; Chen et al. 2010; Wang et al. 2012), *Schizochytrium* sp. (Wang et al. 2011), *Isochrysis galbana* (Sun et al. 2014), *Nostoc carneum* (Hussein et al. 2015), as well as *Scenedesmus quadricauda* and *Chlorella vulgaris* (Mohamed 2008). In all cases, the activities were shown to be dose dependent.

Antibacterial and antifungal properties have also been described, with strong variations in minimum inhibitory concentrations (MIC) from a polymer to another, and from a bacterial strain to another. For EPS from *Gloeocapsa* sp., *Rhodella reticulata* and *Synechocystis* sp. they inhibit the proliferation of *Staphylococcus aureus*, with MIC equal to 0.125, 0.25 and 1.00 mg.mL^{-1} respectively (Najdenski et al. 2013). In addition, EPS from *Gloeocapsa* sp. was the only one inhibiting the growth of the fungus *Candida albicans*, and that of *Synechocystis* sp. inhibiting the growth of *Pseudomonas aeruginosa* (Najdenski et al. 2013). EPS from *Porphyridium marinum* was also found efficient against *Staphylococcus aureus*, (MIC of 0.125 g L^{-1}), *E. coli, Salmonella enteritidis*, and to a lesser extend on *S. aureus* Methicillin Resistant (1 g L^{-1}). Moreover, this EPS is able to prevent Candida albicans biofilm formation (90% inhibition at 31.3 µg.L^{-1}, Gargouch et al. 2021).

Some antitumor activities have also been highlighted. In most cases, the activity has been related to the immunomodulatory effect of polymers. Indeed, the EPS from *Porphyridium cruentum* improves the immune response by stimulating the proliferation of macrophages and lymphocytes. These activities have been found to be greater for small molecules (Mw = 6.55 . 10^4 Da) by comparison with larger ones (Mw = 2.56 . 10^5 Da) (Sun et al. 2012). Indeed, in the study of Gargouch et al. (2021), an antiproliferative effect of EPS from *P. marinum* was noticed on mammary carcinoma cells, with an increasing efficiency with the decrease in molar mass. The homopolysaccharide from *G. impudicum* was also shown to prevent tumor cell growth both in vitro and in vivo (Yim et al. 2005).

The EPS from microalgae can also be used as anti-inflammatory agents. This is the case of the polymer isolated from *Porphyridium* sp. which inhibits the migration of leukocytes to the site of inflammation (Matsui et al. 2003).

Finally, some other biological activities have been demonstrated, such as the specific reduction of the frequency and the intensity of the cough without avoiding the expectoration by the EPS of *Rhodella grisea* (Nosálová et al. 2012), or the anticoagulant activity of the EPS from *Arthrospira platensis* (Majdoub et al. 2009). Additionally, the EPS of *Rhodella violacea, Rhodella maculata, Porphyridium marinum*, and *P. purpureum* have shown an in vitro activity toward *Encephalitozoon cuniculi* microsporidia and in vivo against *Nosema ceranae* (Roussel et al. 2015). This antiparasitic activity can give rise to their use in treatment of nosema, a disease contributing to honey bee mortality.

Despite the numerous studies demonstrating biological activities of EPS from microalgae and cyanobacteria, the mechanisms by which they act remain almost unclear. As these polymers are frequently sulfated, they may act as mimetics of glycosaminoglycans (GAG). These GAG (heparin/heparan sulfate, chondroitin sulfate, dermatan sulfate, keratan sulfate, hyaluronan) constitute a family of polymers with high sulfate content (except hyaluronan) and are major components of the extracellular matrix in animal tissues where they are implied in many biological functions.

Many authors have claimed that activity is directly linked to the sulfate content of the GAGs, as they may interact with proteins such as receptors or growth factors (Hayman et al. 2005; Hintze et al. 2012; Kreuger et al. 2006; Corsuto et al. 2018). Thus, for EPS from *Porphyridium*, Arad and Levy-Ontman (2010) confirmed that there was a linear relationship between the sulfate level in the polysaccharide and its antiviral activity against the herpesvirus. For the specific case of antiviral activities, several mechanisms have been proposed, related to the replication of the viruses during the early stages (Martinez et al. 2005; Kim et al. 2012), or inhibiting the virus penetration (Hayashi et al. 1996a, b) by competing with the virus for the attachment sites (Damonte et al. 2004; Radonic et al. 2010; Rashid et al. 2009; Kanekiyo et al. 2007). However, action mechanism could be different depending on the biological activity tested, and it remains very difficult to rely a structure and a function. Moreover, due to the EPS conformation in solution, the level of sulfation appears not to be always a sufficient parameter to consider, as shown for antiparasitic activity (Roussel et al. 2015). These authors suggested a possible implication of the position of the sulfate groups, as several polysaccharides from algae were tested, and the more active one was not the more sulfated. While testing the antitumor activity of polysaccharides from the seaweed *Gayralia oxysperma*, the same conclusion was obtained (Ropellato et al. 2015). More than the sulfate presence, it seems that the anionic behavior of the polysaccharide could be of importance in some cases, as some polymers exhibiting uronic acids in their structure (but no sulfate groups) were shown to be active, hyaluronic acid being one of the most extensively studied examples. Moreover, implication of some specific monosaccharides in the EPS composition could also been discussed. As an example, and even if this assumption is still under discussion, some biological activities described for the fucans may be in relation to their high content in fucose. Indeed, polysaccharides from 23 green algae species, with similar sulfation levels, have been tested for their anticoagulant activities (Hayakawa et al. 2000). The rhamnose-rich polysaccharides exhibited potent thrombin inhibition effect (even stronger than that of heparin), followed by arabinose-rich polysaccharides and then galactose-rich ones. Alike, a polysaccharide containing rhamnose from *Chlorella pyrenoidosa* presented greater antitumor activity than another containing galactose as the main monosaccharide (Sheng et al. 2007). All these findings suggest also that the operating parameters chosen to improve EPS production yields are of importance depending on the targeted application, as the composition of the polymer could vary as previously discussed. Finally, other parameters should have an implication on biological activities (molecular mass and branching pattern of polysaccharides, type of activity tested, acting mechanism, and so on), explaining the difficulty to establish a clear scheme for the structure-function relationship.

9.5.2 Physicochemical Properties

PS from bacteria or higher plants (including seaweeds) are commonly used as thickeners, stabilizers, and gelling agents in the agrifood industries, pharmaceutical, and cosmetics (Funami 2017). EPS from microalgae and cyanobacteria could present interesting physicochemical properties, making them promising for such applications. The main limiting factor for their exploitation is actually the production costs that are still not competitive with PS from other sources.

Several studies have been conducted on the flow properties of EPS in solution from microalgae and cyanobacteria (De Philippis et al. 1998; De Jesus Raposo et al. 2014) but unfortunately limited to a really low number of species. Concerning microalgae, all works were focused on EPS from red microalgae such as *Rhodella* or *Porphyridium*. As an example, dilute solutions (0.25%, m/v) of EPS produced by red microalgae present viscosities between 23.5 and 27.5 cP for the polysaccharide of *Porphyridium* sp. and equal to 28.7 cP for the polymer of *Rhodella reticulata* (Geresh et al. 2002). Similar results were obtained with the EPS from *Porphyridium cruentum* for which the viscosity was found significantly higher when the EPS was extracted from cultures containing 21 mM of $MgSO_4$ or $MgCl_2$ (De Jesus Raposo et al. 2014). Solutions of these microalgae EPS presented all a non-Newtonian behavior (Geresh and Arad 1991; Sun et al. 2009), with a viscosity decreasing with the increase of shear rate, which is typical of a shear-thinning comportment (Eteshola et al. 1998). This decrease in viscosity with shear rate has been suggested as being linked to the dissociation of the strong hydrogen bonds existing between the chains of EPS (De Jesus Raposo et al. 2013). Partial depolymerization of EPS from *Porphyridium* has been shown to strongly affect viscosity, with, for the lower molecular weight samples, a comportment typical of Newtonian fluids (Sun et al. 2009). This finding confirms that the high viscosity is correlated (at least partly) to the high molecular mass of EPS. Recently, Bernaerts et al. (2018) studied the rheological properties of RPS and BPS (2% w/w) from *Porphyridium cruentum*. The RPS possessed high intrinsic viscosities in comparison with those described for commercial thickeners, such as alginates, locust bean gums, and guar gums, but lower than those of xanthan gums (Imeson 1997, 2011; Rao 2010). A lower intrinsic viscosity was observed for BPS, which is not surprising as the molecular weight of the polymer was found lower. Nevertheless, the value obtained was still in the range of some commercial thickening agents confirming that both fractions could be promising for use in the hydrocolloid market. While studying physicochemical comportment in oscillation mode, some differences were highlighted. For RPS, predominant elastic behavior was observed, since values for the storage modulus (G′) exceeded those of the loss modulus (G″). However, no true gel has been observed (ratio G″/G′ should have been in the order 10^{-2}) but clearly presented a weak gel behavior. In contrast, viscous behavior was observed for BPS, with a constant loss modulus and $G' < G''$ over the whole range of shear strains explored. This fraction can then be assimilated to a viscoelastic fluid (Bernaerts et al. 2018). Furthermore, the rheological properties of EPS from *Porphyridium* were found stable in a wide range of pH values (2–9) and temperatures (30–60 °C) (Arad et al. 2006). However, contrasting observations were reported by different authors when temperatures were further increased. Three gelation mechanisms are well documented for polysaccharides: (i) ionotropic gelation (typically cation-mediated gelation), (ii) cold-set gelation, and (iii) heat-set gelation. Eteshola et al. (1998) observed a strong gelation at temperatures above 60 °C which was reversible upon cooling (thus belonging to the last category), whereas other authors reported the rheological properties to be independent of temperature (Arad et al. 2006; Patel et al. 2013). Regarding the impact of temperature on rheological properties of EPS from *Porphyridium*, attention has also to be paid on the drying process used as postproduction treatment, as Ginzberg et al. (2008) have shown that temperature >90 °C may cause significant alteration to the polymer and thus affecting the intrinsic viscosity and elasticity properties. As these polymers are anionic due to their content in uronic acids and sulfate groups, it has been suggested that the high viscosity values and the weak gel behavior observed can be linked to interaction with cations leading to a partial cation-mediated gelation (De Jesus Raposo et al. 2015; Eteshola et al. 1998). However, no increase in viscosity were observed by Bernaerts et al. (2018) in the presence of different cations.

EPS from *Cyanobacteria* also possess interesting rheological properties. For example, the production of EPS by *Cyanospira capsulata* induced a viscosity increase of the culture medium. After 31 days of batch culture, the viscosity has been measured around 450–500 cP (Vincenzini et al. 1990). The viscometric analysis of supernatants collected at different times of culture have revealed a typical pseudoplastic behavior (Vincenzini et al. 1990). An increase in the culture medium viscosity was also observed for *Synechococcus* sp. The viscosity was mainly due to the EPS, as viscosity measurements results were similar for solutions with or without cells (Phlips et al. 1989). The rheological study has shown that the viscosity decreased when the shear rate increased, reflecting once more a shear-thinning behavior. *Anabaena* sp. EPS also presented this typical property in diluted solutions (0.6% w/w) (Moreno et al. 2000). Similar results were obtained on several EPS isolated from *Cyanothece* even at lower concentration (0.1% (w/w) (De Philippis et al. 1998). For 1% (w/v) EPS solutions from *Nostoc flagelliforme*, apparent viscosity was found intermediate between 0.5% and 1% (w/v) xanthan gum solutions, but the decrease observed with shear rate was faster, leading quickly to low viscosity solutions (two times less than viscosity of the 0.5% xanthan gum solution). The difference in viscosity with xanthan gum was attributed by authors to a lower molecular mass of EPS, which was about 2×10^5 Da for EPS and 2×10^6 Da for xanthan (Han et al. 2014). For EPS from *Nostoc carneum*, shear-thinning behavior was also observed, but the viscosities reported were found lower and requested a 15 $mg.mL^{-1}$ solution to obtain a viscosity about 140 cP (Hussein et al. 2015). The same kind of comportment was described for EPS from *Arthrospira platensis*, requesting a high polymer concentration (5% w/w) to observe a significant viscosity (Chentir et al. 2017). At this concentration, solutions were found to present linear viscoelastic range (G' and G'' are constant and $G' > G''$), indicating a viscoelastic gel-like behavior (Chentir et al. 2017).

In addition to these rheological properties, the anionic nature of the EPS confers some interesting affinity toward cations. Thus, some EPS may present some gelation capacity similar to alginates or pectins PS. The low number of EPS from microalgae and cyanobacteria studied for their physicochemical properties has not allowed yet to identify this kind of gelation ability, but a lot of original comportment remains to be discovered. The chelation ability of EPS will not be developed here but has already been described for several EPS, mainly from cyanobacteria. Such a property may open the way to industrial use of EPS for biosorption and water treatment to remove heavy metals (Urrutia 1997; De Philippis et al. 2007, 2011; Micheletti et al. 2008).

Other than their hydrophilic nature, EPS can also present significant hydrophobic behavior, in relation to ester-linked methyl or acetyl groups, together with peptidic moieties (Flaibani et al. 1989; Shepherd et al. 1995). The presence of both hydrophilic and hydrophobic groups can suggest potential utilization of the polymers as bioflocculants (Han et al. 2014), biosurfactants or bioemulsifiers (Paniagua-Michel et al. 2014). This kind of study is rather scarce, but the use of EPS from *Nostoc flagelliforme* for stabilization of emulsions has been demonstrated, with, depending on the oils used, emulsification activities similar to xanthan gum (Han et al. 2014).

9.5.3 Valorization and Markets

Taking into account their rheological properties and their biological activities, valorization of EPS from microalgae and cyanobacteria can be considered in the areas of food, cosmetics, and pharmaceuticals. Figure 9.2 illustrates the EPS properties and the potential commercial markets. However, only a very few numbers of commercial EPS can actually be found and mainly in the cosmetic industry. Their

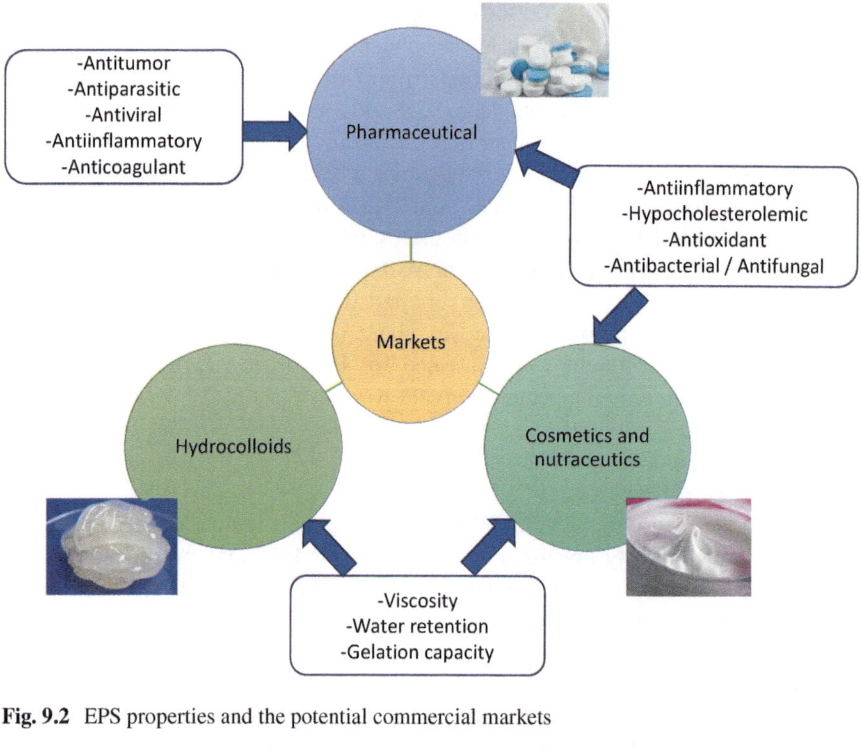

Fig. 9.2 EPS properties and the potential commercial markets

valorization in the other areas (therapeutic, human food and animal feed, nutraceuticals) is very reduced, mainly because of the too high production costs and the lack of information on their structure. The polysaccharide market has a constant growing rate, evaluated between 1% and 3% in 2011 (Bixler and Porse 2011). Indeed, the market could assimilate some polysaccharides from microalgae with original biological or technofunctional properties. Even if some technological bottlenecks subsist, the microalgae technologies (production, harvesting, biorefinery) have been significantly improved in the last decades. Autotrophic production of microalgae could reach the breakeven point with high cell density cultivations and their treatments by specific downstream processes, opening them low value markets as the food industry. If despite the efforts and technical improvements the production cost of EPS remains high, their valorization will be limited to the high value molecule markets such as those of the cosmetic or of the therapy.

The next section deals with a few examples of the use of EPS in various industrial areas, mainly for EPS from red microalgae or from *Arthrospira platensis*.

9.5.3.1 Cosmetic and Nutraceutical Markets

As previously said, this market is the most developed for the use of EPS from microalgae and cyanobacteria, mainly for their biological activities and especially their anti-inflammatory, antimicrobial, and antioxidant activities. They are also described as moisturizing agents, anti-aging, and strengthening of the barrier function of the skin (Potter et al. 2013; Tannin-Spitz et al. 2005). This moisturizing effect is linked to the strong water retention capacity of EPS, related to the fact that EPS is supposed to be produced to support desiccation (Tamaru et al. 2005; Li et al. 2011b). As an example, the moisture retention capacity of EPS extracted from the cyanobacterium *Aphanothece sacrum* is tenfold higher than hyaluronic acid (Okajima et al. 2008, 2009). However, the market is surprisingly mainly occupied with *Porphyridium* EPS, despite the many other EPS that can have interesting applications in this field. An exception is a mixture of polysaccharides extracted from a *Chlorella* strain grown heterotrophically (Coragliotti et al. 2012), patented by the Solazyme company and commercialized with the registered trade name “Alguronic Acid”. Recently, Yanhui (2017) has developed a cosmetic formulation containing EPS from *Porphyridium* allowing to repair and strengthen the skin barrier. Previously, the L’Oréal company had patented a mixture of EPS from *Porphyridium* sp. of a marine bacterium and ulvan (a polysaccharide from the seaweed *Ulva lactuca*), associated with a C-glycoside (Potter et al. 2014). They claimed that this active ingredient can improve the barrier function of the skin, as well as its hydration. This same company has developed another patent based on the use of the EPS to eliminate and/or reduce the dandruff of the scalp, as an alternative to the zinc pyrithione frequently incorporated into shampoo (Potter et al. 2013). The same EPS from *Porphyridium cruentum* have been tested on three enzymes (hyaluronidase, elastase, and collagenase) involved in the degradation of the three respective substrates hyaluronan, Elastan, and collagen, which are key compounds for the appearance of the skin (Diaz Bayona et al. 2012). Even if no activity against collagenase has been noticed, the EPS strongly reduced the activity of the hyaluronidase at concentrations between 0.25 and 2.5 mg.mL^{-1} (96.6 ± 0.3%), as well as the activity of elastase at a concentration of 5 mg.mL^{-1} (46.0 ± 7.1%). These results have been patented by the company Solazyme (Dillon et al. 2007a) for use in cosmetic formulations. EPS from several microalgae (*Phaeodactylum ornatum, Synechococcus, P. cruentum, Glossomastix* sp., *Chrysotila dentata, Pavlova* sp., *Diacronema* sp.) have been patented as anti-aging agent. Indeed, these EPS were shown to stimulate collagen and hyaluronic acid production, with a greater activity while decreasing molecular mass (Goncalves et al. 2022a). Thus, several companies propose these cosmetic formulations or active ingredients, such as Frutarom, Asta technologies, and Yemoja, all three located in Israel, AlgoSource technologies (France), Micoperi Blue Growth (Italy), and Solazyme (USA).

Concerning food and nutraceutical market, the *Porphyridium* EPS is once more the most represented, even if some *Arthrospira* (spirulina) extracts can be found. But in this latter case, as *Arthrospira platensis* is allowed for food consumption in its whole biomass form, the EPS is generally not extracted and purified. Nutraceuticals correspond to biomolecules having a positive effect on health that can be incorporated into food. This market is in constant growth for 20 years as shown the exponential increase in the number of publications on the subject (Bernal et al. 2011).

Arad et al. (1993) have shown that EPS from *Porphyridium cruentum* is not hydrolyzed in the gastrointestinal tract. Therefore, its bioavailability is null, and associated with the high viscosity of EPS solutions, its use as dietary fibers could be of interest. Moreover, Dvir et al. (2009) have demonstrated hypocholesterolemic effect (on rat model) of EPS from *Porphyridium* sp. Feeding rats with this EPS was found to significantly improve levels of total serum cholesterol, serum triglycerides, hepatic cholesterol, HDL/LDL ratios, and increased fecal excretion of neutral sterols and bile acids. The nutraceutical use related to these biological activities has been patented by the company Solazyme in 2007 (Dillon et al. 2007b). The same year, Solazyme company has also developed a nutraceutical composition with an antioxidant activity from the same EPS (Dillon et al. 2007c). This formulation allows to reduce inflammation and prevent the oxidative damage in the tissues of mammals. Finally, Huang et al. (2006) have shown that the EPS from *Porphyridium* could be used to reduce the blood glucose level in mice.

9.5.3.2 Pharmaceutical and Medical Markets

As previously described, EPS from microalgae and cyanobacteria present numerous biological activities that may allow their use as pharmaceutical drugs. However, to our knowledge, such compounds are still not on the market, despite some patents can be found. EPS from *Porphyridium* sp. has been patented as antiviral agent (Arad et al. 1997), whereas EPS from red microalgae can be used for veterinary purpose as antiparasitic agent to prevent infections of honey bees by microsporidia (Michaud et al. 2017). The reason why no EPS are on the market at this time is not only linked to economic reasons but rather to the drastic legislation of therapeutic compounds. Before exploitation of these EPS, biological activities at lab scale should first be confirmed by clinical trials that remain to conduct. A strict control of the production systems and on purity, stability, and safety is required, implying still some process development. But the main scientific lock to their exploitation is certainly the lack of detailed structure information, together with difficulties to identify mechanism of action. The road for commercialization of new efficient drugs from microalgae will still be long and challenging.

The pharmaceutical application that appears to be the faster to develop can be the use of EPS in wound healing bandages. Such hydrocolloid dressings can absorb and retain liquid exudating from the wound, prevent bacteria infection, and help healing. It has been developed and commercialized with polysaccharides such as hyaluronan, gelatin, or alginates (Piacquadio and Nelson 1992; Sweeney et al. 2012). The high water-retention capacity of EPS from microalgae and cyanobacteria, associated with their antibacterial activity, is then promising for such a market.

9.5.3.3 Hydrocolloids

There is neither EPS from microalgae nor cyanobacteria on the hydrocolloids market, despite their original physicochemical properties. A first patent has been recently published, proposing the use of EPS from several microalgae species as "anti-settling" agent for food beverages, due to their weak gel behavior (Goncalves et al. 2022b). However, the reason of the absence of microalgae EPS on this market is probably mostly of economic nature. The hydrocolloid market is mainly driven by seaweed polysaccharides. The annual market for seaweed polysaccharides is around 9600 tons for agars, 26,500 tons for alginate, and 50,000 tons for carrageenans for a market value around 1 billion US dollars (Kraan 2012). However, the market demand is still in expansion. Seaweeds are mostly harvested in the natural environment or cultivated in open ponds or in the sea. But the development cycle of seaweeds takes 1 year, and the harvesting is not possible during the whole year, even with sourcing all over the world. Moreover, the extraction yields and the composition (and then properties) of these seaweed PS are varying with seasons and harvesting location. Probably to cope with the market request, overcome variability, and stabilize sourcing, the global demand for EPS from bacteria (such as xanthan, gellan, or curdlan) has increased these last years. In this context, EPS from microalgae and cyanobacteria could be able to find their place in the future, on the condition to further reduce the production and downstream process costs, making them competitive. Moreover, many EPS from microalgae have still not been characterized for their physicochemical properties, and regarding the diversity of composition and structures, many different behaviors could be expected, this originality opening the way to specific market demand.

9.6 Conclusion

Microalgae and cyanobacteria represent a huge taxonomic diversity as several hundred thousand species may exist. Only a limited number have been identified as EPS producers and even less have been subject of studies for their characterization. Most of these EPS have only been described for their monosaccharide composition, and the complexity of structural studies has until now limited their elucidation. At the moment, we are still at the beginning of the comprehension of factors affecting the EPS production, and the diversity of comportment between strains, associated with still expensive production/extraction processes that limit the cost-effective mass production. Furthermore, metabolic pathways involved in the production are still unresolved, and a better comprehension could open the way to greater production yields. Besides, many biological activities and original physicochemical properties have been described, allowing considering these molecules in several industrial areas. If some of these EPS are already available on the cosmetic market, the introduction in the pharmaceutical market will still require long and hard work to secure production, purity, and safety and elucidate action mechanisms. Finally, a promising valorization area could be the hydrocolloid sector, which is in constant expansion and looking for atypical properties, in answer to more and more complex food processes and elaborated products.

References

Ai XX, Liang JR, Gao YH, Lo SCL, Lee FWF, Chen CP, Luo CS, Du C (2015) MALDI-TOF MS analysis of the extracellular polysaccharides released by the diatom *Thalassiosira pseudonana* under various nutrient conditions. J Appl Phycol 27:673–684

Allan GG, Lewin J, Johnson PG (1972) Marine polymers. IV Diatom polysaccharides. Bot Mar 15:102–108

Allard B, Casadevall E (1990) Carbohydrate composition and characterization of sugars from the green microalga *Botryococcus braunii*. Phytochemistry 29(6):1875–1878

Allard B, Tazi A (1992) Influence of growth status on composition of extracellular polysaccharides from two *Chlamydomonas* species. Phytochemistry 32:41–47

Angelaalincy MJ, Senthilkumar N, Karpagam R, Kumar GG, Ashokkumar B, Varalakshmi P (2017) Enhanced extracellular polysaccharide production and self sustainable electricity generation for PAMFCs by *Scenedesmus* sp. SB1. ACS Omega 2:3754–3765

Arad SM (1988) Production of sulphated polysaccharides from red unicellular algae. In: Stadler T, Mollion J, Verdus MC, Karamanos Y, Morvan H, Christiaen D (eds) Algal biotechnology, pp 65–87

Arad SM, Levy-Ontman O (2010) Red microalgal cell-wall polysaccharides: biotechnological aspects. Curr Opin Biotechnol 21:358–364

Arad SM, Lerental YB, Dubinsky O (1992) Effect of nitrate and sulfate starvation on polysaccharide formation in *Rhodella reticulata*. Bioresour Technol 42:141–148

Arad SM, Keristovesky G, Simon B, Barak Z, Geresh S (1993) Biodegradation of the sulphated polysaccharide of *Porphyridium* by soil bacteria. Phytochemistry 32(2):287–290

Arad SM, Huliheil M, Tal J (1997) Antiviral agents. Patent WO1997000689A1

Arad SM, Rapoport L, Moshkovich A, van Moppes D, Karpasas M, Golan R, Golan Y (2006) Superior biolubricant from a species of red microalga. Langmuir 22:7313–7317

Azeredo J, Henriques M, Sillankorva S, Oliveira R (2003) Extraction of exopolymers from biofilms: the protective effect of glutaraldehyde. Water Sci Technol 47:175–179

Badel S, Callet F, Laroche C, Gardarin C, Petit E, El Alaoui H, Bernardi T, Michaud P (2011) A new tool to detect high viscous exopolymers from microalgae. J Ind Microbiol Biotechnol 38:319–326

Barberousse H, Ruiz G, Gloaguen V, Lombardo RJ, Djediat C, Mascarell G, Castaing JC (2006) Capsular polysaccharides secreted by building façade colonisers: characterisation and adsorption to surfaces. Biofouling 22:361–370

Bellinger BJ, Abdullahi AS, Gretz MR, Underwood GJC (2005) Biofilm polymers: relationship between carbohydrate polymers from estuarine mudflats and unialgal cultures of benthic diatoms. Aquat Microb Ecol 38:169–180

Belnap J, Lange OL (eds) (2013) Biological soil crusts: structure, function and management. Springer, Berlin

Bergman B (1986) Glyoxylate induced changes in the carbon and nitrogen metabolism of the cyanobacterium *Anabaena cylindrica*. Plant Physiol 80:698–701

Bernaerts TMM, Kyomugasho C, Van Looveren N, Gheysen L, Foubert I, Hendrickx ME, Van Loey AM (2018) Molecular and rheological characterization of different cell wall fractions of *Porphyridium cruentum*. Carbohydr Polym 195:542–550

Bernal J, Mendiola JA, Ibáñez E, Cifuentes A (2011) Advanced analysis of nutraceuticals. J Pharm Biomed Anal 55:758–774

Bixler HJ, Porse H (2011) A decade of change in the seaweed hydrocolloids industry. J Appl Phycol 23:321–335

Boonchai R, Kaewsuk J, Seo G (2014) Effect of nutrient starvation on nutrient uptake and extracellular polymeric substance for microalgae cultivation and separation Desal. Water Treat 55(2):360–367

Borjas Esqueda A, Gardarin C, Laroche C (2022) Exploring the diversity of red microalgae for exopolysaccharide production. Mar Drugs 20:246–269

Borowitzka MA (2013) High-value products from microalgae – their development and commercialisation. J Appl Phycol 25:743–756

Brányiková I, Marsalková B, Doucha J, Brányik T, Bisová K, Zachleder V (2011) Microalgae-novel highly efficient starch producers. Biotechnol Bioeng 108:766–776

Brown MR (1991) The amino acid and sugar composition of 16 species of microalgae used in mariculture. J Exp Mar Biol Ecol 145:79–99

Brown MJ, Lester JN (1980) Comparison of bacterial extracellular polymer extraction methods. Appl Environ Microbiol 40:179–185

Capek P, Matulová M, Combourieu B (2008) The extracellular proteoglycan produced by *Rhodella grisea*. Int J Biol Macromol 43:390–393

Chen LZ, Li DH, Song LR, Hu CX, Wang GH, Liu YD (2006) Effects of salt stress on carbohydrate metabolism in desert soil alga *Microcoleus vaginatus*. J Integr Plant Biol 48:914–919

Chen B, You W, Huang J, Yu Y, Chen W (2010) Isolation and antioxidant property of the extracellular polysaccharide from *Rhodella reticulata*. World J Microbiol Biotechnol 26:833–840

Chen L, Rossi F, Deng S, Liu Y, Wang G, Adessi A, De Philippis R (2014) Macromolecular and chemical features of the excreted extracellular polysaccharides in induced biological crusts of different ages. Soil Biol Biochem 78:1–9

Cheng YS, Labavitch JM, Vabdergheynst JS (2015) Elevated CO_2 concentration impacts cell wall polysaccharide composition of green microalgae of the genus *Chlorella*. Lett Appl Microbiol 60:1–7

Chentir I, Hamdi M, Doumandji A, HadjSadok A, Ben Ouada H, Nasri M, Jridi M (2017) Enhancement of extracellular polymeric substances (EPS) production in *Spirulina* (*Arthrospira* sp.) by two-step cultivation process and partial characterization of their polysaccharidic moiety. Int J Biol Macromol 105:1412–1420

Chiovitti A, Molino P, Crawford SA, Ten R, Spurck T, Wetherbee R (2004) The glucans extracted with warm water from diatoms are mainly derived from intracellular chrysolaminaran and not extracellular polysaccharides. Eur J Phycol 39:117–128

Chiovitti A, Harper RE, Willis A, Bacic A, Mulvaney P, Wetherbee R (2005) Variations in the substituted 3-linked mannans closely associated with the silicified walls of diatoms. J Phycol 41:1154–1161

Cloete TE, Jacobs L, Brozel VS (1998) The chemical control of biofouling in industrial water systems. Biodegradation 9:23–37

Colica G, Li H, Rossi F, Li D, Liu Y, De Philippis R (2014) Microbial secreted exopolysaccharides affect the hydrological behavior of induced biological crusts in desert soils. Soil Biol Biochem 68:62–72

Comte S, Guibaud G, Baudu M (2006) Relations between extraction protocols for activated sludge extracellular polymeric substances (EPS) complexation properties. Part I. Comparison of the efficiency of eight EPS extraction methods. Enzym Microb Technol 38:237–245

Coragliotti A, Franklin S, Day AG, Decker SM (2012) Microalgal polysaccharide compositions US Patent 2012/0202768

Corsuto L, Rother S, Koehler L, Bedini E, Moeller S, Schnabelrauch M, Hintze V, Schiraldi C, Scharnweber D (2018) Sulfation degree not origin of chondroitin sulfate derivatives modulates keratinocyte response. Carbohydr Polym 191:53–64

Damonte EB, Matulewicz MC, Cerezo AS (2004) Sulfated seaweed polysaccharides as antiviral agents. Curr Med Chem 11(18):2399–2419

De Brouwer JFC, Stal LJ (2004) Does warm-water extraction of benthic diatoms yield extracellular polymeric substances or does it extract intracellular chrysolaminaran? Eur J Phycol 39:129–131

De Brouwer JFC, Wolfstein K, Stal LJ (2002) Physical characterization and diel dynamics of different fractions of extracellular polysaccharides in an axenic culture of a benthic diatom. Eur J Phycol 37:37–44

De Jesus Raposo MF, de Morais RMSC, de Morais AMMB (2013) Bioactivity and applications of sulphated polysaccharides from marine microalgae. Mar Drugs 11:233–252

De Jesus Raposo MF, de Morais AMMB, de Morais RMSC (2014) Influence of sulphate on the composition and antibacterial and antiviral properties of the exopolysaccharide from *Porphyridium cruentum*. Life Sci 101:56–63

De Jesus Raposo MF, de Morais AMMB, de Morais RMSC (2015) Bioactivity and applications of polysaccharides from marine microalgae. In: Ramawat K, Mérillon JM (eds) Polysaccharides, pp 1683–1727

De Philippis R, Vincenzini M (1998) Exocellular polysaccharides from cyanobacteria and their possible applications. FEMS Microbiol Rev 22:151–175

De Philippis R, Sili C, Tassinato G, Vincenzini M, Materassi R (1991) Effects of growth conditions on exopolysaccharide production by *Cyanospira capsulata*. Bioresour Technol 38:101–104

De Philippis R, Margheri MC, Pelosi E, Ventura S (1993) Exopolysaccharide production by a unicellular cyanobacterium isolated from a hypersaline habitat. J Appl Phycol 5:387–394

De Philippis R, Sili C, Vincenzini M (1996) Response of an exopolysaccharide-producing heterocystous cyanobacterium to changes in metabolic carbon flux. J Appl Phycol 8:275–281

De Philippis R, Margheri MC, Materassi R, Vincenzini M (1998) Potential of unicellular cyanobacteria from saline environments as exopolysaccharide producers. Appl Environ Microbiol 64:1130–1132

De Philippis R, Paperi R, Sili C (2007) Heavy metal sorption by released polysaccharides and whole cultures of two exopolysaccharide-producing cyanobacteria. Biodegradation 18(2):181–187

De Philippis R, Colica G, Micheletti E (2011) Exopolysaccharide-producing cyanobacteria in heavy metal removal from water: molecular basis and practical applicability of the biosorption process. Appl Microbiol Biotechnol 92:697–708

Deamici KM, de Morais MG, Santos LO, Koenraad M, Gardarin C, Costa JAV, Laroche C (2021) Static Magnetic Fields effects on polysaccharides production by different microalgae strains. Appl Sci 11:5299–5311

Decamp A, Martineau E, Grizeau D, Pruvost J, Gonçalves O (2023) Effects of the salinity on the biosynthesis of the polysaccharides of the marine microalgae *Porphyridium cruentum*. Algal Res 71:103089

Delattre C, Pierre G, Laroche C, Michaud P (2016) Production, extraction and characterization of microalgal and cyanobacterial exopolysaccharides. Biotechnol Adv 34:1159–1179

Di Pippo F, Ellwood NTW, Gismondi A, Bruno L, Rossi F, Magni P, De Philippis R (2013) Characterization of exopolysaccharides produced by seven biofilm-forming cyanobacterial strains for biotechnological applications. J Appl Phycol 25:1697–1708

Diaz Bayona KC, Gallón SMN, Lara DR, Colorado JR (2012) Activity of sulfated polysaccharides from microalgae *Porphyridium cruentum* over degenerative mechanisms of the skin. Int J Sci Adv Technol 2:85–92

Dillon H, Somanchi A, Zaman A, Rao K, Wolfson J (2007a) Methods and compositions for improving the health and appearance of skin. Patent US20070166266A1

Dillon H, Somanchi A, Rao K (2007b) Methods and compositions for cholesterol reduction in mammals. Patent US20070167396A1

Dillon H, Somanchi A, Rao K (2007c) Methods and compositions for reducing inflammation and preventing oxidative damage. Patent US20070167398A1

Dimitriu S (ed) (2004) Polysaccharides, structural diversity and functional versatility. Dekker M, New York

Domozych CR, Plante K, Blais P, Paliulis L, Domozych DS (1993) Mucilage processing and secretion in the green alga *Closterium*. I. Cytology and biochemistry. J Phycol 29:650–659

Domozych DS, Kort S, Benton S, Yu T (2005) The extracellular polymeric substance of the green alga *Penium margaritaceum* and its role in biofilm formation. Biofilms 2:129–144

Dubinsky O, Lerental YB, Christiaen D, Glaser R, Barak Z, Arad SM (1988) Production and characterization of polysaccharides in the unicellular red alga *Rhodella reticulata*. In: Stadler T, Mollion J, Verdus MC, Karamanos Y, Morvan H, Christiaen D (eds) Algal biotechnology. Elsevier, London, pp 451–461

Dubois M, Gilles KA, Hamilton JK, Rebers PA, Smith F (1956) Colorimetric method for determination of sugars and related substances. Anal Chem 28:350–356

Dvir I, Stark AH, Chayoth R, Madar Z, Arad SM (2009) Hypocholesterolemic effects of nutraceuticals produced from the red microalga *Porphyridium* sp. in rats. Nutrients 1:156–167

Eteshola E, Karpasas M, Arad SM, Gottlieb M (1998) Red microalga exopolysaccharides: 2. Study of the rheology, morphology and thermal gelation of aqueous preparations. Acta Polym 49:549–556

Farges B, Laroche C, Cornet J-F, Dussap C-G (2009) Spectral kinetic modelling and long-term behavior assessment of *Arthrospira platensis* growth in photobioreactor under red (620 nm) light illumination. Biotechnol Prog 25(1):151–162

Fattom A, Shilo M (1984) *Phormidium* J-1 bioflocculant: production and activity. Arch Microbiol 139:421–426

Fattom A, Shilo M (1985) Production of emulcyan by *Phormidium* J-1: its activity and function. FEMS Microbiol Ecol 31:3–9

Filali-Mouhim R, Cornet J-F, Fontane T, Fournet B, Dubertret G (1993) Production, isolation and preliminary characterization of the exopolysaccharide of the cyanobacterium *Spirulina platensis*. Biotechnol Lett 15:567–572

Fischer D, Schlösser UG, Pohl P (1997) Exopolysaccharide production by cyanobacteria grown in closed photobioreactors and immobilized using white cotton towelling. J Appl Phycol 9:205–213

Flaibani A, Olsen Y, Painter TJ (1989) Polysaccharides in desert reclamation: compositions of exocellular proteoglycan complexes produced by filamentous blue-green and unicellular green edaphic algae. Carbohydr Res 190:235–248

Forni C, Telo FR, Caiola MG (1997) Comparative analysis of the polysaccharides produced by different species of *Microcystis* (Chroococcales, Cyanophyta). Phycologia 36:181–185

Freire-Nordi CS, Vieira AAH, Nakaie CR, Nascimento OR (2006) Effect of polysaccharide capsule of the microalgae *Staurastrum iversenii* var. *americanum* on diffusion of charged and uncharged molecules, using EPR technique. Braz J Phys 36(1A):75–82

Fresnedo O, Serra JL (1992) Effect of nitrogen starvation on the biochemistry of *Phormidium laminosum* (Cyanophyceae). J Phycol 28:786–793

Frølund B, Palmgren R, Keiding K, Nielsen PH (1996) Extraction of extracellular polymers from activated sludge using a cation exchange resin. Water Res 30:1749–1758

Funami T (2017) In vivo and rheological approaches for characterizing food oral processing and usefulness of polysaccharides as texture modifiers – a review. Food Hydrocoll 68:2–14

Gaignard C, Macao V, Gardarin C, Rihouey C, Picton L, Michaud P, Laroche C (2018) The red microalga *Flintiella sanguinaria* as a new exopolysaccharide producer. J Appl Phycol 30:2803–2814

Gaignard C, Gargouch N, Dubessay P, Delattre C, Pierre G, Laroche C, Fendri I, Abdelkafi S, Michaud P (2019a) New horizons in culture and valorization of red microalgae. Biotechnol Adv 37:193–222

Gaignard C, Laroche C, Pierre G, Dubessay P, Delattre C, Gardarin C, Gourvil P, Probert I, Dubuffet A, Michaud P (2019b) Screening of microalgae: investigation of new exopolysaccharide producers. Algal Res 44:101711

Gantar M, Rowell P, Kerby NW, Sutherland IW (1995) Role of extracellular polysaccharide in the colonization of wheat (*Triticum vulgare* L.) roots by N2-fixing cyanobacteria. Biol Frtil Soils 19:41–48

Gargouch N, Elleuch F, Karkouch I, Tabbene O, Pichon C, Gardarin C, Rihouey C, Picton L, Abdelkafi S, Fendri I, Laroche C (2021) Potential of exopolysaccharide from Porphyridium marinum to contend with bacterial proliferation, biofilm formation, and breast cancer. Mar Drugs 19:66–84

Ge H, Xia L, Zhou X, Zhang D, Hu C (2014a) Effects of light intensity on components and topographical structures of extracellular polysaccharides from the cyanobacteria *Nostoc* sp. J Microbiol 52:179–183

Ge H, Zhang J, Zhou X, Xia L, Hu C (2014b) Effects of light intensity on components and topographical structures of extracellular polymeric substances from *Microcoleus vaginatus* (Cyanophyceae). Phycologia 53:167–173

Geresh S, Arad SM (1991) The extracellular polysaccharides of the red microalgae: chemistry and rheology. Bioresour Technol 38:195–201

Geresh S, Lupescu N, Arad S (1992) Fractionation and partial characterization of the sulphated polysaccharide of *Porphyridium*. Phytochemistry 31(12):4181–4186

Geresh S, Mamontov A, Weinstein J (2002) Sulfation of extracellular polysaccharides of red microalgae: preparation, characterization and properties. J Biochem Biophys Methods 50:179–187

Ginzberg A, Korin E, Arad SM (2008) Effect of drying on the biological activities of a red microalga polysaccharide. Biotechnol Bioeng 99:411–420

Giroldo D, Vieira AAH, Paulsen BS (2003) Relative increase of deoxy sugars during microbial degradation of an extracellular polysaccharide released by a tropical freshwater *Thalassiosira* sp. (Bacillariophyceae). J Phycol 39:1109–1115

Giroldo D, Vieira AAH, Paulsen BS (2005) Microbial degradation of extracellular polysaccharides released by a tropical strain of *Staurastrum orbiculare* (Zygnematophyceae). Phycologia 44:671–677

Gloaguen V, Ruiz G, Morvan H, Mouradi-Givernaud A, Maes E, Krausz P, Strecker G (2004) The extracellular polysaccharide of *Porphyridium* sp.: an NMR study of lithium – resistant oligosaccharidic fragments. Carbohydr Res 339:97–103

Goncalves O, Pruvost J, Massé A, Decamp A, Probert I, Maugard T, Arnaudin I, Bridiau N, Toucheteau C, Le Cerf D, Picton L, Dulong V, Rihouey C, Gaignard C, Laroche C, Pierre G, Delattre C, Dubessay P, Michaud P (2022a) Novel depolymerized exopolysaccharides, derived from microalgae, method for preparing same, and uses thereof in cosmetics for delaying the delaying the effect of skin ageing. International Patent n°WO2022184746

Goncalves O, Pruvost J, Massé A, Decamp A, Probert I, Gaignard C, Laroche C, Pierre G, Delattre C, Dubessay P, Michaud P, Le Cerf D, Picton L, Dulong V, Rihouey C, Maugard T, Arnaudin I, Bridiau N, Toucheteau C (2022b) Use of microalgal exopolysaccharides as texturing agents. International Patent n° WO2022184745

Gonzalez Bautista E, Laroche C (2021) *Arthrospira platensis* as a feasible feedstock for bioethanol production. Appl Sci 11:6756–6767

Gonzalez-Fernandez C, Ballestros M (2012) Linking microalgae and cyanobacteria culture conditions and key-enzymes for carbohydrate accumulation. Biotechnol Adv 30:1655–1661

Gouvêa SP, Vieira AAH, Lombardi AT (2005) Copper and cadmium complexation by high molecular weight materials of dominant microalgae and of water from a eutrophic reservoir. Chemosphere 60:1332–1339

Guerrini F, Cangini M, Boni L, Trost P, Pistocchi R (2000) Metabolic responses of the diatom *Achnanthes brevipes* (Bacillariophyceae) to nutrient limitation. J Phycol 36(5):882–890

Gügi B, Le Costaouec T, Burel C, Lerouge P, Helbert W, Bardor M (2015) Diatom-specific oligosaccharide and polysaccharide structures help to unravel biosynthetic capabilities in Diatoms. Mar Drugs 13:5993–6018

Han P-P, Sun Y, Wu X-Y, Yuan Y-J, Dai Y-J, Jia S-R (2014) Emulsifying, flocculating, and physicochemical properties of exopolysaccharide produced by cyanobacterium *Nostoc flagelliforme*. Appl Biochem Biotechnol 172:36–49

Han P-P, Shen S-G, Wang H-Y, Sun Y, Dai Y-J, Jia S-R (2015) Comparative metabolomic analysis of the effects of light quality on polysaccharide production of cyanobacterium *Nostoc flagelliforme*. Algal Res 9:143–150

Harun R, Danquah MK (2011) Influence of acid pre-treatment on microalgal biomass for bioethanol production. Process Biochem 46:304–309

Hasui M, Matsuda M, Okutani K, Shigeta S (1995) In vitro antiviral activities of sulfated polysaccharides from a marine microalga (*Cochlodinium polykrikoides*) against human immunodeficiency virus and other enveloped viruses. Int J Biol Macromol 17:293–297

Hayashi T, Hayashi K, Maeda M, Kojima I (1996a) Calcium spirulan, an inhibitor of enveloped virus replication, from a blue-green alga *Spirulina platensis*. J Nat Prod 59:83–87

Hayashi K, Hayashi T, Kojima I (1996b) A natural sulphated polysaccharide, calcium spirulan, isolated from *Spirulina platensis*: *in vitro* and *ex vivo* evaluation of anti-herpes simplex virus and anti-human immunodeficiency virus. AIDS Res Hum Retrovir 12:1463–1471

Hayman JR, Southern TR, Nash TE (2005) Role of sulfated glycans in adherence of the microsporidian Encephalitozoon intestinalis to host cells in vitro. Inf Immun 73:841–848

Hayakawa Y, Hayashi T, Lee JB, Srisomporn P, Maeda M, Ozawa T, Sakuragawa N (2000) Inhibition of thrombin by sulfated polysaccharides isolated from green algae. Biochim Biophys Acta 1543:86–94

Helm RF, Huang Z, Edwards D, Leeson H, Peery W, Potts M (2000) Structural characterization of the released polysaccharide of desiccation-tolerant *Nostoc commune* DRH-1. J Bacteriol 182:974–982

Herrero M, Martin-Alvarez PJ, Senoráns FJ, Cifuentes A, Ibánez E (2005) Optimization of accelerated solvent extraction of antioxidants from *Spirulina platensis* microalga. Food Chem 93:417–423

Hintze V, Miron A, Moeller S, Schnabelrauch M, Wiesmann HP, Worch H, Scharnweber D (2012) Sulfated hyaluronan and chondroitin sulfate derivatives interact differently with human transforming growth factor-β1 (TGF-β1). Acta Biomater 8:2144–2152

Hokputsa S, Hu C, Smestad Paulsen B, Harding SE (2003) A physico-chemical comparative study on extracellular carbohydrate polymers from five desert algae. Carbohydr Polym 54:27–32

Hu C, Liu Y, Paulsen BS, Petersen D, Klaveness D (2003) Extracellular carbohydrate polymers from five desert soil algae with different cohesion in the stabilization of fine sand grain. Carbohydr Polym 54:33–42

Huang Z, Liu Y, Paulsen BS, Klaveness D (1998) Studies on polysaccharides from three edible species of *Nostoc* (Cyanobacteria) with different colony morphologies: comparison of monosaccharide compositions and viscosities of polysaccharides from field colonies and suspension cultures. J Phycol 34:962–968

Huang J, Liu L, Yu Y, Lin W, Chen B, Li M, Rao X (2006) Reduction in the blood glucose level of exopolysaccharide of *Porphyridium cruentum* in alloxan-induced diabetic mice. J Fujian Norm Univ 22:77–80

Huheihel M, Ishanu V, Tal J, Arad SM (2001) Antiviral effect of red microalgal polysaccharides on Herpes simplex and Varicella zoster viruses. J Appl Phycol 13:127–134

Huheihel M, Ishanu V, Tal J, Arad SM (2002) Activity of *Porphyridium* sp. polysaccharide against herpes simplex viruses in vitro and in vivo. J Biochem Biophys Methods 50:189–200

Hussein MH, Abou-ElWafa GS, Shaaban-Dessuuki SA, Hassan NI (2015) Characterization and antioxidant activity of exopolysaccharide secreted by *Nostoc carneum*. Int J Pharmacol 11:432–439

Imeson A (1997) Thickening and gelling agents for food, 2nd edn. Springer

Imeson A (2011) Food stabilisers, thickeners and gelling agents. Wiley

Iqbal M, Zafar SI (1993) Effects of photon flux density, CO_2, aeration rate, and inoculum density on growth and extracellular polysaccharide production by *Porphyridium cruentum*. Folia Microbiol 38:509–514

Jhaveri JH, Murthy ZVP (2016) A comprehensive review on anti-fouling nanocomposite membranes for pressure driven membrane separation processes. Desalination 379:137–154

Jin C, Yu Z, Peng S, Feng K, Zhang L, Zhou X (2018) The characterization and comparison of exopolysaccharides from two benthic diatoms with different biofilm formation abilities. An Acad Brasil de Ciênc 90:1503–1519

Kanekiyo K, Hayashi K, Takenaka H, Lee JB, Hayashi T (2007) Anti-herpes simplex virus target of an acidic polysaccharide, Nostoflan, from the edible blue-green alga *Nostoc flagelliforme*. Biol Pharmac Bull 30:1573–1575

Kiemle SN, Domozych DS, Gretz MR (2007) The extracellular polymeric substances of desmids (Conjugatophyceae, Streptophyta): chemistry, structural analyses and implications in wetland biofilms. Phycologia 46:617–627

Kim M, Yim JH, Kim S-Y, Kim HS, Lee WG, Kim SJ, Kang PS, Lee CK (2012) In vitro inhibition of influenza A virus infection by marine microalga-derived polysaccharide p-KG03. Antivir Res 93:253–259

Kreuger J, Spillmann D, Li JP, Lindahl U (2006) Interactions between heparan sulfate and proteins: the concept of specificity. J Cell Biol 174:323–327

Kraan S (2012) Algal polysaccharides, novel applications and outlook. In: Carbohydrates – Comprehensive Studies on Glycobiology and Glycotechnology, Chang C-F (Ed). InTech, Croatia, pp 489–532

Kroen WK, Rayburn WR (1984) Influence of growth status and nutrients on extracellular polysaccharide synthesis by the soil alga *Chlamydomonas mexicana* (Chlorophyceae). J Phycol 20:253–257

Kumar AS, Mody K, Jha B (2007) Bacterial exopolysaccharides – a perception. J Basic Microbiol 47:103–117

Lama L, Nicolaus B, Calandrelli V, Manca MC, Romano I, Gambacorta A (1996) Effect of growth conditions on endo- and exopolymer biosynthesis in *Anabaena cylindrica* 10C. Phytochemistry 42:655–659

Laroche C (2022) Exopolysaccharides from microalgae and cyanobacteria: diversity of strains, production strategies, and valorizations. Mar Drugs 20:336–379

Leandro SM, Gil MC, Delgadillo I (2003) Partial characterisation of exopolysaccharides exudated by planktonic diatoms maintained in batch cultures. Acta Oecol 24:S49–S55

Lee J-B, Hayashi T, Hayashi K, Sankawa U, Maeda M, Nemoto T, Nakanishi H (1998) Further purification and structural analysis of calcium Spirulan from *Spirulina platensis*. J Nat Prod 61:1101–1104

Levy-Ontman O, Arad SM, Harvey DJ, Parsons TB, Fairbanks A, Tekoah Y (2011) Unique N-glycan moieties of the 66-kDa cell wall glycoprotein from the red microalga *Porphyridium* sp. J Biol Chem 286(24):21340–21352

Li S-Y, Lellouche J-P, Shabtai Y, Arad SM (2001) Fixed carbon partitioning in the red microalga *Porphyridium* sp. (rhodophyta). J Phycol 37(2):289–297

Li S-Y, Shabtai Y, Arad SM (2002) Floridoside as a carbon precursor for the synthesis of cell-wall polysaccharides in the red microalga *Porphyridium* sp. (rhodophyta). J Phycol 38(5):931–938

Li L, Gao YT, Dai Y, Yang YL, Wang XM (2007) Scavenging effects of *spirulina* and polysaccharides *Spirulina platensis* on active oxygens and its antioxidation in vitro. Chem Bioeng 24:55–57

Li H, Li Z, Xiong S, Zhang H, Li N, Zhou S, Liu Y, Huang Z (2011a) Pilot-scale isolation of bioactive extracellular polymeric substances from cell-free media of mass microalgal cultures using tangential-flow ultrafiltration. Process Biochem 46:1104–1109

Li H, Xu J, Liu Y, Ai S, Qin F, Li Z, Zhang H, Huang Z (2011b) Antioxidant and moisture-retention activities of the polysaccharide from *Nostoc commune*. Carbohydr Polym 83:1821–1827

Liu H, Fang HP (2002) Extraction of extracellular polymeric substances (EPS) of sludges. J Biotechnol 95:249–256

Liu B, Sun Z, Ma X, Yang B, Jiang Y, Wei D, Chen F (2015) Mutation breeding of extracellular polysaccharide-producing microalga *Crypthecodinium cohnii* by a novel mutagenesis with atmospheric and room temperature plasma. Int J Mol Sci 16:8201–8212

Lopes DC, Maurer JBB, Stevan-Hancke FR, de Oliveira Proença LA, Zawadzki-Baggio SF (2012) Chemical analysis of exopolysaccharide fractions and lipid compounds of the microalga *Heterosigma akashiwo* grown in vitro. Bot Mar 55:565–580

Lourenço SO, Barbarino E, Marquez UML, Aidar E (1998) Distribution of intracellular nitrogen in marine microalgae: basis for the calculation of specific nitrogen-to-proteins conversion factors. J Phycol 34:798–811

Lupi FM, Fernandes HML, Sá-Correia I, Novais JM (1991) Temperature profiles of cellular growth and exopolysaccharide synthesis by *Botryococus braunii* Kütz UC 58. J Appl Phycol 3:35–42

Lupi FM, Fernandes HML, Tomé MM, Sá-Correia I, Novais JM (1994) Influence of nitrogen source and photoperiod on exopolysaccharide synthesis by the microalga *Botryococcus braunii* UC 58. Enzyme Microb Technol 16:546–550

Magaletti E, Urbani R, Sist P, Ferrari CR, Cicero AM (2004) Abundance and chemical characterization of extracellular carbohydrates released by the marine diatom *Cylindrotheca fusiformis* under N- and P-limitation. Eur J Phycol 39:133–142

Mager DM, Thomas AD (2011) Extracellular polysaccharides from cyanobacterial soil crusts: a review of their role in dryland soil processes. J Arid Environ 75:91–97

Majdoub H, Mansour MB, Chaubet F, Roudesli MS, Maaroufi RM (2009) Anticoagulant activity of a sulfated polysaccharide from the green alga *Arthrospira platensis*. Biochim Biophys Acta Gen Subj 1790:1377–1381

Mancuso Nichols CA, Nairn KM, Glattauer V, Blackburn SI, Ramshaw JAM, Graham LD (2009) Screening microalgal cultures in search of microbial exopolysaccharides with potential as adhesives. J Adhes 85(2–3):97–125

Marcati A, Ursu AV, Laroche C, Soanen N, Marchal L, Jubeau S, Djelveh G, Michaud P (2014) Extraction and fractionation of polysaccharides and B-phycoerythrin from the microalga *Porphyridium cruentum* by membrane technology. Algal Res 5:258–263

Martinez MJA, Del Olmo LMB, Benito PB (2005) Antiviral activities from natural sources. In: Atta-ur-Rahman (ed) Studies in natural products chemistry, vol 30. Elsevier B.V, London, pp 393–418

Matsui MS, Muizzuddin N, Arad SM, Marenus K (2003) Sulfated polysaccharides from red microalgae have antiinflammatory properties *in vitro* and *in vivo*. Appl Biochem Biotechnol 104:13–22

McDonnell G, Russell AD (1999) Antiseptics and disinfectants: activity, action and resistance. Clin Microbiol Rev 12(1):147–179

Mehta VB, Vaidya BS (1978) Cellular and extracellular polysaccharides of the bluegreen alga *Nostoc*. J Exp Bot 29:1423–1430

Michaud P, Laroche C, Villay A, Roussel M, Diogon M, El Alaoui H, Delbac F (2017) Sulphated polysaccharide composition. Patent US9827277B2

Micheletti E, Colica G, Viti C, Tamagnini P, De Philippis R (2008) Selectivity in the heavy metal removal by exopolysaccharide-producing cyanobacteria. J Appl Microbiol 105:88–94

Mishra A, Jha B (2009) Isolation and characterization of extracellular polymeric substances from micro-algae *Dunaliella salina* under salt stress. Bioresour Technol 100:3382–3386

Mohamed ZA (2008) Polysaccharides as a protective response against microcystin-induced oxidative stress in *Chlorella vulgaris* and *Scenedesmus quadricauda* and their possible significance in the aquatic ecosystem. Ecotoxicology 17:504–516

Molina Grima E, Belarbi EH, Acién Fernández FG, Robles Medina A, Chisti Y (2003) Recovery of microalgal biomass and metabolites: process options and economics. Biotechnol Adv 20:491–515

Monsigny M, Petit C, Roche AC (1988) Colorimetric determination of neutral sugars by a resorcinol sulfuric acid micromethod. Anal Biochem 175:525–530

Moore BG, Tischer RG (1964) Extracellular polysaccharides of algae: effects on life-support systems. Science 145:586–587

Moreno J, Vargas MA, Olivares H, Rivas J, Guerrero MG (1998) Exopolysaccharide production by the cyanobacterium *Anabaena* sp. ATCC 33047 in batch and continuous culture. J Biotechnol 60:175–182

Moreno J, Vargas MA, Madiedo JM, Muñoz J, Rivas J, Guerrero MG (2000) Chemical and rheological properties of an extracellular polysaccharide produced by the cyanobacterium *Anabaena* sp. ATCC 33047. Biotechnol Bioeng 67:283–290

Mugnai B, Rossi F, Felde VJMNL, Colesie C, Budel B, Peth S, Kaplan A, De Philippis R (2018) The potential of the cyanobacterium *Leptolyngbya ohadii* as inoculum for stabilizing bare sandy substrates. Soil Biol Biochem 127:318–328

Myklestad S, Haug A, Larsen B (1972) Production of carbohydrates by the marine diatom *Chaetoceros affinis* var. willei (Gran) Hustedt. II. Preliminary investigation of the extracellular polysaccharide. J Exp Mar Biol Ecol 9:137–144

Myklestad S, Holm-Hansen O, Vårum KM, Volcani BE (1989) Rate of release of extracellular amino acids and carbohydrates from the marine diatom *Chaetoceros affinis*. J Plankton Res 11:763–773

Najdenski HM, Gigova LG, Iliev II, Pilarski PS, Lukavský J, Tsvetkova IV, Ninova MS, Kussovski VK (2013) Antibacterial and antifungal activities of selected microalgae and cyanobacteria. Int J Food Sci Technol 48:1533–1540

Nicolaus B, Panico A, Lama L, Romano I, Manca MC, De Giulio A, Gambacorta A (1999) Chemical composition and production of exopolysaccharides from representative members of heterocystous and non-heterocystous cyanobacteria. Phytochemistry 52:639–647

Norsker NH, Barbosa MJ, Vermuë MH, Wijffels RH (2011) Microalgal production – a close look at the economics. Biotechnol Adv 29:24–27

Nosálová G, Capek P, Matáková T, Nosál S, Flešková D, Jureček L (2012) Antitussive activity of an extracellular Rhodella grisea proteoglycan on the mechanically induced cough reflex. Carbohydr Polym 87(1):752–756

Okajima MK, Bamba T, Kaneso Y, Hirata K, Fukusaki E, Kajiyama SI, Kaneko T (2008) Supergiant ampholytic sugar chains with imbalanced charge ratio form saline ultra-absorbent hydrogels. Macromolecules 41:4061–4064

Okajima MK, Miyazato S, Kaneko T (2009) Cyanobacterial megamolecule sacran efficiently forms LC gels with very heavy metal ions. Langmuir 25:8526–8531

Otero A, Vincenzini M (2004) *Nostoc* (cyanophyceae) goes nude: extracellular polysaccharides serve as a sink for reducing power under unbalanced c/n metabolism. J Phycol 40:74–81

Ozturk S, Aslim B (2010) Modification of exopolysaccharide composition and production by three cyanobacterial isolates under salt stress. Environ Sci Pollut Res 17:595–602

Palmer J, Flint S, Brooks J (2007) Bacterial cell attachment, the beginning of a biofilm. J Ind Microbiol Biotechnol 34:577–588

Paniagua-Michel J, Olmos-Soto J, Morales-Guerrero ER (2014) Algal and microbial exopolysaccharides: new insights as biosurfactants and bioemulsifiers. Adv Food Nutr Res 73:221–257

Panoff JM, Priem B, Morvan H, Joset F (1988) Sulphated exopolysaccharides produced by two unicellular strains of Cyanobacteria, *Synechocystis* PCC 6803 and 6714. Arch Microbiol 150:558–563

Parikh A, Madamwar D (2006) Partial characterization of extracellular polysaccharides from Cyanobacteria. Bioresour Technol 97:1822–1827

Patel AK, Laroche C, Marcati A, Ursu AV, Jubeau S, Marchal L, Petit E, Djelveh G, Michaud P (2013) Separation and fractionation of exopolysaccharides from *Porphyridium cruentum*. Bioresour Technol 145:345–350

Paulsen BS, Vieira AAH (1994) Structure of the capsular and extracellular polysaccharides produced by the desmid *Spondylosium panduriforme* (Chlorophyta). J Phycol 30:638–641

Paulsen BS, Aslaksen T, Freire-Nordi CS, Vieira AAH (1998) Extracellular polysaccharides from *Ankistrodesmus Densus* (Chlorophyceae). J Phycol 34:638–641

Percival E, Foyle RAJ (1979) The extracellular polysaccharides of *Porphyridium cruentum* and *Porphyridium aerugineum*. Carbohydr Res 72:165–176

Pereira S, Zille A, Micheletti E, Moradas-Ferreira P, De Philippis R, Tamagnini P (2009) Complexity of cyanobacterial exopolysaccharides: composition, structures, inducing factors and putative genes involved in their biosynthesis and assembly. FEMS Microbiol Rev 33:917–941

Pereira S, Mota R, Santos CL, De Philippis R, Tamagnini P (2013) Assembly and export of extracellular polymeric substances (EPS) in Cyanobacteria: a phylogenomic approach. Adv Bot Res 65:235–279

Phelippé M, Goncalves O, Thouand G, Cogne G, Laroche C (2019) Characterization of the polysaccharides chemical diversity of the cyanobacteria *Arthrospira platensis*. Algal Res 38:101426

Phlips EJ, Zeman C, Hansen P (1989) Growth, photosynthesis, nitrogen fixation and carbohydrate production by a unicellular cyanobacterium *Synechococcus* sp. (Cyanophyta). J Appl Phycol 1:137–145

Piacquadio D, Nelson DB (1992) Alginates: a "new" dressing alternative. Dermatol Surg 18(11):992–995

Potter A, Thibaut S, Ribaut C (2013) Utilisation de polysaccharides sulfatés comme agent antipelliculaire. Patent WO2013093307A1

Potter A, Ghibaudo M, Baltenneck C (2014) Association de polysaccharides sulfatés et de C-glycoside et leurs utilisations. Patent WO2014174188A1

Radonic A, Thulke S, Achenbach J, Kurth A, Vreemann A, König T, Walter C, Possinger K, Nitsche A (2010) Anionic polysaccharides from phototrophic microorganisms exhibit antiviral activities to Vaccinia virus. J Antivir Antiretrovir 2(4):51–55

Ramus J (1981) The capture and transduction of light energy. In: Lobban CS, Wynne MJ (eds) The biology of seaweeds. Blackwell Scientific Publications, Boston, pp 458–492

Ramus J (1972) The production of extracellular polysaccharide by the unicellular red alga Porphyridium aeruginum. J Phycol 8:97–11

Rao MA (2010) Rheology of fluid and semisolid foods: principles and applications, 2nd edn. Springer. 481p

Rashid ZM, Lahaye E, Defer D, Douzenel P, Perrin B, Bourgougnon N, Sire O (2009) Isolation of a sulfated polysaccharide from a recently discovered sponge species (*Celtodoryx girardae*) and determination of its anti-herpetic activity. Int J Biol Macromol 44:286–293

Raungsomboon S, Chidthaisong A, Bunnag B, Inthorn D, Harvey NW (2006) Production, composition and Pb 2+ adsorption characteristics of capsular polysaccharides extracted from a cyanobacterium *Gloeocapsa gelatinosa*. Water Res 40:3759–3766

Razaghi A, Godhe A, Albers E (2014) Effects of nitrogen on growth and carbohydrate formation in *Porphyridium cruentum*. Cent Eur J Biol 9:156–162

Rechter S, König T, Auerochs S, Thulke S, Walter H, Dörnenburg H, Walter C, Marschall M (2006) Antiviral activity of *Arthrospira*-derived spirulan-like substances. Antivir Res 72:197–206
Reddy KJ, Soper BW, Tang J, Bradley RL (1996) Phenotypic variation in exopolysaccharide production in the marine, aerobic nitrogen-fixing unicellular cyanobacterium Cyanothece sp. W J Microbiol Biotechnol 12: 311–318
Richert L, Golubic S, Guédès RL, Ratiskol J, Payri C, Guezennec J (2005) Characterization of exopolysaccharides produced by cyanobacteria isolated from Polynesian microbial mats. Curr Microbiol 51:379–384
Rickman M, Pellegrino J, Davis R (2012) Fouling phenomena during membrane filtration of microalgae. J Memb Sci 423–424:33–42
Ropellato J, Carvalho MM, Ferreira LG, Noseda MD, Zuconelli CR, Gonçalves AG, Ducatti DRB, Kenski JCN, Nasato PL, Winnischofer SMB, Duarte MER (2015) Sulfated heterorhamnans from the green seaweed *Gayralia oxysperma*: partial depolymerization, chemical structure and antitumor activity. Carbohydr Polym 117:476–485
Rossi F, De Philippis R (2016) Exocellular polysaccharides in microalgae and cyanobacteria: chemical features, role and enzymes and genes involved in their biosynthesis. In: Borowitzka MA, Beardall J, Raven JA (eds) Developments in applied phycology 6 – the physiology of microalgae. Springer, Cham, pp 565–590
Rossi N, Jaouen P, Legentilhomme P, Petit I (2004) Harvesting of cyanobacterium *Arthrospira platensis* using organic filtration membranes. Food Bioprod Process 82:244–250
Rossi N, Petit I, Jaouen P, Legentilhomme P, Derouiniot M (2005) Harvesting of cyanobacterium *Arthrospira platensis* using inorganic filtration membranes. Sep Sci Technol 40:3033–3050
Rossi F, Micheletti E, Bruno L, Adhikary SP, Albertano P, De Philippis R (2012) Characteristics and role of the exocellular polysaccharides produced by five cyanobacteria isolated from phototrophic biofilms growing on stone monuments. Biofouling 28:215–224
Roussel M, Villay A, Delbac F, Michaud P, Laroche C, Roriz D, El Alaoui H, Diogon M (2015) Antimicrosporidian activity of sulphated polysaccharides from algae and their potential to control honeybee nosemosis. Carbohydr Polym 133:213–220
Roux JM (1996) Production of polysaccharide slime by microbial mats in the hypersaline environment of a Western Australian solar salt field. Int J Salt Lake Res 5:103–130
Sangar VK, Dugan PR (1972) Polysaccharide produced by *Anacystis nidulans*: its ecological implication. Appl Microbiol 24:732–734
Sathya A, Thirunavukkarasu A, Nithya R, Nandan A, Sakthishobana K, Kola AK, Sivashankar R, Tuan HA, Deepanraj B (2023) Microalgal biofuel production: potential challenges and prospective research. Fuel 332(2):126199
Senni K, Pereira J, Gueniche F, Delbarre-Ladrat C, Sinquin C, Ratiskol J, Godeau G, Fisher AM, Helley D, Colliec-Jouault S (2011) Marine polysaccharides: a source of bioactive molecules for cell therapy and tissue engineering. Mar Drugs 9:1664–1681
Shepherd R, Rockey J, Sutherland IW, Roller S (1995) Novel bioemulsifiers from microorganisms for use in foods. J Biotechnol 40:207–217
Sheng J, Yu F, Xin Z, Zhao L, Zhu X, Hu Q (2007) Preparation, identification and their antitumor activities in vitro of polysaccharides from Chlorella pyrenoidosa. Food Chem 105:533–539
Shnyukova EI, Zolotareva EK (2015) Diatom exopolysaccharides: a review. Int J Algae 17:50–67
Shnyukova EI, Zolotareva EK (2017) Ecological role of exopolysaccharides of Bacillariophyta: a review. Algologia 27:22–44
Singh S, Arad SM, Richmond A (2000) Extracellular polysaccharide production in outdoor mass cultures of *Porphyridium* sp. in flat plate glass reactors. J Appl Phycol 12:269–275
Soanen N, Da Silva E, Gardarin C, Michaud P, Laroche C (2016) Improvement of exopolysaccharide production by *Porphyridium marinum*. Bioresour Technol 213:231–238
Staats N, De Winder B, Stal L, Mur L (1999) Isolation and characterization of extracellular polysaccharides from the epipelic diatoms *Cylindrotheca closterium* and *Navicula salinarum*. Eur J Phycol 34(2):161–169

Sudo H, Burgess JG, Takemasa H, Nakamura N, Matsunaga T (1995) Sulfated exopolysaccharide production by the halophilic cyanobacterium *Aphanocapsa halophytia*. Curr Microbiol 30:219–222

Sun L, Wang C, Shi Q, Ma C (2009) Preparation of different molecular weight polysaccharides from *Porphyridium cruentum* and their antioxidant activities. Int J Biol Macromol 45:42–47

Sun L, Wang L, Zhou Y (2012) Immunomodulation and antitumor activities of different-molecular-weight polysaccharides from *Porphyridium cruentum*. Carbohydr Polym 87:1206–1210

Sun Y, Wang H, Guo G, Pu Y, Yan B (2014) The isolation and antioxidant activity of polysaccharides from the marine microalgae *Isochrysis galbana*. Carbohydr Polym 113:22–31

Sweeney IR, Miraftab M, Collyer G (2012) A critical review of modern and emerging absorbent dressings used to treat exuding wounds. Int Wound J 9(6):601–612

Takahashi E, Ledauphin J, Goux D, Orvain F (2009) Optimizing extraction of extracellular polymeric substances (EPS) from benthic diatoms: comparison of the efficiency of six EPS extraction methods. Austr J Mar Freshw Res 60:1201–1210

Tamaru Y, Takani Y, Yoshida T, Sakamoto T (2005) Crucial role of extracellular polysaccharides in desiccation and freezing tolerance in the terrestrial cyanobacterium *Nostoc commune*. Appl Environ Microbiol 71:7327–7333

Tannin-Spitz T, Bergman M, van-Moppes D, Grossman S, Arad SM (2005) Antioxidant activity of the polysaccharide of the red microalga *Porphyridium* sp. J Appl Phycol 17:215–222

Tease B, Jürgens UJ, Golecki JR, Heinrich UR, Rippka R, Weckesser J (1991) Fine-structural and chemical analyses on inner and outer sheath of the cyanobacterium *Gloeothece* sp. PCC 6909. Anton van Leeuwen 59:27–34

Tesson B, Hildebrand M (2013) Characterization and location of insoluble organic matrices associated with diatom cell walls: insight into their roles during cell wall formation. PLoS One 8(4):e61675

Tischer RG, Davis EB (1971) The effect of various nitrogen sources upon the production of extracellular polysaccharide by the blue-green alga *Anabaena flos-aquae* A-37. J Exp Bot 22:546–551

Tischer RG, Moore BG (1964) An extracellular polysaccharide produced by *Palmella mucosa* Kütz. Archiv für Mikrobiol 49:158–166

Underwood GJC, Paterson DM (2003) The importance of extracellular carbohydrate production by marine epipelic diatoms. Adv Bot Res 40:184–240

Underwood GJC, Boulcott M, Raines CA, Waldron K (2004) Environmental effects on exopolymer production by marine benthic diatoms: dynamics, changes in composition, and pathways of production. J Phycol 40:293–304

Urbani R, Magaletti E, Sist P, Cicero AM (2005) Extracellular carbohydrates released by the marine diatoms *Cylindrotheca closterium*, *Thalassiosira pseudonana* and *Skeletonema costatum*: effect of P-depletion and growth status. Sci Total Environ 353:300–306

Urrutia MM (1997) General bacteria sorption processes. In: Wase J, Forster C (eds) Biosorbents for metal ions. Taylor & Francis, London, pp 39–66

Villay A, Laroche C, Roriz D, El Alaoui H, Delbac F, Michaud P (2013) Optimisation of culture parameters for exopolysaccharides production by the microalga *Rhodella violacea*. Bioresour Technol 146:732–735

Vincenzini M, De Philippis R, Sili C, Materassi R (1990) A novel exopolysaccharide from a filamentous cyanobacterium: production, chemical characterization and theological properties. In: Dawes EA (ed) Novel biodegradable microbial polymers. Kluwer Academic Publishers, Dordrecht, pp 295–310

Vincenzini M, De Philippis R, Sili C, Materassi R (1993) Stability of molecular and rheological properties of the exopolysaccharide produced by *Cyanospira capsulata* cultivated under different growth conditions. J Appl Phycol 5:539–541

Volk RB, Venzke K, Blaschek W (2006) Structural investigation of a polysaccharide released by the cyanobacterium *Nostoc insulare*. J Appl Phycol 19:255–262

Wang SY, Jiang Y, Meng C, OuYang YH, Lin XZ (2011) Preparation of extracellular polysaccharide from fermentation liquor of marine microalgae *Schizochytrium* and study on the bioactivities. J Fuzhou Univ (Nat Sci Ed) 39:786–791
Wang L, Sun LQ, Zhao XH (2012) Physicochemical property analysis of polysaccharides from marine microalgae and their antioxidation, hygroscopicity and moisture retention activities. Fine Chem 1:20–25
Wustman BA, Gretz MR, Hoagland KD (1997) Extracellular matrix assembly in diatoms (Bacillariophyceae). 1. A model of adhesives based on chemical characterization and localization of polysaccharides from the marine diatom *Achnanthes longipes* and other diatoms. Plant Physiol 113:1059–1069
Yang A, Liu Y, Ge J, Wang W, Chen Y, Montagnes D (2010) Aggregate formation and polysaccharide content of *Chlorella pyrenoidosa* Chick (Chlorophyta) in response to simulated nutrient stress. Bioresour Technol 101:8336–8341
Yanhui C (2017) Cosmetic composition with functions of repairing and strengthening skin barrier and application. Patent CN107412042A
Ye Y, Clech PL, Chen V, Fane AG (2005) Evolution of fouling during crossflow filtration of model EPS solutions. J Memb Sci 264:190–199
Yim JH, Kim SJ, Ahn SH, Lee HK (2003) Optimal conditions for the production of sulfated polysaccharide by marine microalga *Gyrodinium impudicum* strain KG03. Biomol Eng 20:273–280
Yim JH, Son E, Pyo S, Lee HK (2005) Novel sulfated polysaccharide derived from red-tide microalga *Gyrodinium impudicum* strain KG03 with immunostimulating activity *in vivo*. Mar Biotechnol 7:331–338
You T, Barnett SM (2004) Effect of light quality on production of extracellular polysaccharides and growth rate of *Porphyridium cruentum*. Biochem Eng J 19:251–258
Zhang SJ, Santschi PH (2009) Application of cross-flow ultrafiltration for isolating exopolymeric substances from a marine diatom (*Amphora* sp.). Limnol Oceanogr Methods 7:419–429
Zhang S, Xu C, Santschi PH (2008) Chemical composition and 234Th (IV) binding of extracellular polymeric substances (EPS) produced by the marine diatom *Amphora* sp. Mar Chem 112:81–92

Chapter 10
Challenges and Perspectives for Alternative Media in Microalgae Cultivation: Main Substrates and Approaches and Potential of Alternative Derived Biomasses as Source of Bioactives

F. A. Silva, V. P. B. Medeiros, W. K. A. Costa, R. L. dos Santos, T. Somera, T. C. Pimentel, and M. Magnani

Abstract This chapter presents an overview of the utilization of food industrial waste/byproducts for cultivating freshwater microalgae biomasses, highlighting their main characteristics and potential application. The main industrial byproducts used in microalgae cultivation were dairy wastewater, cheese whey, starch wastewater, fruit peels or discarded fruits, oil mill waste, tofu, brewery and restaurant food wastes, and fish protein hydrolysate. The main microalgae species evaluated were *Chlorella, Porphyridium, Ascochloris, Desmodesmus, Haematococcus, Lagerheimia, Monoraphidium, Scenedesmus, Crypthecodinium, Chlamydomonas,* and *Spirulina.* Industrial byproduct utilization may increase microalgae growth rate and biomass productivity, resulting in biomasses with higher concentrations of bioactive compounds, such as carotenoids, phycoerythrin, phycocyanin, beta-carotene, lutein, zeaxanthin, astaxanthin, docosahexaenoic acid, omega fatty acids, and exopolysaccharides compared with the synthetic medium. This strategy can reduce the use of freshwater and growth medium for microalgae cultivation, reducing production costs. Besides, microalgae cultivation removes nutrients from wastewater, reduces possible environmental impacts, and improves the industry's carbon foot-

F. A. Silva · V. P. B. Medeiros · W. K. A. Costa · R. L. Santos · M. Magnani (✉)
Laboratory of Microbial Processes in Foods, Department of Food Engineering, Technology Center, Federal University of Paraíba, João Pessoa, PB, Brazil

T. Somera
Department of Food Science and Technology, State University of Londrina, Londrina, PR, Brazil

T. C. Pimentel
Department of Food Science and Technology, State University of Londrina, Londrina, PR, Brazil

Federal Institute of Paraná, Paranavaí, Brazil

J. A. V. Costa et al. (eds.), *Microalgal Bioengineering*,
https://doi.org/10.1007/978-3-031-61253-4_10

print. Optimizing pretreatment and cultivation parameters may contribute to an increased concentration of bioactive compounds, such as using different pretreatment methods (maceration, aqueous extraction, heat-assisted extraction), control culture conditions (light time and intensity, pH, temperature), and optimizing byproduct concentration and promote nutritional manipulations in the culture medium (nutrient supplementation/privation), among others.

Keywords Wastewater · Microalgae biomass · Cultivation medium · Bioactive compounds

10.1 Introduction

Microalgae constitute a group of photosynthetic microorganisms that are single celled, fast growing, and capable of accumulating various compounds (Dolganyuk et al. 2020; Wong et al. 2022). More than 200,000 species of microalgae are known, with an ample distribution among blue-green (*Cyanophyta*), golden-brown (*Chrysophyta*), and green (*Chlorophyta*) species. However, only some microalgae species present commercial significance, such as *Schizochytrium, Tetraselmis, Haematococcus, Dunaliella, Chaetoceros, Crypthecodinium, Skeletonema, Botryococcus, Phaeodactylum, Porphyridium, Spirulina, Chlorella, Isochrysis, Nannochloris,* and *Nitzschia* (Barkia et al. 2019; Niccolai et al. 2019; Sathasivam et al. 2019).

Microalgae have stood out recently due to their prospective application in various food matrices. Their utilization may generate unique results for new products developed by the industry, as their biomass is the source of diverse bioactive compounds contributing to improving the products' nutritional content and potential health functionality (Niccolai et al. 2019; Wang et al. 2020; Wong et al. 2022). Furthermore, the microalgal biotechnology industry has attracted considerable attention due to the development and application of biorefinery, considering that microalgae species can produce several substances, including chlorophylls, carotenoids, carbohydrates, lipids, proteins, nucleic acids, and nutraceuticals, among others (Mutanda et al. 2020). Finally, extracts and compounds from microalgae have been used for diverse finalities since they are classified as safe substances for human consumption (GRAS), according to the Food and Drug Administration (FDA 2017). In this way, they are commonly used in food processing, cosmetics, medicines, supplements, biofuel, and fertilizers (Bernaerts et al. 2019; Medeiros et al. 2022; Wong et al. 2022).

However, the costs involved in the production and processing stages of microalgae and extracted products are generally high, limiting their use (Dolganyuk et al. 2020; Medeiros et al. 2022; Mu et al. 2019; Wong et al. 2022).

The microalgae's biochemical composition can be significantly influenced by the culture parameters, such as luminosity (quality and quantity), temperature, and nutritional composition of the medium. Furthermore, microalgae have high metabolic flexibility and adaptation to various cultivation conditions. In this sense, improvements in microalgae production have been suggested (Hemalatha et al. 2019; Medeiros et al. 2022; Chandrasekhar et al. 2022). However, to accomplish this kind of cultivation enlargement, the productivity of industrial cultivation systems must go along with the ability to manage biologically valuable compound biosynthesis in intensive culture growth conditions (Fasolin et al. 2019; Mu et al. 2019; Wang et al. 2020).

The transformation of industrial waste and side streams to produce sustainable energy and added-value compounds has been stimulated to design a circular/sustainable economy (Dragone et al. 2020). In addition, transforming discarded substrates into valuable feed or food for humans, animals, and plants is an environmentally friendly and healthy business (Sharif et al. 2021; Alves et al. 2023). Nowadays, industrial waste pollutes the environment or is used to manufacture low-value-added byproducts such as biofuels and biogas. Nevertheless, high-quality products with different production methods have been developed (Alves et al. 2023). Wastewater and other industrial byproducts may be employed as freshwater substitutes or supplements in microalgal cultivation, which may decrease the cost of biomass production. Moreover, it may break the restrictions with ample applications such as pollution remediation (Azam et al. 2020; Bhatia et al. 2021).

Considering these new demands, this chapter presents an overview of the utilization of food industrial waste/byproducts for cultivating freshwater microalgae biomasses, highlighting their main characteristics and potential application.

10.2 Cultivation of Microalgae Using Nonconventional Media

Different types of industrial wastes can be used for microalgae cultivation, and the most used are presented in Table 10.1. Figure 10.1 summarizes the main findings of using industrial wastes in microalgae cultivation

10.2.1 Dairy Processing Waste

The dairy industry has an environmental issue due to properly disposing of large amounts of waste (whey, sludge, and wastewater) from milk processing to produce various dairy products (Galan et al. 2022). Microalgae cultivation is an eco-friendly possibility for dairy waste management and the simultaneous valorization of these residues. This strategy can reduce the use of freshwater and growth medium for microalgae cultivation, reducing production costs and resulting in a lipid-rich biomass that can be employed to develop high-value byproducts (Bonett et al. 2020; Verma et al. 2022). Besides, microalgae cultivation removes nutrients from dairy

Table 10.1 Main types of food waste used as alternative media in microalgae cultivation

	Waste substrate	Microalgae specie	Biomass production	Nutritional composition	Bioactive compounds	References
Dairy processing waste	Milk processing wastewater	*Chlorella vulgaris*	Microalgae were cultivated for three consecutive cycles of 4 weeks each and had differences in the bioavailable amounts of biomass (400.38 mg L^{-1} in the first cycle, 407.12 mg L^{-1} in the second, and 421.56 mg L^{-1} in the third) and its effective productivity (19.06, 19.39, and 20.08 mg L^{-1} day^{-1} in the first, second, and third cycles, respectively)	During the three cultivation cycles, biomass had higher levels of sugars (44.3–48.52%) and lipids (38.7–41.2%)	Carotenoid production was 7.76 mg g^{-1} in the first cycle, 9.89 mg g^{-1} in the second, and 8.65 mg g^{-1} in the third The lipid profile was composed of C16:0 (palmitic acid: 39.98%), C18:1 (oleic acid: 21.08%), C18:2 (linoleic acid: 13.31%), and C18:1 (α linolenic acid: 10.39%)	Verma et al. (2022)
	Cheese whey	*Porphyridium purpureum*	Biomass productivities in medium with concentrations of cheese whey correspondents to lactose amounts of 0, 2.5, 5, 7.5, and 10 g L^{-1} were, respectively, 0.44, 1.36, 1.46, 1.58, and 1.70 g L^{-1} under continuous illumination and 0.32, 1.05, 1.58, 1.89, and 1.40 g L^{-1} under a day-night cycle	Not reported	Approximate amounts of the pigments phycoerythrin (continuous light, 5.8, 13.0, 5.5, 2.5, and 1.5 mg g^{-1}; day-night cycle, 7.9, 6.5, 1.5, 1.0, 0.8 mg g^{-1}), phycocyanin (continuous light and day-night cycle: <1 mg g^{-1}), and carotenoids (continuous light and day-night cycle: <0.2 mg g^{-1})	Galan et al. (2022)
	Raw dairy wastewater	*Ascochloris* sp.	Biomass yield at day 17 was 1.46 g L^{-1} in the BG11 medium and 1.54, 1.62, 1.71, and 1.74 g L^{-1} in medium with 20, 50, 80, and 100% of raw dairy wastewater (RDW), respectively	Lipid contents at day 17 were 10.67% in BG11 medium and 31.72% in RDW under indoor cultivation conditions	The lipid profile of medium with RDW presented omega fatty acids C18:3 (linolenic acid: 9.8%), C18:2 (linoleic acid: 5.3%), C18:1 (oleic acid: 1.7%), C20:5 (eicosapentaenoic acid—EPA: 0.3%) and C22:6 (docosahexaenoic acid—DHA: 6.6%)	Kumar et al. (2019)
	Cheese whey	*Chlorella vulgaris*	The total amount of biomass in medium with cheese whey cultivated under mixotrophic conditions was 2.59 g L^{-1}, double that of the autotrophic control group	Not reported	The protein composition of the biomass presented high amounts of essential amino acids (lysine, 2.4; histidine, 0.87; phenylalanine, 2.26; leucine, 3.47; isoleucine, 1.78; methionine, 0.53; valine, 2.6; and threonine, 2.34)	Salati et al. (2017)
	Cheese whey	*Desmodesmus* sp.	Biomass yields for cheese whey were 8.63 and 9.97 g L^{-1} and for BG11 were 12.87 and 11.76 g L^{-1}, respectively, under controlled and environmental conditions	Contents of carbohydrates for cheese whey (0.68 and 2.20 g L^{-1}) and BG11 (3.84 and 2.66 g L^{-1}) and lipids for cheese whey (4.00 and 2.37 g L^{-1}) and BG11 2.83 and 1.00 g L^{-1}), respectively under controlled and environmental conditions	Carotenoid contents were 0.25 and 0.50 μg mL^{-1} for cheese whey and 0.06 and 0.12 μg mL^{-1} for BG11, respectively, under controlled and environmental conditions	Bonett et al. (2020)
	Scotta cheese whey	*Chlorella protothecoides*	Biomass amounts were 3.60 g L^{-1} and 3.17 g L^{-1} in medium with ovine scotta and mixed scotta, respectively, and approximately 1.2 g L^{-1} for control in BG11 medium	Not reported	Contents of carotenoids (2.70, 2.31, and 6.38 mg g_{cdw}^{-1}), beta-carotene (128.24, 73.61, and 335.71 μg g_{cdw}^{-1}), lutein/zeaxantin (598.29, 532.49, and 1178.48 μg g_{cdw}^{-1}), and astaxanthin (73.73, 28.85, and 96.54 μg g_{cdw}^{-1}) in medium with ovine scotta, mixed scotta, and control, respectively	Ribeiro et al. (2017)

Starch processing waste	Cassava wastewater (CW)	*Haema-tococcus pluvialis*	Cell growth in CW from Jaguaruna (JAG) at 2.5, 5.0, and 10% was approximately 5.0 x 10^4, 2.0 x 10^4 and 3.5 x 10^4 cel mL^{-1}, respectively	Not reported	Contents of astaxanthin for CW-JAG were respectively 106.10, 123.76, and 97.42 μg mL^{-1} at concentrations of 2.5, 5.0, and 10%	Rodrigues et al. (2021)
	Corn starch wastewater	*Chlorella sorokiniana* FACHB-275	Biomass concentrations were approximately 1.1, 1.3, and 1.2 g L^{-1} under continuous illumination at 1200, 2200, and 3200 Lux, respectively	Not reported	Lutein contents were approximately 9.5, 11.0, and 6.0 mg L^{-1} under continuous illumination at 1200, 2200, and 3200 Lux, respectively	Zheng et al. (2022)
Fruit and vegetable waste	Red onion peels (ROP)	*Chlorella vulgaris*	Biomass concentrations were 1.40 and 1.44 g L^{-1} in medium add of ROP and approximately 0.25 and 0.35 g L^{-1} in control at 2 and 4% of *C. vulgaris* inoculum	Lipid contents were approximately 20% in medium add of ROP and 22% in control	Not reported	Supar-maniam et al. (2023)
	Biocompost of discarded fruits and vegetables (HBM)	*Lagerheimia longiseta, Monora-phidium contortum,* and *Scenedesmus quadricauda*	Biomass yields of *L. longiseta* (380 and 430 mg L^{-1}), *M. contortum* (248 and 291 mg L^{-1}), and *S. quadricauda* (506 and 515 mg L^{-1}) in HBM and standard medium, respectively	Contents of lipids, proteins, and carbohydrates, respectively, for *L. longiseta* in HBM (19.38, 55.76, and 11.98 g 100 g^{-1}) and standard medium (17.61, 55.45, and 20.23 g.100 g^{-1}), *M. contortum* in HBM (24.21, 49.08, and 16.97 g.100 g^{-1}) and standard medium (24.09, 53.58, and 14.49 g 100 g^{-1}), and *S. quadricauda* in HBM (10.62, 59.15, and 18.08 g 100 g^{-1}) and standard medium (14.01, 44.96, and 16.13 g 100 g^{-1})	Carotenoid contents for *L. longiseta* (7.05 and 6.86 g kg^{-1}), *M. contortum* (12.38 and 12.14 g kg^{-1}), and *S. quadricauda* (1.51 and 1.47 g kg^{-1}) and total phenolic compounds for *L. longiseta* (13.33 and 10.56 g EAG Kg^{-1}), *M. contortum* (5.08 and 10.32 g EAG.Kg^{-1}), and *S. quadricauda* (3.19 and 11.78 g EAG Kg^{-1}) in HBM and standard medium, respectively	Medeiros et al. (2020)
	Date syrup	*Crypthe-codinium cohnii*	Biomass concentration of 28 g.L^{-1} for optimized date syrup under combined stress cultivation	Not reported	Docosahexaenoic acid (DHA 22:6 ω-3) content of 0.2 g DHA.g^{-1} biomass	Rumiani et al. (2018)

(continued)

Table 10.1 (continued)

	Waste substrate	Microalgae specie	Biomass production	Nutritional composition	Bioactive compounds	References
Oil mill waste	Palm oil mill effluent (POME)	*Chlorella vulgaris, Chlorella sorokiniana* UKM3,*Chlamydomonas* sp. UKM6, and *Scenedesmus* sp. UKM9	Biomass yields for *C. vulgaris* (1.21, 1.15, and 1.58 g.L^{-1}), *C. sorokiniana* (0.95, 1.23, and 1.87 g. L^{-1}), *Chlamydomonas* sp. (0.97, 188, and 1.68 g. L^{-1}), and *Scenedesmus* sp. (0.43, 1.04, and 1.46 g L^{-1}) at 10, 20, and 30% of POME, respectively	Contents of lipids, proteins, and carbohydrates, respectively, for *C. vulgaris* (1.9–71.9, 8.9–3.1, and 6.1–8.5%), *C. sorokiniana* (4.1–49.7, 19.2–36.7, and 7.3–8.6%), *Chlamydomonas* sp. (8.9–20.8, 9.6–20.0, and 4.9–9.9%) and *Scenedesmus* sp. (8.6–24.4, 7.6–24.9, and 5.5–10.7%)	Carotenoid yields of approximately 5.0, 5.5, 4.5, and 5.8 mg.g^{-1} for *C. vulgaris, C. sorokiniana, Chlamydomonas* sp., and *Scenedesmus* sp., respectively	Japar et al. (2021)
	POME	*Haematococcus pluvialis* and *Chromochloris zofingiensis*	Biomass yields of approximately 0.30, 0.45, and 0.50 g.L^{-1} for *H. pluvialis* and 0.45, 0.35, and 0.30 g.L^{-1} for *C. zofingiensis* at 2.5, 5.0, and 7.5% of POME, respectively	Not reported	Contents of astaxanthin for *H. pluvialis* (9.25, 14.62, and 22.43 mg L^{-1}) and *C. zofingiensis* (2.71, 2.32, and 2.67 mg L^{-1}) at 2.5, 5.0, and 7.5% of POME, respectively	Fernando et al. (2021)
	POME	*Arthrospira platensis*	Biomass productivity in 50% POME was 69.59 and 33.64 mg.L^{-1}.day^{-1} in cultivation under semicontinuous and batch systems, respectively	Not reported	C-phycocyanin concentration in 50% POME was 13.80 and 22.69 mg L^{-1} in cultivation under semicontinuous and batch systems, respectively	Nur et al. (2019b)
	POME	*Phaeodactylum tricornutum*	Biomass yield at 25 °C was 0.29 g L^{-1} in standard f/2 medium, 0.11 g L^{-1} in POME alone, and 0.17–0.31 g L^{-1} in POME with different nutrient additions	Carbohydrate contents of 48.49, 52.68, and 38.95–49.42% for standard medium, POME alone, and POME with different nutrient additions, respectively	Exopolysaccharide contents of 101.25, 100.52, and 94.64–131.18 mg L^{-1} for standard medium, POME alone, and POME with different nutrient additions, respectively	Nur et al. (2019a)
	POME	*Chlorella vulgaris*	Dry cell weight of *C. vulgaris* at day 10 was 0.2981, 0.3487, 0.3741, and 0.4273 g L^{-1} for 1:1, 1:2, 1:3, and 1:4 dilution ratio of POME and 0.3409 g L^{-1} in BG11 medium	Lipids contents of 21% for 1:4 POME and 18.5% for BG11	The lipid profile in 1:4 POME presented omega fatty acids C18:3 (linolenic acid: 1.86%), C18:2 (linoleic acid: 24.22%), and C18:1 (oleic acid: 28.53%)	Kumaran et al. (2023)

Other waste substrates	Tofu wastewater (TW)	*Chlorella pyrenoidosa*	Biomass concentrations were 2.01, 3.66, and 5.30 $g.L^{-1}$ for TW, while for BG11, they were 0.54, 1.86, and 2.56 $g.L^{-1}$, respectively, under autotrophic, heterotrophic, and mixotrophic conditions	Contents of lipids for TW (17.5, 23.4, and 23.2%) and BG11 (18.3, 26.0, and 25.1%), and proteins for TW (34.7, 22.7, and 30.3%) and BG11 (40.2, 26.8, and 33.6%), respectively under autotrophic, heterotrophic, and mixotrophic conditions	Not reported	Wang et al. (2018)
	Brewery waste (BW)	*Spirulina platensis* and *Spirulina maxima*	Biomass yields for *Spirulina* sp. in BW (503–511 mg L^{-1}) and control (504–510 mg L^{-1})	Contents of lipids, proteins, and carbohydrates, respectively, for *Spirulina* sp. in BW (12.3–13.9, 53.3–55.9, and 16.9–17.2 g.100 g^{-1}) and control (12.0–13.9, 49.8–50.1, and 12.8–15.0 g.100 g^{-1})	Exopolysaccharide contents for *Spirulina* sp. in BW (~190 mg L^{-1}) and control (~110 mg L^{-1})	Varandas et al. (2022)
	Fish protein hydrolysate (FPH)	*Spirulina*	Biomass productivity was 126, 131, 113, and 116 mg L^{-1} d^{-1} in medium with different concentrations (0.3, 0.5, 0.7, and 1%) of FPH and 94 $mg.L^{-1}.day^{-1}$ in control	Approximate contents of lipids (3–8%), proteins (25–70%), and carbohydrates (5–50%) for FPH and lipids (9%), proteins (60%), and carbohydrates (9%) for control	Highest contents of carotenoids for FPH (~2.8 mg g^{-1}) and control (~3.3 mg g^{-1}) and phycocyanin for FPH (0.488 g g^{-1}) and control (0.245 g g^{-1})	Shanthi et al. (2021)
	Restaurant food waste	*Chlorella pyrenoidosa*	Biomass yields in nutrient sufficient batch (0.77 g g^{-1}) and under nutrient limitation (0.26–0.63 g g^{-1})	Contents of lipids (104 and 264–317 mg g^{-1}), proteins (281 and 157–179 mg g^{-1}), and carbohydrates (352 and 391–619 mg g^{-1}), respectively, in nutrient sufficient batch and under nutrient limitation	The lipid profile presented omega fatty acids C18:3 (linolenic acid: 15 and 25–35 mg g^{-1}), C18:2 (linoleic acid: 31 and 58–92 mg g^{-1}) and C18:1 (oleic acid: 17 and 48–76 mg g^{-1})	Pleissner et al. (2017)

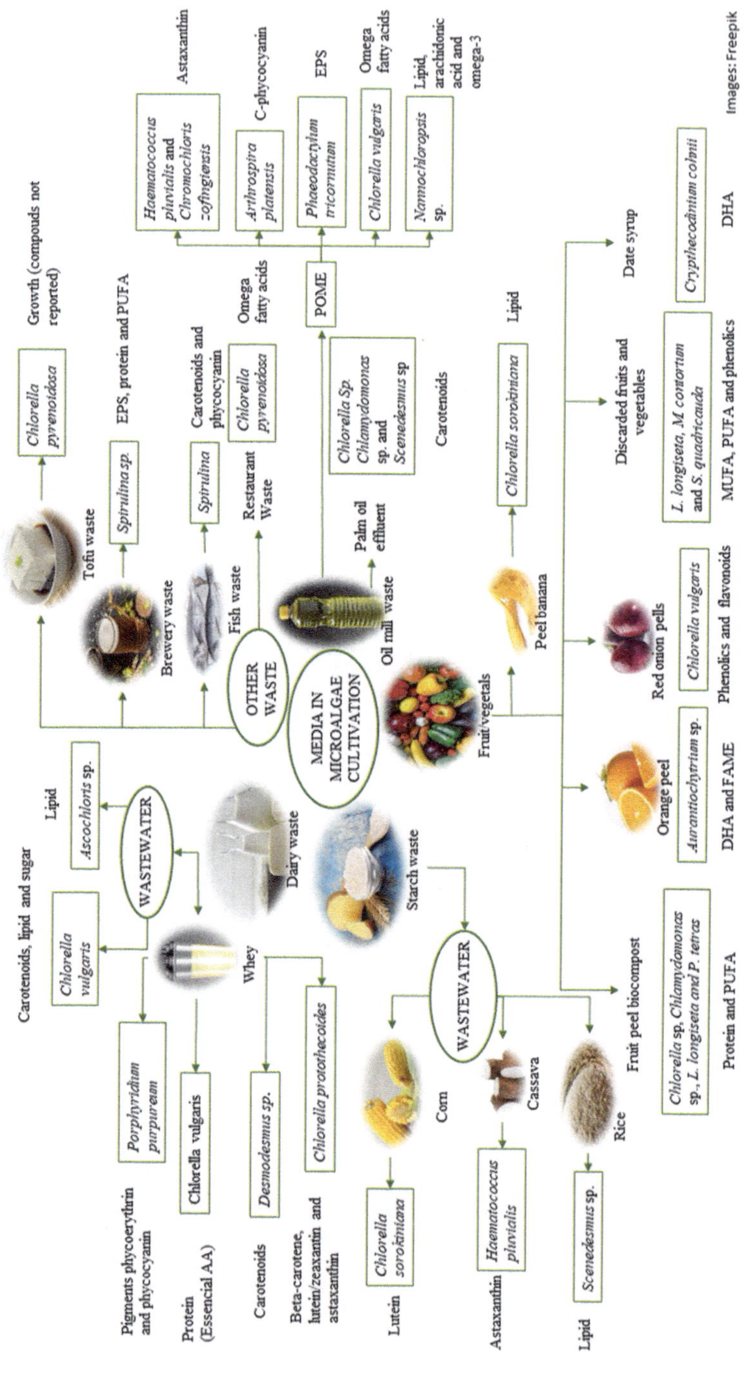

Fig. 10.1 Industrial wastes in microalgae cultivation

wastewater, reduces possible environmental impacts, and improves the dairy industry's carbon footprint (Biswas et al. 2021; Galan et al. 2022).

Dairy wastewater is a good source of nitrogen, phosphorus, and potassium, essential nutrients in the medium for microalgae growth. Thus, it allows for achieving an optimal growth rate without adding nutrient supplements. Furthermore, this residue is also rich in lactose, a carbon source that can be used for microalgae cultivation (Kumar et al. 2019; Verma et al. 2022). Kumar et al. (2019) used raw dairy wastewater (RDW) to cultivate *Ascochloris* sp. ADW007 in indoor bench-scale and outdoor pilot-scale conditions. The indoor experiments used a mixed growth media composed of BG11 (synthetic medium) and RDW at 20, 50, 80, and 100%, with cultivation conditions of 8:16 h light/dark cycles at 25 °C. Still, for the outdoor studies (environmental conditions), only RDW (100%) was used without dilutions and supplementation of nutrients. As a result, the biomass production achieved was higher in RDW compared with BG11 medium, regardless of indoor or outdoor cultivation conditions. Concerning indoor cultivation, increased RDW concentration promoted increases in both microalgae growth rate and biomass productivity. However, outdoor cultivation presented significantly higher biomass rates than indoor experiments, which is interesting for scaleup production of microalgae with low costs. The biomasses from RDW showed higher lipid content. This study demonstrated the feasibility of using dairy wastewater to obtain lipid-rich biomass.

Cheese whey and ricotta cheese whey (scotta) are other dairy discarded substrates that can be used as a medium for microalgae cultivation once they are good sources of carbon consisting mainly of lactose (Bonett et al. 2020; D'Imporzano et al. 2018). Biomass production may increase proportionally with the lactose concentration in the dairy waste used in the microalgae medium growth (Galan et al. 2022).

Bonett et al. (2020) used controlled conditions of light (12 h light/dark) and temperature (27 °C) for the cultivation of *Desmodesmus* sp. L2B Bold in cheese whey and BG11 medium. The biomass productivity using cheese whey was higher under environmental conditions than under controlled conditions, while for the BG11 medium, the higher biomass production occurred under controlled conditions. In addition, the use of environmental conditions for cultivating *Desmodesmus* sp. L2B Bold in cheese whey also resulted in twice more carotenoids than under controlled conditions. Besides, the values obtained in cheese whey were higher than in the BG11 medium under both evaluated conditions. These results highlight the cheese whey's viability in the microalgae cultivation process, which exhibits economic advantages due to no need for an additional concentration process commonly employed for this byproduct and better biomass yield without controlled conditions.

Galan et al. (2022) cultivated *Porphyridium purpureum* in a microalgae growth medium prepared from the dilution of different concentrations of cheese whey, calculated as the amount of lactose (0, 2.5, 5, 7.5, and 10 g L^{-1}), in a modified optimum growth medium. A previous deproteinization process was realized in cheese whey to avoid inhibiting microalgae growth. The experiments were carried out at high-light continuous illumination or day-night cycle, at a temperature of 25 °C, with 100 rpm agitation. The results showed that the increase of lactose concentration in the growth medium promoted increases in biomass production regardless of the illumination option. The work also evaluated the production of pigments, like phycobiliproteins (phycoerythrin and phycocyanin) and carotenoids, which are recognized as bioactive compounds. These pigments had higher contents in the medium with small lactose concentrations (2.5 g L^{-1}). However, it differs according to the illumination option. Phycobiliproteins presented a more pronounced increase under continuous illumination, while carotenoid content was higher in the case of the day-night cycle of illumination over the equivalent samples. This carotenoid behavior probably occurs due to the light stress caused by continuous illumination, affecting its production by the microalgae. This study demonstrated that higher bioactive compound concentration may be produced by optimizing the culture medium composition and illumination type.

Ribeiro et al. (2017) used ovine and mixed scotta to cultivate *Chlorella protothecoides* under mixotrophic conditions where BG11 was partially replaced by 30% of these alternative media or by the same concentration of pure lactose, glucose, and galactose. Cultivation also was carried out under autotrophic conditions using only BG11 medium as a control. The cultivation conditions used were 16:8 h light/dark cycles at 25 °C. The addition of scotta in the medium significantly increased biomass production compared with the autotrophic culture and pure sugars. This result demonstrates that other nutrients contained in scotta seem to stimulate the use of the sugars by *C. protothecoides* to support its growth. Both mixotrophic and autotrophic modes of growth had high amounts of carotenoids (astaxanthin, lutein/zeaxanthin, and β-carotene). Still, considering these compound's concentrations for the medium's unit volume, the mixotrophic culture presented higher values. Besides, the use of stress factors in the cultivation process, like salt and light, can control carotenogenesis and improve the efficiency of the production of pigment. This study also corroborates the importance of optimizing the diverse cultivation parameters to improve the concentration of bioactive compounds.

Another interesting microalga application is for protein production. Salati et al. (2017) used cheese whey, white wine lees, and glycerol for *Chlorella vulgaris* cultivation. They obtained the best results for biomass productivity using cheese whey, besides having high protein content and especially rich in essential amino acids. In the last few years, another important application exploited for microalgae is the industrial production of enzymes. Concerning the use of dairy wastes as enzyme producers, Bentahar et al. (2019) described that adding acid whey permeate in the cultivation medium for *Tetradesmus obliquus* increases the biomass and β-galactosidase enzyme production. Some advantages of microalgae use as enzyme producers are the reduced costs that can be achieved, especially using alternative substrates as cultivation medium, in addition to the other valuable byproducts that can be synthesized (Brasil et al. 2017).

10.2.2 Starch Processing Waste

Starch is recognized as one of the natural polymers existing in greater amounts on the planet (Jangyubol et al. 2018). Besides, this polymer is also cheap and renewable, making it an interesting substrate in fermentation industries (Tan et al. 2019). The industrial processing of starch is extensively employed; however, the generated wastewater could contribute to negative environmental impacts (Radhakrishnan et al. 2021).

Cassava starch is one of the world's most important staple crops. Its use as raw material for the industrial sector has increased due to its low price, renewability, and biodegradability (Carvalho et al. 2018; Jangyubol et al. 2018). The processing of cassava generates wastewater with highly suspended solids (lipids and nonsoluble carbohydrates—starch and cellulose fibers) but also rich in soluble solids (carbohydrates, nitrate, proteins, phosphate, and potassium) (Carvalho et al. 2018; Rodrigues et al. 2021). In the same way, corn is a very important cereal for the human diet, widely cultivated worldwide. Therefore, it is also extensively used for starch production, which results in large amounts of corn starch wastewater (Zheng et al. 2022).

Starch wastewater has been described as an alternative carbon source for microalgae cultivation. Its use may contribute to reducing the limitation of microalgae culture caused by adding an extra carbon source in the cultivation medium (Wang et al. 2018). This approach may result in less water spent and chemicals added for cultivation and wastewater purification (Tan et al. 2019).

Starch wastewater is rich in nutrients (e.g., carbon, nitrogen, phosphorus, and potassium), representing a good culture medium for microalgae cultivation. Nevertheless, starch wastewater nutrients are mostly in complex forms that are difficult to assimilate by microalgae. A pretreatment process is usually required before microalgae cultivation, aiming to degrade the solid organic substances and improve nutrient availability (Tan et al. 2019). Among these processes, anaerobic digestion, enzymolysis, acidification, and oxidation can be cited. However, the enzymatic process is highlighted due to its good characteristics related to more selective and efficient reactions, with moderate conditions and no inhibition products, besides the environmental benefits (Zheng et al. 2022).

Rodrigues et al. (2021) cultivated *Haematococcus pluvialis* in a microalgae growth medium prepared by adding 2.5%, 5.0%, and 10% (v/v) of cassava wastewater (CW) in the BBM medium. CW samples were collected in three cities (Imaruí, Jaguaruna, and Paulo Lopes) of Santa Catarina, Brazil. Cultivation also was realized using only the BBM medium as a control. The experiments were conducted under 12 h light/dark cycle, temperature of 23 °C, and constant agitation at 100 rpm. Adding CW in the culture medium positively affected microalgal growth and astaxanthin accumulation. The medium with CW from Jaguaruna (JAG) had the best results, with the highest biomass yield (3.33 g L^{-1}) at 2.5% and higher astaxanthin accumulation at 5%. These results probably occur due to differences in the physicochemical composition of samples cultivated in different locations. In this way, the higher nutrient availability of CW-JAG stimulated algae growth and astaxanthin production. This study demonstrated the importance of byproduct composition and concentration concerning the concentration of the bioactive compounds.

Zheng et al. (2022) used corn starch wastewater (CSW) pretreated by enzymolysis to cultivate *Chlorella sorokiniana.* First, different concentrations (2.5, 5.0, 7.5, and 10% v/v) of the pretreated CSW under 24 h of continuous illumination and a temperature of 28 °C were evaluated. Then, the optimized concentration (7.5%), according to the maximum biomass and lutein production, was used to evaluate the effect of different light/dark cycles (24/0, 18/6, 12/12, 6/18, and 0/24 h) and light intensity (1200, 2200, and 3200 Lux). Finally, cultivation in a medium with no pretreated CSW was used as a control. The results obtained in the study showed that under longer photoperiod (24 h) and medium light intensity (2200), the culture had the highest biomass concentration and lutein production, highlighting the impact of light conditions on microalgae growth and lutein biosynthesis.

Other studies have shown that microalgae cultivation using different types of starch wastes like rice mill effluent (Keerthana et al. 2020) and different pure and mixed starch wastewater (Tan et al. 2019) could result in high biomass production and lipid contents. Furthermore, strategies like the co-culture of microalgae with microorganisms that produce amylolytic enzymes are described in the literature to facilitate the assimilation of nutrients from starch wastes for microalgae growth at low costs without complicating the cultivation process. Wang et al. (2018) cultivated *Chlorella pyrenoidosa* in co-culture with immobilized *Saccharomycopsis fibuligera* using soluble starch as a carbon source. Using starch in co-culture conditions for microalgae growth had better results in biomass concentration than the control using glucose as a carbon source. This probably occurs because the hydrolyses of starch by the amylolytic *S. fibuligera* promote a controlled release of glucose in the medium, helping to avoid inhibiting microalgae growth in the concentration limit of glucose.

Microalgae are considered an interesting choice for commercial and sustainable production of high-value compounds like carotenoids (e.g., lutein and astaxanthin). These compounds are natural pigments with strong antioxidant properties and numerous beneficial effects on human health. Besides, it is very important in industry sectors like nutrition, pharmaceutical, and cosmetics. However, despite the relevance of natural carotenoid production, the costs are still much higher than for chemical production (Zarekarizi et al. 2023; Zheng et al. 2022). Therefore, microalgae cultivation in different starch wastes has the potential to be explored for low-cost biosynthesis of these compounds (Rodrigues et al. 2021; Zheng et al. 2022).

10.2.3 Fruits and Vegetable Waste

The fruit and vegetable industry is a sector of great growth in the world agricultural market, responsible for commercializing various food products such as juices, alcoholic beverages, jellies, jams, vinegar, and dehydrated products, among others (Osorio et al. 2021). In 2020, world fruit production exceeded about 900 million tons, and consequently, fibrous byproducts were produced, called bagasse, constituted of peels, pulp, stems, and seeds (Iqbal et al. 2021). Similarly, vegetable processing generates many byproducts (leaves, peels, stems, and bagasse) from inadequate raw materials during processing or production (Kim et al. 2022). As a result, fruits and vegetables are estimated to contribute 40–50% of the annual food waste production (Psaki et al. 2023).

In general, the improper disposal of tons of bagasse directly into landfills leads to various environmental problems (Osorio et al. 2021; Ray et al. 2023), mainly due to the high biological oxygen demand (BOD), chemical oxygen demand (COD), and biodegradable organic contents that can promote microbial decomposition and result in environmental pollution (Iqbal et al. 2021). Therefore, properly managing this byproduct is challenging for agribusinesses (Osorio et al. 2021).

Fruit and vegetable wastes present several nutritional and bioactive constituents that include polysaccharides, proteins, lipids, polyphenols, vitamins, pigments, dietary fibers, lipids, and micronutrients (Jun et al. 2022) that can act as a substrate for the growth of several microorganisms (Psaki et al. 2023; Hasanin and Hashem 2020) and for obtaining high value-added products through biotechnological approaches. Furthermore, considering the nutritional composition and the presence of various bioactive compounds, the use of microalgae as a biological resource for treating fruit and vegetable waste has also been reported.

The microalgae species and the byproduct concentration and type are important parameters to be considered. In this perspective, Tan et al. (2021) used different concentrations (10, 15%, 20%, and 25%) of tropical fruit residues (papaya, pineapple, and mango) diluted in water as culture media for the growth of *C. vulgaris* and *H. pluvialis* species in comparison to basal basic medium (control). The experiments were conducted at room temperature with illumination composed of white fluorescent lamps in the light-dark cycle under constant stirring. After 7 days of cultivation, *C. vulgaris* showed higher biomass yield in the culture added with 20% papaya (4.13 g L^{-1}), pineapple (5.33 g L^{-1}), and mango (4.60 g L^{-1}) when compared with the control treatment (2.23 g L^{-1}). On the contrary, *H. pluvialis* showed higher productivity and biomass yield when grown in media at 5% and 10% concentrations with fruit residue. The differences may be associated with different cell metabolism between microalgae species (Tan et al. 2021).

Similarly, the use of fruit peels (mango, papaya, and pineapple) as growth media for *Anabaena cylindrica* was evaluated at different concentrations (5%, 10%, and 20%) diluted in distilled water compared with BG11 medium (positive control) and deionized water (negative control). The microalgae were grown under fluorescent lamp illumination (16:8 h light-dark cycle) at a controlled temperature (22–24 °C) and with gentle agitation. According to the results, the media prepared with the evaluated fruit peels promoted satisfactory growth of *A. cylindrica*. Furthermore, the parameters of biomass concentration and productivity, as well as the growth rate cells, were higher in the cultivation of *A. cylindrica* in a medium made with fruit peel than controls (Hwang et al. 2021). The highest biomass production occurred in the medium with 20% papaya (933.33 mg L^{-1}), while the cultivation made with

20% mango peel was lower (100.00 mg L^{-1}). Furthermore, a 10% pineapple peel medium yielded the highest cell growth rate (Hwang et al. 2021).

An alternative medium based on fruits may be used for cultivating microalgae with an improved nutritional profile. Medeiros et al. (2020) investigated the cultivation of four microalgae isolated from freshwater environments (*Lagerheimia longiseta*, *Monoraphidium contortum*, and *Scenedesmus quadricauda*) in a medium made with biocompost of discarded fruits and vegetables (HBM), to evaluate the nutritional and bioactive constituents present in the biomass. Cultures were conducted under controlled conditions [25 ± 1 °C, illumination with 40 W fluorescent lamps (4.5 ± 0.3 K lux) under a 12 h photoperiod, with continuous aeration (0.12 L h^{-1})]. The control treatment using synthetic medium (WC) was used for all species. The research results indicated that the HBM medium promoted satisfactory cell growth, especially for *S. quadricauda*, which obtained the highest biomass yield (506 mg L^{-1}), daily biomass productivity (33.73 mg L^{-1} day^{-1}), growth speed (1.27 *k*), and maximum cell yield (17.78 × 10^5 mL^{-1}). Furthermore, the HBM medium improved the antioxidant activity and the levels of MUFA and PUFA. In addition, biomass obtained in an alternative medium exhibited high protein content, total phenolic concentration, and better health indices, indicated by a decrease in atherogenic, thrombogenic, and hypercholesterolemic saturated fatty acids indices and an increase in desired fatty acids. In this way, the use of HBM for microalgal cultivation was efficient and resulted in the production of biomass sources of valuable nutrients that could be applied to various foods (Medeiros et al. 2020).

Similarly, Calixto et al. (2016) determined the biomass production and chemical composition of four species of microalgae from freshwater environments in northeastern Brazil (*Chlorella* sp., *Chlamydomonas* sp., *Lagerheimia longiseta*, and *Pediastrum tetras*) grown on alternative media, including fruit/horticultural biocompost (HBM), in comparison with the synthetic media WC and Zarrouk. Cultures were conducted in media with constant aeration, under controlled temperature conditions (25 ± 1 °C) and illumination (12 h photoperiod, under 150 μmol photons cm^{-2} s^{-1}). The results demonstrated that the fruit/horticultural biocompost promoted greater *Chlorella* sp., *Chlamydomonas* sp. and *Lagerheimia longiseta* growth (1.4X greater) than the control. In addition, the microalgae *Pediastrum tetras* showed a higher growth rate and biomass production in the HBM medium. Regarding the biochemical composition, the microalgae *Chlamydomonas* sp. and *L. longiseta* presented high concentrations of proteins, while all species produced high content of lipids. Generally, the highest concentrations of PUFA and lowest saturated fatty acid concentration were identified in the cultures performed in HBM, which demonstrates the viability of the culture and production of biomass source of important nutrients to take advantage of the fruit and vegetable residues and minimize the costs with synthetic media (Calixto et al. 2016).

In another study, orange peel waste obtained from a squeezing juice factory was evaluated as a nutrient source for cultivating *Aurantiochytrium* sp. KRS101. For this purpose, 200 g of orange peel waste was added with 1 L of water (500 mL distilled water and 500 mL natural seawater). After simple aqueous extraction, the solid residues were removed. Three nitrogen sources (1.5 g L^{-1} of sodium nitrate, 0.994 g L^{-1} of ammonium chloride, and 0.53 g L^{-1} of urea) were evaluated as a replacement for the N source (yeast extract) in the basal medium. Cultures were conducted at 28 °C with 120 rpm agitation for 3 days (Park et al. 2018). Supplementation and pH adjustment was required. The medium prepared with orange peel waste supplemented with nitrate ($NaNO_3$) and glucose at pH 5.5 promoted a 2.5X fold increase in docosahexaenoic acid (DHA) synthesis and fatty acid methyl ester yield (1.97 g L^{-1}), respectively (Park et al. 2018). These results demonstrate that some byproducts should be supplemented with limiting compounds to improve biomass production and its biochemical composition.

With a similar objective, Rumiani et al. (2018) evaluated date syrup added with yeast extract and sea salt as a carbon source for *Crypthecodinium cohnii* growth and DHA production. The cultivations were carried out at 25 °C, in orbital agitation (250–350 rpm), with aeration in approximately 40% of natural air. Cultivation was divided into four distinct steps to increase the synthesis of DHA. According to the exposed results, the date syrup presented high concentrations of carbohydrates (glucose and fructose), favorable to microalgae cultivation. The highest biomass production (6.65 g L^{-1}) was observed in the treatment that included 17 g L^{-1} of date syrup, 2 g L^{-1} of yeast extract, pH 7, and 20 g L^{-1} of sea salt, while the lowest biomass production (0.57 g L^{-1}) was using 17 g L^{-1} of syrup, 20 g L^{-1} of sea salt, and in the absence of yeast extract. Regarding DHA production, it was seen that the stress caused by decreased nitrogen and temperature resulted in increased DHA production (5.65 g L^{-1}) in contrast to the treatment in the absence of stress (2.9 g L^{-1}). Therefore, date syrup can be used in large-scale cultivation to cheapen the cost with synthetic media and produce biomass and derived bioactive compounds (Rumiani et al. 2018).

The research conducted by Malakar et al. (2022) evaluated the use of banana peel waste as substrate for microalgae (*Chlorella sorokiniana* KMBM_I and *Chlorella sorokiniana* KMBM_K) culture medium compared with control in BG11 medium, after using various pretreatment and enzymatic hydrolysis techniques. The experiments were conducted in a shaker incubator, with 5000 lx florescent lights, at a controlled temperature (25 ± 2 °C and 16:8 h light-dark cycle). Given the results presented, autoclaving the residue followed by enzymatic hydrolysis resulted in the highest concentration of glucose (13.04 ± 1.56 g L^{-1}) compared with the other treatments and therefore was used for process optimization. Higher biomass yield of *Chlorella sorokiniana* KMBM_K was obtained with banana peel waste compared with the control medium, while lipid content increased for both *Chlorella sorokiniana* KMBM_I and *Chlorella sorokiniana* KMBM_biomass. The authors did not perform analyses of the biomass' other nutritional and bioactive constituents. However, they reinforced the possibility of using this waste as a substrate for microalgae growth as an economically viable, sustainable, and promising alternative for biomass production. In addition, their derived co-products have the potential to be applied in the food industry (Malakar et al. 2022).

Finally, red onion peels (ROP) were also used to promote the growth of *C. vulgaris* and improve the cost-effectiveness of microalgae. The ROP extracts were obtained using heat-assisted extraction (water bath) and maceration. The researchers also optimized ROP extraction for microalgae cultivation through a response surface model employing different variables. The culture was conducted in a closed system, with continuous aeration and illumination composed of fluorescent lamps of luminous intensity of 60–70 μ mol/m^2 at an ambient temperature of 25–28 °C for 14 days (Suparmaniam et al. 2023). The findings reveal that a low concentration of inoculum (2 and 4% v/v) under optimal extraction conditions could promote a high biomass yield of 1.76 g L^{-1} without negatively affecting lipid production. Furthermore, the bioactivity of ROP extracted using the mentioned techniques was also investigated. According to the results, the extract obtained by heat-assisted extraction exhibited higher values of total phenolic content (9.1094 ± 0.748 mg GAE g^{-1} dry weight), total flavonoid content (3.1307 ± 0.210 mg), and antioxidant activity (78.65 ± 3.421% and 73.68 ± 9.608% for DPPH and ABTS inhibition, respectively). These results may be associated with preserving antioxidant metabolites (flavonoids and phenolic compounds) during heat-assisted extraction (Suparmaniam et al. 2023).

Taken together, the use of fruit and vegetable residues as a nutrient source for the cultivation of microalgal species is promising. Optimizations in waste pretreatment methods (maceration, aqueous extraction, heat-assisted extraction), as well as nutritional manipulations in the culture medium (nutrient supplementation/privation) and in culture conditions (temperature and pH), have promoted satisfactory changes in the nutritional composition of biomasses, comparable or higher to synthetic media.

Although some research has attested to the efficiency and feasibility of using culture media made from fruit and vegetable residues, little has been discussed about the composition of the resulting biomass and its potential applications. Therefore, it is necessary to explore the use of new sources of residues, the bioprospecting of new species, and the evaluation of the nutritional and bioactive potential of biomass and/or derived co-products and their possible applications in the food industry.

10.2.4 Oil Mill Waste

The recent expansion of the oilseed agroindustry has promoted the generation of residues and effluents, especially oil mill effluents from palm oil processing. Considered a dark brown viscous liquid with high turbidity, the palm oil mill effluent (POME) consists of approximately 95–96% by weight of water and 4–5% total solids (Cheng et al. 2021). The main nutrients present in POME are complex substrates such as amino acids, carbohydrates (hemicellulose and simple sugars), organic acids, and inorganic nutrients (calcium, magnesium, phosphorus, potassium, and sodium) (Li et al. 2019).

The high amount of easily degradable organic substances reflects the high COD, BOD, and total suspended solids content (TSS) that can contaminate soils, surface, and underground water bodies when not treated correctly (Khadaroo et al. 2020). Thus, proper disposal and management of POME are important to avoid harmful environmental impacts. The main conventional treatment methods are physical, biological, and chemical (Ahmada et al. 2023).

In this perspective, research has described using oil mill effluents as a culture medium for microalgae. Furthermore, due to the variety of nutritional constituents present in POME, it has been considered a beneficial substrate for microalgae cultivation and producing high-added-value compounds (Fernando et al. 2021). However, the dark color of raw POME can negatively interfere with light penetration. For this reason, some authors have suggested diluting the medium and/or adding substrates associated with the optimization of the cultivation process to enhance the production of biomass and derived compounds (Kumaran et al. 2023).

Using POME in a growth medium may improve the fatty acid profile or EPS concentration in microalgae biomass, but the medium composition is an important parameter to be evaluated. The effect of different concentrations of POME was investigated in the cultivation of *Nannochloropsis* sp. The experiments were conducted at POME concentrations of 20, 40, 60, and 80% (v/v), diluted in seawater (POME + SW) or in Walne's medium (POME + WM), under controlled temperature conditions (23 ± 10.5 °C), pH (7.0 ± 0.4), and light intensity (100 μmol m^{-2} s^{-1}) with 24 h photoperiod and CO_2 aeration for 14 days. The results showed that *Nannochloropsis* sp. showed a higher growth rate (0.39 d^{-1}), maximum cell density (7.93 × 10^7 cells mL^{-1}), and lipid content (61.60%) in 60% POME + WM compared with control. There was a predominance of saturated fatty acids (35.49%), and methyl palmitate (20.85%) was the major fatty acid. Furthermore, a higher arachidonic acid and omega-3 content and an improved omega n-6/n-3 were observed. *Nannochloropsis* sp. did not show significant growth in POME + SW medium, possibly due to the medium's absence of Mg, Na, and Ca minerals (Resdi et al. 2021).

In another study, the diatom *Phaeodactylum tricornutum* showed satisfactory cell growth in POME culture, especially at the concentration of 30% POME + 70% v/v filtered natural oligotrophic seawater. Adding nitrate to the culture medium promoted increased growth rate and final biomass yield, while phosphate stimulated growth only in the low-temperature treatment. Moreover, the interaction of the parameters of temperature (25 °C), salinity (2.6%), and addition of urea (100 mg L^{-1}) resulted in the highest synthesis of EPS (140 mg L^{-1}) compared with the treatment with POME alone at the same concentration (Nur et al. 2019a).

Using POME in a growth medium may also improve astaxanthin production in the microalgae biomass. However, the effect may depend on the microalgae specie. Fernando et al. (2021) carried out the cultivation of freshwater microalgae (*H. pluvialis* and *C. zofingiensis*) in an alternative medium prepared with POME at concentrations of 2.5%, 5.0%, and 7.5% under a 12/12 h light/dark cycle with continuous aeration and temperature (28 ± 2 °C) for 14 days. *H. pluvialis* showed greater adaptability and biomass yield when cultivated in a high concentration of POME (0.52 ± 0.01 g L^{-1}) compared with *C. zofingiensis* (0.48 ± 0 0.06 g L^{-1}). Although the biomass concentration was higher in Bold's basal medium (BBM), after the 14th day of cultivation and increased lighting, the species *H. pluvialis* accumulated higher concentrations of astaxanthin (22.43 ± 1.68 mg L^{-1}) in cultivation with the addition of 7.5% POME. The biomass from this cultivation presented the content of metals within the allowed standards and, therefore, could be used in the food industry or for astaxanthin extraction.

POME effect on bioactive compound content may depend on its concentration, dilution with water, and interaction with other cultivation parameters. Nur et al. (2019b) cultivated *A. platensis* in different concentrations of POME to determine the best concentration for biomass and C-phycocyanin (C-PC) synthesis. In addition, the effect of the association of light intensity, nitrogen, salinity, POME concentration, and form of semicontinuous cultivation with the addition of distinct nutrients (urea or phosphorus) was evaluated. The findings indicated that biomass synthesis and C-PC concentration were dependent on POME content, where the highest growth and highest C-PC synthesis (4.08 ± 1.3 mg L^{-1} day^{-1}) was observed in semicontinuous cultivation (175 μmol of photons m^{-2} s^{-1}) with 50% POME in the first treatment and second treatment with 100% when compared with synthetic Zarrouk medium in batch cultivation. Furthermore, it was seen that the addition of urea at the optimal salinity positively impacted C-PC production. These results demonstrate the importance of evaluating the chemical composition of the byproduct used and, if necessary, supplementing it with limiting compounds.

Chlorella sp. utilization was investigated in a culture medium prepared with POME diluted in distilled water at different concentrations (1:1, 1:2, 1:3, and 1:4). The results showed that the highest biomass yield (0.42 g L^{-1}) and lipid production was identified in POME in the 1:4 dilution; in addition, cultivation in these conditions promoted a reduction in BOD, COD, total nitrogen, and phosphate values. The predominantly identified fatty acids were saturated (palmitic acid), monounsaturated (oleic acid), and polyunsaturated (linoleic acid), which have potential applications in food supplements and the food industry due to their health claims (Kumaran et al. 2023).

Improvements in microalgae biomass composition may be found using POME in the culture medium. In one study, indigenous microalgae species *Scenedesmus* sp. and *Chlorella* sp. were cultivated in BBM, maintained at 25 ± 2 °C, and constantly illuminated in an integrated operating system involving POME and CO_2. The results show the synthesis of biomass sources of nutrients such as proteins (48.6–57.36%), carbohydrates (2.06–10.58%), and lipids (17.19–19.29%) (Hariza et al. 2019). Bhuyar et al. (2021) used POME as a growth medium for cultivating *Chlorella* sp., which provided a biomass source of lipids 66% (w/w), especially in saturated fatty acids (palmitic acid, myristic acid, and octadecanoic acid) and PUFA (eicosapentaenoic acid and linoleic acid).

Strategies may be used to improve the POME effect, such as acclimation. The effects of acclimation of microalgal species *Chlorella vulgaris*, *Chlorella sorokiniana* UKM3, *Chlamydomonas* sp. UKM6, and *Scenedesmus* sp. UKM9 grown in anaerobic digested POME and in BBM medium (control) on the biochemical composition and pigment synthesis of microalgae were investigated by Japar et al.

(2021). Cultures were developed under controlled temperature conditions at 25 °C, air injection at 1 L min^{-1}, and continuous illumination with a light intensity of 190 μmol m^{-2} s^{-1}. Three steps were employed during acclimation. The first was the determination of the concentration of 10% (v/v), 20% (v/v), and 30% (v/v) microalgal inoculum to be added to the POME medium. The inoculum that presented the highest growth rate and biomass production was used in the second step, which sought to evaluate the growth rate and biomass production in anaerobic digested POME without additional substrates or dilutions. The third and final stage comprised the evaluation of the impact of acclimatization on the biochemical composition of the biomasses. Based on the results obtained, all cultured species showed higher biomass yield when 20% (v/v) and 30% (v/v) inoculum were employed in BBM and anaerobic digested POME, respectively. Furthermore, all cultured microalgae showed higher maximum cell growth rates (μmax) in BBM medium than in POME. This fact can be justified by several nutrients in the synthetic medium. The microalga *C. vulgaris* showed the highest biomass synthesis (144.9% from 0.98 ± 0.34 to 2.40 ± 0.05 g L^{-1}), followed by *Scenedesmus* sp. UKM9 with an increase of 58.3% from 1.03 ± 0.01 to 1.63 ± 0.07 g L^{-1}. In general, the synthesis of biomass, lipids, chlorophylls, and carotenoids was increased in *C. vulgaris* and *C. sorokiniana* UKM3 species, reinforcing the hypothesis that acclimation tolerance is associated with the metabolism of each species. Therefore, the findings suggest that acclimatization reduces microalgae cultivation time and increases biomass synthesis without the need to add/modify substrates present in the medium (Japar et al. 2021).

Taken together, the results denote the feasibility of using POME as an alternative medium for cultivating various freshwater and/or marine microalgal species, capable of growing and producing biomasses that are sources of nutrients and bioactive compounds, especially polyunsaturated acids, exopolysaccharides, and pigments. First, however, research is needed to explore the potential application and functionality of biomass and co-products derived from these cultures to expand the commercial applications of this valuable natural resource and promote the valorization of the effluents used to reduce the cost of large-scale cultivation, especially with synthetic culture media as well as mitigate the environmental impacts of improper disposal of these effluents.

10.2.5 Other Waste Substrates

Replacing standard medium with food and processing wastewaters has aroused growing interest as a low-cost alternative for reducing microalgae cultivation costs. Furthermore, these growing media from food sources highlights when compared with other industrial alternative media used for microalgae cultivation because they usually do not present toxic compounds to microalgae growth inhibition. Besides, various usable macro- and micronutrients (e.g., carbon, nitrogen, and phosphorus) were converted into high-value biomass such as pigment, fatty acid, amino acid, and polysaccharide (Pleissner et al. 2017; Song et al. 2020b; Wang et al. 2018).

In addition to those already mentioned, other important alternative media for microalgae cultivation are soy processing waste (Elystia et al. 2023; Shen et al. 2020; Song et al. 2020a; Wang et al. 2018), brewery processing waste (Song et al. 2020b; Varandas et al. 2022), meat processing waste (Azam et al. 2020; Hu et al. 2019), fish and seafood waste (Forghani et al. 2022; Shanthi et al. 2021), and kitchen food wastes (Pleissner et al. 2017; Sharma et al. 2023).

Soy processing wastes include mostly soybean and tofu wastewater. These nutrient-rich food-discarded substrates can increase biomass and microalgae growth (Arsad et al. 2020; Qiu et al. 2019). Elystia et al. (2023) cultivated *Chlorella* sp. in a medium with tofu wastewater (TW) at 40, 60, 80, and 100%, achieving the highest microalgae growth at 40% TW. The presence of decomposing bacteria in nonsterile TS could have broken down organic material into macronutrients and inorganic minerals in the form of ions more easily used by microalgae for growth. Moreover, TW concentrations higher than 40% could have inhibited microalgae growth due to the excess nutrients and decreased photosynthesis caused by the medium's turbidity. Wang et al. (2018) also used raw TW for *Chlorella pyrenoidosa* cultivation under different culture modes. After simple pH adjustments of the TW, better microalgae growth was obtained compared with the control (BG11 medium) at autotrophic, heterotrophic, and mixotrophic conditions. In the same way, Song et al. (2020a) concluded that pH adjustment of soybean wastewater reduces ammonia toxicity and improves *Chlorella* sp. L38 growth.

Brewery waste has also been used as a microalgal growth medium. Varandas et al. (2022) evaluated the growth and biomass production of *Spirulina platensis* and *Spirulina maxima* in an elaborated medium with partial addition of a biocompound extract obtained from malt residue at different concentrations (100%, 50%, and 25%) in substitution of the medium synthetic (Zarrouk). After 13 days of cultivation in an alternative medium, the highest growth rates were observed in *S. platensis* and *S. maxima* grown at a concentration of 50%, similar to the control. In addition, cultivation in an alternative medium promoted a greater accumulation of proteins in the evaluated species. Similarly, the contents of PUFAs (linoleic and alpha-linolenic acid), polysaccharides, and EPS were higher in both species compared with the control.

The slaughter industry is the most significant contributor to water pollution because it contains high concentrations of contaminants like fats, suspended solids, chemical oxygen demand, chlorides, and nitrates and is rich in nitrogen and phosphorous (Azam et al. 2020; Hu et al. 2019). These amounts of organic matter make the slaughter a favorable condition for microalgae growth and, alternatively, produce higher biomass content.

Azam et al. (2020) assessed the impact of variation in the concentration of slaughterhouse wastewater on the growth of *C. pyrenoidosa* for high biomass production for 15 days. A direct relation was found between pollution load and nutrient load of slaughterhouse wastewater with the biochemical profile of *C. pyrenoidosa*. Hu et al. (2019) selected *C. protothecoides*, *S. obliquus*, and *C. vulgaris* microalgae species for organic matter and pollutants removal from the meat wastewater while cultivating for microalgae biomass. As a result, there was an increase in biomass production and significant removal of total phosphorus, nitrogen, and COD.

Considering the fish and seafood waste, Forghani et al. (2022) evaluated the wasted shrimp processing waters as *Chlorella sorokiniana* and *Auxenochlorella protothecoides* growth media obtaining protein-rich biomass with high concentrations of essential amino acids. Besides, it highlights the input as an alternative to microalgal biorefinery comprising chemical, physical, and microbial processes. Shanthi et al. (2021) used fish protein hydrolysate (FPH) to cultivate *Spirulina* as an organic nitrogen source. Results showed that the fish waste in the form of FPH could be potential organic nitrogen for cultivating *Spirulina* with high-quality pigment.

Still, tons of food waste (FW) are generated annually from households, food services, and other retail services. These wastes are still rich in nutrients such as starch, proteins, and lipids, but the composition varies according to the source. The mixture of FW from different sources can be realized to solve an unbalanced composition and achieve adequate carbon and nitrogen rates that improve microalgae growth (Pleissner et al. 2017; Sharma et al. 2023). Pleissner et al. (2017) used a hydrolysate obtained after enzymatic digestion of a mixed restaurant FW (noodles, rice, meat, eggs, bread, cake, vegetables, and sauce) as a growth medium for *Chlorella pyrenoidosa* cultivation. The obtained biomass yield was higher in nutrient-sufficient batch culture than in cultivation at different dilution rates. A decrease in biomass yield occurred due to the nutrient limitations caused by FW dilution. This also reduced the carbohydrate content but, on the other hand, increased lipid concentration. Thus, limiting nutrients like nitrogen and phosphorus can manipulate biomass composition.

10.3 Concluding Remarks, Main Challenges, and Future Perspectives

Industrial byproduct utilization may increase microalgae growth rate and biomass productivity, resulting in biomasses with higher concentrations of bioactive compounds, such as carotenoids, phycoerythrin, phycocyanin, beta-carotene, lutein, zeaxanthin, astaxanthin, docosahexaenoic acid, omega fatty acids, and exopolysaccharides compared with the synthetic medium. This strategy can reduce the use of freshwater and growth medium for microalgae cultivation, reducing production costs. Besides, microalgae cultivation removes nutrients from wastewater, reduces possible environmental impacts, and improves the industry's carbon footprint.

Optimizing pretreatment and cultivation parameters may contribute to an increased concentration of bioactive compounds in the microalgae biomass, such as using different pretreatment methods (maceration, aqueous extraction, heat-assisted extraction), control culture conditions (light time and intensity, pH, temperature), byproduct concentration, and nutritional manipulations in the culture medium (nutrient supplementation/privation), among others. These results open opportunities for new studies with emergent technologies for the pretreatment of industrial wastes and manipulation of culture medium with low-cost substances (besides waste) to replace totally the synthetic medium with no detrimental effects on productivity.

However, there is a need to standardize different waste substrates regarding general composition, genetically engineered microalgal strains, tackle process scalability issues, control wastewater toxicity, and establish a waste transportation chain. The main gaps in the studies are as follows: (1) most of them are related to biodiesel production, which may be considered a low-value-added byproduct, (2) the produced biomass is not generally evaluated for toxicity, precluding its utilization for food and pharmacological products, (3) many studies do not determine the bioactive compounds and, mainly, the biological activity, and (4) the correlation between the concentration and composition of the alternative medium and the biomass composition has been scarcely explored. These are areas with great perspectives for future studies.

References

Ahmada I, Ibrahima NNB, Abdullaha N, Koji I, Mohamada SE, Khoo KS, Cheah WY, Ling CT, Show PL (2023) Bioremediation strategies of palm oil mill effluent and landfill leachate using microalgae cultivation: an approach contributing towards environmental sustainability. Chin J Chem 34:1–15. https://doi.org/10.1016/j.cclet.2022.107854

Alves SC, Díaz-Ruiz E, Lisboa B, Sharma M, Mussato SI, Thakur VK, Kalaskar DM, Gupta VK, Chandel AK (2023) Microbial meat: a sustainable vegan protein source produced from agri-waste to feed the world. Food Res Int 166:112596. https://doi.org/10.1016/j.foodres.2023.112596

Arsad S, Luthfiana A, Sari LA, Suherman SP, Cahyani D, Nadhira T, Yulinda EN, Musa M, Lusiana ED, Prasetiya FS (2020) Utilization of tofu wastewater as *Chlorella pyrenoidosa* growth medium. AACL Bioflux 13(5):2878–2885

Azam R, Kothari R, Singh HM, Ahmad S, Ashokkumar V, Tyagi VV (2020) Production of algal biomass for its biochemical profile using slaughterhouse wastewater for treatment under axenic conditions. Bioresour Technol 306:123116. https://doi.org/10.1016/j.biortech.2020.123116

Barkia I, Saari N, Manning SR (2019) Microalgae for high-value products towards human health and nutrition. Mar Drugs 17:304. https://doi.org/10.3390/md17050304

Bentahar J, Doyen A, Beaulieu L, Deschênes JS (2019) Acid whey permeate: An alternative growth medium for microalgae *Tetradesmus obliquus* and production of β-galactosidase. Algal Res 41:101559. https://doi.org/10.1016/j.algal.2019.101559

Bernaerts TMM, Gheysen L, Foubert I, Hendrickx ME, Van Loey AM (2019) The potential of microalgae and their biopolymers as structuring ingredients in food: a review. Biotechnol Adv 37(8):107419. https://doi.org/10.1016/j.biotechadv.2019.107419

Bhatia SK, Mehariya S, Bhatia RK, Kumar M, Pugazhendhi A, Awasthi MK, Atabani AE, Kumar G, Kim W, Seo SO, Yang YH (2021) Wastewater based microalgal biorefinery for bioenergy production: progress and challenges. Sci Total Environ 751:141599. https://doi.org/10.1016/j.scitotenv.2020.141599

Bhuyar P, Sundararaju S, Rahim MHA, Ramaraj R, Maniam GP, Govindan N (2021) Microalgae cultivation using palm oil mill effluent as growth medium for lipid production with the effect of CO_2 supply and light intensity. Biomass Convers Biorefinery 11:1555–1563

Biswas T, Bhushan S, Prajapati SK, Chaudhuri SR (2021) An eco-friendly strategy for dairy wastewater remediation with high lipid microalgae-bacterial biomass production. J Environ Manag 286:112196. https://doi.org/10.1016/j.jenvman.112196

Bonett JBA, Geraldino BS, Cardoso PG, Coelho FF, Duarte WF (2020) Isolation of freshwater microalgae and outdoor cultivation using cheese whey as substrate. Biocatal Agric Biotechnol 29:101799. https://doi.org/10.1016/j.bcab.2020.101799

Brasil BSAF, Siqueira FG, Salum TFC, Zanette CM, Spier MR (2017) Microalgae and cyanobacteria as enzyme biofactories. Algal Res 25:76–89. https://doi.org/10.1016/j.algal.04.035

Calixto CD, Santana JKS, Lira EB, Sassi PGP, Rosenhaim R, Sassi CFC, Conceição MM, Sassi R (2016) Biochemical compositions and fatty acid profiles in four species of microalgae cultivated on household sewage and agro-industrial residues. Bioresour Technol 221:458–446. https://doi.org/10.1016/j.biortech.2016.09.066

Carvalho JC, Borghetti IA, Cartas LC, Woiciechowski AL, Soccol VT, Soccol CR (2018) Biorefinery integration of microalgae production into cassava processing industry: potential and perspectives. Bioresour Technol 247:1165–1172. https://doi.org/10.1016/j.biortech.2017.09.213

Chandrasekhar K, Raj T, Ramanaiah SV, Kumar G, Banu JR, Varjani S, Sharma P, Pandey A, Kumar S, Kim S-H (2022) Algae biorefinery: a promising approach to promote microalgae industry and waste utilization. J Biotechnol 345:1–16. https://doi.org/10.1016/j.jbiotec.2021.12.008

Cheng YW, Chong CC, Lam MK, Leong WH, Chuah LF, Yusup S, Setiabudi HD, Tang Y, Lim JW (2021) Identification of microbial inhibitions and mitigation strategies towards cleaner bioconversions of palm oil mill effluent (POME): a review. J Clean Prod 280:1–23. https://doi.org/10.1016/j.jclepro.2020.124346

D'Imporzano G, Veronesi D, Salati S, Adani F (2018) Carbon and nutrient recovery in the cultivation of *Chlorella vulgaris*: a life cycle assessment approach to comparing environmental performance. J Clean Prod 194:685e694. https://doi.org/10.1016/j.jclepro.2018.05.174

Dolganyuk V, Belova D, Babich O, Prosekov A, Ivanova S, Katserov D, Patyukov N, Sukhikh S (2020) Microalgae: a promising source of valuable bioproducts. Biomol Ther 10(8):1153. https://doi.org/10.3390/biom10081153

Dragone G, Kerssemakers AAJ, Driessen JLSP, Yamakawa CK, Brumano LP, Mussatto SI (2020) Innovation and strategic orientations for the development of advanced biorefineries. Bioresour Technol 302:122847. https://doi.org/10.1016/j.biortech.2020.122847

Elystia S, Nasution FHM, Sasmita A (2023) Rotary Algae Biofilm Reactor (RABR) using microalgae *Chlorella* sp. for tofu wastewater treatment. Mater Today Proc. https://doi.org/10.1016/j.matpr.2023.03.206

Fasolin LH, Pereira RN, Pinheiro AC, Martins JT, Andrade CCP, Ramos OL, Vicente AA (2019) Emergent food proteins – towards sustainability, health and innovation. Food Res Int 125:108586. https://doi.org/10.1016/j.foodres.2019

FDA – Food and Drug Administration (2017) Regulatory framework for substances intended for use in human food or animal food on the basis of the generally recognized as safe (GRAS) Provision of the Federal Food, Drug, and Cosmetic Act: Guidance for Industry. November, 28

Fernando JSR, Premaratne M, Dinalankara DDMS, Perera GLNJ, Ariyadasa TU (2021) Cultivation of microalgae in palm oil mill effluent (POME) for astaxanthin production and simultaneous phycoremediation. J Environ Chem Eng 9:1–13. https://doi.org/10.1016/j.jece.2021.105375

Forghani B, Mayers JJ, Albers E, Undeland I (2022) Cultivation of microalgae – *Chlorella sorokiniana* and *Auxenochlorella protothecoides* – in shrimp boiling water residues. Algal Res 65:102753. https://doi.org/10.1016/j.algal.2022.102753

Galan AM, Vlaicu A, Vintila ACN, Udrescu MC, Cerchezan G, Frone AN, Vasilievici G, Paulenco A (2022) Microalgae Strain *Porphyridium purpureum* for nutrient reduction in dairy wastewaters. Sustain For 14:8545. https://doi.org/10.3390/su14148545

Hariza HB, Takriffa MS, Yasinb NHM, Ba-Abbadc MM, Hakimi NIMN (2019) Potential of the microalgae-based integrated wastewater treatment and CO_2 fixation system to treat Palm Oil Mill Effluent (POME) by indigenous microalgae: *Scenedesmus* sp. and *Chlorella* sp. J Water Process Eng 9:1–13. https://doi.org/10.1016/j.jwpe.2019.100907

Hasanin SM, Hashem AH (2020) Eco-friendly, economic fungal universal medium from watermelon peel waste. J Microbiol Methods 168:1–9. https://doi.org/10.1016/j.mimet.2019.105802

Hemalatha M, Sravan JS, Min B, Venkata Mohan S (2019) Microalgae-biorefinery with cascading resource recovery design associated to dairy wastewater treatment. Bioresour Technol 284:424–429. https://doi.org/10.1016/j.biortech.2019.03.106

Hu X, Meneses YE, Stratton J, Wang B (2019) Acclimation of consortium of micro-algae help removal of organic pollutants from meat processing wastewater. J Clean Prod 214:95–102. https://doi.org/10.1016/j.jclepro.2018.12.255

Hwang TY, Qi AH, Kin CM, Shing WL, Djearamane S, Thirunavukkarasu C (2021) Potential of fruit peel waste in growing Cyanobacteria *Anabaena cylindrica*. Int J Technol 12:711–719. https://doi.org/10.14716/ijtech.v12i4.4852

Iqbal A, Schulz P, Rizvi SSH (2021) Valorization of bioactive compounds in fruit pomace from agro-fruit industries: present Insights and future challenges. Food Biosci 44:1–15. https://doi.org/10.1016/j.fbio.2021.101384

Jangyubol K, Kasemwong K, Charoenrat T, Chittapun S (2018) Magnetic–cationic cassava starch composite for harvesting *Chlorella* sp. TISTR8236. Algal Res 35:561–568. https://doi.org/10.1016/j.algal.2018.09.027

Japar SA, Takriff MS, Yasin NHM (2021) Microalgae acclimatization in industrial wastewater and its effect on growth and primary metabolite composition. Algal Res 53:1–17. https://doi.org/10.1016/j.algal.2020.102163

Jun Y, Yifan W, Qiongyin W, Shuo Z, Meizhen W, Huajun F, Jun J, Xiaopeng Q, Yanfeng Z, Tin C (2022) Generation of fruit and vegetable wastes in the farmers' market and its influencing factors: a case study from Hangzhou, China. Waste Manag 154:331–339. https://doi.org/10.1016/j.wasman.2022.10.023

Keerthana S, Sekar S, Kumar SD, Santhanam P, Divya M, Krishnaveni N, Arthikha R, Kim M-K (2020) *Scenedesmus pecsensis* cultivation in rice mill effluent using commercial scale nutrient sources. Bioresour Technol Rep 9:100379. https://doi.org/10.1016/j.biteb.2019.100379

Khadaroo SNBA, Grassia P, Gouwanda D, Poh PE (2020) The impact of thermal pretreatment on various solid-liquid ratios of palm oil mill effluent (POME) for enhanced thermophilic anaerobic digestion performance. J Clean Prod 261:1–10. https://doi.org/10.1016/j.jclepro.2020.121159

Kim S, Ishizawa H, Inoue D, Toyama T, Yu J, Mori K, Ike M, Lee T (2022) Microalgal transformation of food processing byproducts into functional food ingredientes. Bioresour Technol 334:1–13. https://doi.org/10.1016/j.biortech.2021.126324

Kumar AK, Sharma S, Shah E, Parikh BS, Patel A, Dixit G, Gupta S, Divecha JM (2019) Cultivation of *Ascochloris* sp. ADW007-enriched microalga in raw dairy wastewater for enhanced biomass and lipid productivity. Int J Environ Sci Technol 16:943–954. https://doi.org/10.1007/s13762-018-1712-0

Kumaran M, Palanisamy KM, Bhuyar P, Maniam GP, Rahim MG, Govindan N (2023) Agriculture of microalgae *Chlorella vulgaris* for polyunsaturated fatty acids (PUFAs) production employing palm oil mill effluents (POME) for future food, wastewater, and energy nexus. Energy Nexus 9:1–7. https://doi.org/10.1016/j.nexus.2022.100169

Li K, Liu Q, Fang F, Luo R, Lu Q, Zhou W, Huo S, Cheng P, Liu J, Addy M, Chen P, Chen D, Ruan R (2019) Microalgae-based wastewater treatment for nutrients recovery: a review. Bioresour Technol 290:1–16. https://doi.org/10.1016/j.biortech.2019.121934

Malakar B, Das D, Mohanty K (2022) Evaluation of banana peel hydrolysate as alternate and cheaper growth medium for growth of microalgae *Chlorella sorokiniana*. Biomass Convers Biorefinery 9:1–11. https://doi.org/10.1007/s13399-022-03106-8

Medeiros VPB, Pimentel TC, Varandas RCR, Santos AS, Pedrosa GTS, Sassi CFC, Conceição MM, Magnani M (2020) Exploiting the use of agro-industrial residues from fruit and vegetables as alternative microalgae culture medium. Food Res Int 137:1–17. https://doi.org/10.1016/j.foodres.2020.109722

Medeiros VPB, Costa WKA, Silva RT, Pimentel TC, Magnani M (2022) Microalgae as source of functional ingredients in new-generation foods: challenges, technological effects, biological activity, and regulatory issues. Crit Rev Food Sci Nutr 62:4929–4950. https://doi.org/10.1080/10408398.2021.1879729

Mu N, Mehar JG, Mudliar SN, Shekh AY (2019) Recent advances in microalgal bioactives for food, feed, and healthcare products: commercial potential, market space, and sustainability. Compr Rev Food Sci Food Saf 6:1882–1897. https://doi.org/10.1111/1541-4337.12500

Mutanda T, Naidoo D, Bwapwa JK, Anandraj A (2020) Biotechnological applications of microalgal oleaginous compounds: current trends on microalgal bioprocessing of products. Front Energy Res 8:598803

Niccolai A, Venturi M, Galli V, Pini N, Rodolfi L, Biondi N, D'Ottavio M, Batista AP, Raymundo A, Granchi L, Tredici MR (2019) Development of new microalgae based sourdough "crostini": functional effects of *Arthrospiraplatensi* (spirulina) addition. Nat Sci Rep 9:19433. https://doi.org/10.1038/s41598-019-55840-1

Nur MMA, Swaminathan MK, Boelen P, Buma AGJ (2019a) Sulfated exopolysaccharide production and nutrient removal by the marine diatom Phaeodactylum tricornutum growing on palm oil mill effluent. J Appl Phycol 31:2335–2234. https://doi.org/10.1007/s10811-019-01780-2

Nur MMA, Garcia MG, Boelen P, Buma AGJ (2019b) Enhancement of C-phycocyanin productivity by *Arthrospira platensis* when growing on palm oil mill effluent in a two stage semicontinuous cultivation mode. J Appl Phycol 31:2855–2867. https://doi.org/10.1007/s10811-019-01806-9

Osorio LLDR, Flórez-López E, Grande-Tova CD (2021) The potential of selected agri-food loss and waste to contribute to a circular economy: applications in the food, cosmetic and pharmaceutical industries. Molecules 26:1–42. https://doi.org/10.3390/molecules26020515

Park HK, Moona M, Shina SE, Cho JM, Suh WI, Chang YK, Lee B (2018) Economical DHA (Docosahexaenoic acid) production from *Aurantiochytrium* sp. KRS101 using orange peel extract and low cost nitrogen sources. Algal Res 29:71–71. https://doi.org/10.1016/j.algal.2017.11.017

Pleissner D, Lau KY, Lin CSK (2017) Utilization of food waste in continuous flow cultures of the heterotrophic microalga *Chlorella pyrenoidosa* for saturated and unsaturated fatty acids production. J Clean Prod 142:1417e1424. https://doi.org/10.1016/j.jclepro.2016.11.165

Psaki O, Athanasoulia II, Giannoulis A, Briassoulis D, Koutinas A, Ladakis D (2023) Fermentation development using fruit waste derived mixed sugars for poly(3- hydroxybutyrate) production and property evaluation. Bioresour Technol 23:1–38. https://doi.org/10.1016/j.biortech.2023.129077

Qiu Y, Zu Y, Song C, Xie M, Qi Y, Kansha Y, Kitamura Y (2019) Soybean processing wastewater purification via *Chlorella* L166 and L38 with potential value-added ingredients production. Bioresour Technol Rep 7:100195. https://doi.org/10.1016/j.biteb.2019.100195

Radhakrishnan R, Banerjee S, Banerjee S, Singh V, Das D (2021) Sustainable approach for the treatment of poultry manure and starchy wastewater by integrating dark fermentation and microalgal cultivation. J Mater Cycles Waste Manag 23:790–803. https://doi.org/10.1007/s10163-021-01173-z

Ray A, Dubey KK, Marathe SJ, Singhal R (2023) Supercritical fluid extraction of bioactives from fruit waste and its therapeutic potential. Food Biosci 52:1–15. https://doi.org/10.1016/j.fbio.2023.102418

Resdi R, Lim JS, Idris A (2021) Batch kinetics of nutrients removal for palm oil mill effluent and recovery of lipid by *Nannochloropsis* sp. J Water Process Eng 40:1–9. https://doi.org/10.1016/j.jwpe.2020.101767

Ribeiro JES, Martini M, Altomonte I, Salari F, Nardoni S, Sorce C, Silva FLH, Andreucci A (2017) Production of *Chlorella protothecoides* biomass, chlorophyll and carotenoids using the dairy industry byproduct scotta as a substrate. Biocatal Agric Biotechnol 11:207–213. https://doi.org/10.1016/j.bcab.2017.07.007

Rodrigues OHC, Itokazu AG, Rorig L, Maraschin M, Correa RG, Pimentel-Almeida W, Moresco R (2021) Int Biodeterior Biodegrad 163:105269. https://doi.org/10.1016/j.ibiod.2021.105269

Rumiani LA, Jalili H, Amrane A (2018) Enhanced docosahexaenoic acid production by *Crypthecodinium cohnii* under combined stress in two-stage cultivation with date syrup based médium. Algal Res 34:75–81. https://doi.org/10.1016/j.algal.2018.07.010

Salati S, D'Imporzano G, Menin B, Veronesi D, Scaglia B, Abbruscato P, Mariani P, Adani F (2017) Mixotrophic cultivation of *Chlorella* for local protein production using agro-food byproducts. J Bioresour Technol 230:82–89. https://doi.org/10.1016/j.biortech.2017.01.030

Sathasivam R, Radhakrishnan R, Hashem A, Abd_Allah EF (2019) Microalgae metabolites: A rich source for food and medicine. Saudi J Biol Sci 26:709–722. https://doi.org/10.1016/j.sjbs.2017.11.003

Shanthi G, Premalatha M, Anantharaman N (2021) Potential utilization of fish waste for the sustainable production of microalgae rich in renewable protein and phycocyanin-*Arthrospira platensis/Spirulina*. J Clean Prod 294:126106. https://doi.org/10.1016/j.jclepro.2021.126106

Sharif M, Zafar MH, Aqib AI, Saeed M, Farag MR, Alagawany M (2021) Single cell protein: Sources, mechanism of production, nutritional value and its uses in aquaculture nutrition. Aquac 531:735885. https://doi.org/10.1016/j.aquaculture.2020.735885

Sharma S, Show PL, Aminabhavi TM, Sevda S, Garlapati VK (2023) Valorization of environmental-burden waste towards microalgal metabolites production. Environ Res 227:115320. https://doi.org/10.1016/j.envres.2023.115320

Shen X-F, Gao L-J, Zhou S-B, Huang J-L, Wu C-Z, Qin Q-W, Zeng RJ (2020) High fatty acid productivity from *Scenedesmus obliquus* in heterotrophic cultivation with glucose and soybean processing wastewater via nitrogen and phosphorus regulation. Sci Total Environ 708:134596. https://doi.org/10.1016/j.scitotenv.2019.134596

Song C, Han X, Qiu Y, Liu Z, Li S, Kitamura Y (2020a) Microalgae carbon fixation integrated with organic matters recycling from soybean wastewater: Effect of pH on the performance of hybrid system. Chemosphere 248:126094. https://doi.org/10.1016/j.chemosphere.2020.126094

Song C, Hu X, Liu Z, Li S, Kitamura Y (2020b) Combination of brewery wastewater purification and CO_2 fixation with potential value-added ingredients production via different microalgae strains cultivation. J Clean Prod 268:122332. https://doi.org/10.1016/j.jclepro.122332

Suparmaniam U, Lam MK, Rawindran H, Lim WJ, Pa'ee KF, Yong KTL, Tan IS, Ching BLF, Show PL, Lee TK (2023) Optimizing extraction of antioxidative biostimulant from waste onion peels for microalgae cultivation via response surface model. Energy Convers Manag 286:1–13. https://doi.org/10.1016/j.enconman.2023.117023

Tan X-B, Zhao X-C, Yang L-B (2019) Strategies for enhanced biomass and lipid production by *Chlorella pyrenoidosa* culture in starch processing wastewater. J Clean Prod 236:117671. https://doi.org/10.1016/j.jclepro.2019.117671

Tan YH, Khoo YJ, Chai MK, Wong LS (2021) Tropical fruit wastes as an organic nutrient sources for the cultivation of *Chlorella vulgaris* and *Haematococcus pluvialis*. Nat Environ Pollut Technol 20:613–618. https://doi.org/10.46488/NEPT.2021.v20i02.018

Varandas RCV, Pereira AC, Araújo VBS, Almeida PM, Nonato NS, Costa MHJ, Pina LCC, Hamdan VPT, Sassi CFC, Conceição MM (2022) Utilização do resíduo de malte da indústria cervejeira como meio alternativo para o cultivo da *Spirulina platensis* e *Spirulina maxima*. Res Soc Dev 11:1–16. https://doi.org/10.33448/rsd-v11i16.38249

Verma R, Suthar S, Chand N, Mutiyar PK (2022) Phycoremediation of milk processing wastewater and lipid-rich biomass production using *Chlorella vulgaris* under continuous batch system. Sci Total Environ 833:155110. https://doi.org/10.1016/j.scitotenv.2022.155110

Wang S-K, Wang X, Tian Y-T, Tao H-H, Sun X-S (2018) Direct utilization of starch for heterotrophic cultivation of *Chlorella pyrenoidosa* by co-culture with immobilized *Saccharomycopsis fibuligera*. Algal Res 33:406–411. https://doi.org/10.1016/j.algal.2018.06.019

Wang A, Yan K, Chu D, Nazer M, Lin NT, Samaranayake E, Chang J (2020) Microalgae as a mainstream food ingredient: demand and supply perspective. In: Alam MA, Xu JL, Wang Z (eds) Microalgae for food, health and high value products. Springer Nature, Singapore, pp 29–80. https://doi.org/10.1007/978-981-15-0169-2_2

Wong JF, Hong HJ, Foo SC, Yap MKK, Tan JW (2022) A review on current and future advancements for commercialized microalgae species. Food Sci Hum Welness 11:1156–1170. https://doi.org/10.1016/j.fshw.2022.04.007

Zarekarizi A, Hoffmann L, Burritt DJ (2023) The potential of manipulating light for the commercial production of carotenoids from algae. Algal Res 71:103047. https://doi.org/10.1016/j.algal.2023.103047

Zheng H, Wang Y, Li S, Wu Q, Feng X, Zheng Y, Leong YL, Lee D-J, Chang J-S (2022) Lutein production by microalgae using corn starch wastewater pretreated with rapid enzymatic hydrolysis. Bioresour Technol 352:126940. https://doi.org/10.1016/j.biortech.2022.126940

Chapter 11
Lipids and Fatty Acids from Microalgae

Alexsandra Valério, Ignacio Vieitez, Ângelo Paggi Matos, and J. Vladimir Oliveira

Abstract This chapter examines the lipid and fatty acid composition of some marine and freshwater microalgae species. We also report some general growth conditions that can be modulated during algal growth for maximum lipid production. Some relevant works in the field of lipid extraction at laboratory scale are summarized, including chemically organic solvents, supercritical methods, and ionic liquids. The algal lipid extracted by organic solvents can be converted into biodiesel through a process similar to that used for any other vegetable oil. Alternatively, supercritical fluid extraction can be used to extract polyunsaturated fatty acids with food grade appeal. The choice of the extraction method will determine the final utilization of lipid and fatty acids towards food and feed application or biodiesel production.

Keywords Algal biomass · Lipid metabolism · Extraction methods · Unsaturated fatty acids · Bioproducts · Biodiesel

11.1 Introduction

Microalgae are photosynthetic microorganisms able to synthesize a great variety of intracellular compounds including lipids and fatty acids. Their rapid growth rate and biomass productivity enable microalgae to serve as an effective carbon sequestration platform with fast accumulation of lipid in their biomass (up to 75% of dry

A. Valério · J. V. Oliveira (✉)
Department of Chemical and Food Engineering, Federal University of Santa Catarina, Florianópolis, SC, Brazil

I. Vieitez
Departamento de Ciencia y Tecnologia de Alimentos, Facultad de Química, Universidad de la República, Montevideo, Uruguay

Â. P. Matos
Algae R&D Centre, School of Veterinary and Life Sciences, Murdoch University, Murdoch, WA, Australia

J. A. V. Costa et al. (eds.), *Microalgal Bioengineering*,
https://doi.org/10.1007/978-3-031-61253-4_11

cell mass). Many microalgal strains can be cultivated on non-arable land in a saline water medium, and their mass farming does not compete with traditional food-plant-based production (Draaisma et al. 2013; Vigani et al. 2015).

Microalgae are currently considered one of the most promising alternative sources for biodiesel. There are, however, various technological and economic hurdles, which have to be overcome before industrial-scale production of microalgae biodiesel can take place. The selection and successful outdoor large-scale cultivation of a robust microalgal strain, which has optimum neutral lipid content, possesses an elevated growth rate, and is virtually immune toward invasion or contamination by local microbes, remain a major upstream challenge (Halim et al. 2012; Chisti 2008; Costa and Morais 2011).

Current commercial production of biodiesel or fatty acid methyl ester (FAME) involves alkaline-catalyzed transesterification of triglycerides found in oleaginous food crops with methanol. However, cultivation of these food crops for biodiesel (mainly rapeseed in Europe and soybean in the United States) is no longer sustainable as it requires substantial arable land and consumes large amounts of freshwater (Halim et al. 2012; Ratledge and Cohen 2008; Chisti 2007).

The composition of microalgae lipids is quantitatively different from that of common vegetable oils. In general, microalgae biomass differs from traditional biomass feedstocks in several respects, such as the chemistry of cell wall and plasma membrane, the presence of a large amount of water, and the small size of the cells (Harwood 1998; Cronshaw et al. 1958; Scholz et al. 2014). Furthermore, enhanced lipids in algae can be manipulated through managing some factors during cultivation process, including temperature, light intensity, salt concentrations, pH, nutrients, etc. (Moheimani et al. 2015; Byrd et al. 2017; Lari et al. 2016).

Extraction of lipids from microalgae at laboratory scale has been attempted by chemically organic solvents and supercritical fluids using carbon dioxide eventually combined with co-solvents, for example, ethanol. More recently, "green solvents" have been utilized to extract lipids from microalgae, including subcritical *n*-butane (Feller et al. 2018), ionic liquids (Young et al. 2010), and switchable solvents (Samorì et al. 2010).

This chapter address the main lipid content and fatty acid composition of several marine and freshwater microalgae species. We also review some important factors during algal cultivation that affect lipid metabolism. The core of the chapter is devoted to analyzing the current technologies available for laboratory-scale microalgae lipid extraction, with special attention to the use of organic solvent extraction, supercritical fluid extraction, and some emerging technologies, including subcritical fluid extraction and ionic liquids. The final section presents the biodiesel production from microalgae with an exploratory and simplified economic analysis.

11.2 Lipids from Microalgae

Lipids and fatty acids are constituents of all plant cells in which act as function of membrane components, storage products, metabolites, and source of energy (Moheimani et al. 2015). Lipids can be defined as any biological molecule, which is soluble in an organic solvent, and commonly called as total lipids. Lipids present in microalgae can be classified as (Pohl and Zurheide 1982) follows: (*i*) neutral lipids (NLs) including triacylglycerols (TAGs), (*ii*) phospholipids (PLs), and (*iii*) glycolipids (GLs). TAGs are usually regarded as energy storage products, whereas PLs and GLs are structural lipids, contained mainly in the cell membranes.

Nile red, a lipid-soluble fluorescent dye with chemical structure (9-(diethyl amino) benzo [a] phenoaxazin-5(5H)-one), has been frequently employed to stain intracellular lipid droplets in microalgae. In acetone solvent and resuspended in 20% dimethyl sulfoxide (DMSO), a strong yellow-gold emission from intracellular lipid globules will fluoresce after staining with Nile red. Microscopic observation is, for instance, available when visualize in a scanning confocal microscope coupled with 530 nm excitation wavelengths and 575 nm emissions filters (Feller 2017) (Fig. 11.1).

There are also some types of neutral lipids that do not contain fatty acids, such as hydrocarbons, sterols, phenolic compounds, and pigments including carotenoids and chlorophylls. These lipids fractions are soluble in organic solvents and thus are fitting in the definition of lipids. Many of them are considered high-value products, such as carotenoids, while triglycerides are desirable for biodiesel production (Chisti 2007; Bredda et al. 2018; Mata et al. 2010; Costa and Morais 2011).

Microalgae lipid content is largely affected by several factors during its growth, such as medium composition (Chen et al. 2015), salinity (Matos et al. 2015; Ishika et al. 2017a), temperature (Colla et al. 2007), illumination intensity (Sun et al. 2017), and light/dark cycle (Simionato et al. 2011; Matos et al. 2017). The lipid content of microalgae normally varies between 1% and 40%, but managing the microalgal cultivation with nitrogen starvation, lipid content as high as 50% of dry weight can be reached (James et al. 2011; Bonnefond et al. 2017). In the case of *Botryococcus braunii*, one of the most scrutinized microalgal lipid producers, a total of 75% (w/w) of hydrocarbons in cell mass was reported (Ruangsomboon 2015). Lipid content of a handful marine and freshwater microalgae species is shown in Table 11.1.

11.3 Fatty Acids in Microalgae

A fatty acid molecule consists of a hydrophilic carboxylate group attached to one end of a hydrophobic hydrocarbon chain. Fatty acids are constituents of lipid molecules (both neutral and polar) and designated based on the total number of carbon atoms, double bonds along the hydrocarbon chain (Moheimani et al. 2015). Saturated fatty acids (SFAs) have no double bond, while unsaturated fatty acids consist of at least one double bond. When the carboxylate end of the fatty acid molecule is bonded to an uncharged head group (e.g., glycerol), a neutral lipid molecule is formed (e.g., triacylglycerol). On the other hand, the association of a fatty acid molecule with a charged head group (e.g., glycerol and phosphate complex) forms a polar lipid molecule (e.g., phospholipid) (Medina et al. 1998; Halim et al. 2012).

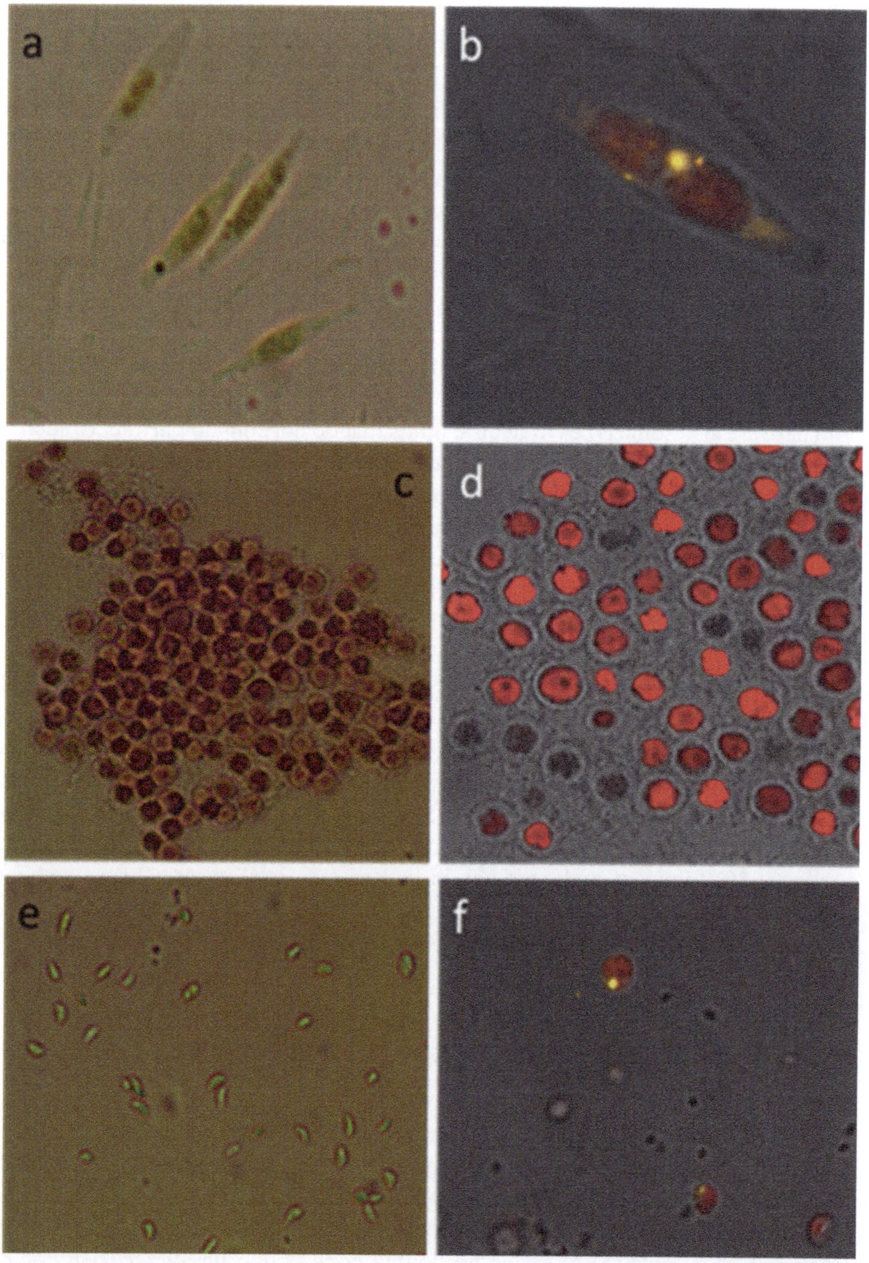

Fig. 11.1 Micrograph images of three marine microalgae. *Phaeodactylum tricornutum* viewed under optical microscope (**a**) and under scanning confocal microscopy (**b**) showing oil droplets with gold color after Nile red staining. Red color from chlorophyll autofluorescence. *Porphyridium cruentum* viewed under optical microscope (**c**) and under scanning confocal microscopy (**d**), where the red fluorescence appearance comes from chlorophyll after UV emission. *Nannochloropsis oculata* viewed under optical microscope (**e**) and under scanning confocal microscopy (**f**), with yellow intracellular lipid bodies stained with Nile red. (Source: Feller 2017)

Table 11.1 Lipid content of some traditional microalgae expressed as percentage on a dry matter

Microalgae	Lipid content	Microalgae	Lipid content
Phaeodactylum tricornutum	18–35	*Chlorella vulgaris*	10–35
Thalassiosira weissflogii	5–20	*Chlorella pyrenoidosa*	10–17
Skeletonema costatum	13–50	*Chlorella protothecoides*	14–28
Chaetoceros muelleri	12–23	*Chlorella minutissima*	14–35
Dunaliella tertiolecta	11–16	*Nostoc commune*	18–22
Dunaliella salina	6–35	*Synechocystis* sp.	12–15
Nannochloropsis oculate	20–35	*Spirulina platensis*	4–10
Nannochloropsis gaditana	10–30	*Spirulina maxima*	6–7
Scenedesmus obliquus	14–28	*Pavlova lutheri*	30–35
Chlamydomonas reinhardtii	18–25	*Pavlova salina*	12–30
Tetraselmis suecica	21–28	*Porphyridium cruentum*	4–12
Neochloris oleoabundans	26–38	*Schizochytrium limacinum*	40–45
Botryococcus braunii	50–75	*Isochrysis galbana*	30–35

Sources: Sajjadi et al. (2018), Menegazzo and Fonseca (2019) and Enamala et al. (2018)

Algal lipids are typically composed of fatty acids having carbon numbers in the range of C_{12}-C_{22}. In general, SFAs in algae are mainly composed of myristic ($C_{14:0}$) and palmitic ($C_{16:0}$) acids. Oleic acid ($C_{18:1}$) is the major monounsaturated fatty acid (MUFA) usually found in algae and is frequently included in daily human diet as a part of animal fats and vegetables oils (Matos 2017; Koller et al. 2014).

Table 11.2 provides an insight into the main composition of fatty acids in microalgae, with the subsequent paragraphs exemplifying the use of long-chain fatty acids for human consumption. Figure 11.2 shows the chemical structures of selected high-value fatty acids.

11.3.1 Polyunsaturated Fatty Acids and Food Applications

Many algal oils contain special polyunsaturated fatty acids (PUFAs) with high market prices, such as omega-3 (ω-3) including α-linolenic (ALA, $C_{18:3}$), eicosapentaenoic (EPA, $C_{20:5}$), and docosahexaenoic (DHA, $C_{22:6}$) acids and omega-6 (ω-6) including linoleic (LA, $C_{18:2}$), ɣ-linolenic (GLA, $C_{18:3}$), and arachidonic (ARA, $C_{20:4}$) acids (Medina et al. 1998). The proportion of PUFA differs greatly depending on the species (Table 11.3). For instance, *Crypthecodinium cohnii* microalgae possess only DHA and no practically other PUFA; *Nannochloropsis oculata* and *Phaeodactylum tricornutum* contain predominantly EPA, whereas *Porphyridium cruentum* contains predominantly ARA (Sun et al. 2017; Batista et al. 2013; Koller et al. 2014).

Microalgae such as *Crypthecodinium cohnii* and *Schizochytrium limacinum* are exceptional producers of DHA and became the major source alternative for production of ω-3 PUFAs, while AA can be synthesized by *Mortierella alpina* fungus, although marine microalgae *Porphyridium cruentum* is also ARA producer. The

Table 11.2 Main fatty acids from microalgae belonging to different groups. Values reported as percentage (%) of total fatty acids

Microalgal strains	C16:0	C18:1ω9	C18:2ω6	C18:3ω3	C18:3ω6	C20:4ω6	C20:5ω3	C22:6ω3	References
Chlorophyta									
Botryococcus braunii	34.8	12.0	2.4	–	2.5	0.1	0.1	–	Ruangsomboon (2015)
Botryococcus terribilis	16.1	27.3	12.4	16.9	–	2.3	7.1	6.6	Hegedűs et al. (2015)
Chlamydomonas reinhardtii	19.3	15.0	10.5	3.1	7.0	–	–	–	James et al. (2011)
Chlorella vulgaris	27.5	1.8	14.7	40.2	3.9	0.4	–	–	Matos et al. (2016)
Dunaliella salina	23.0	9.5	5.5	2.1	37.5	–	–	–	Yao et al. (2016)
Dunaliella tertiolecta	21.0	9.8	15.8	27.6	–	–	–	–	Rizwan et al. (2017)
Haematococcus pluvialis	16.7	31.2	22.0	11.1	1.3	0.8	1.6	–	Batista et al. (2013)
Scenedesmus sp.	16.5	27.1	13.0	22.6	–	–	–	–	Tibbetts et al. (2015a)
Tetraselmis suecica	31.3	43.0	4.7	6.5	–	0.2	1.2	–	Bondioli et al. (2012)
Cyanobacteria									
Arthrospira platensis	45.9	0.6	2.0	1.7	25.8	0.5	0.5	0.8	Matos et al. (2016)
Nostoc flagelliforme	21.2	22.6	15.0	19.3	–	–	–	–	Liu et al. (2005)
Anabaena affinis	30.3	2.5	6.2	35.6	–	–	–	–	Li and Watanabe (2001)
Aphanizomenon flos-aquae	15.0	8.4	8.1	33.8	–	–	0.6	–	Syrpas et al. (2018)
Synechocystis sp.	52.0	4.2	13.7	4.7	20.4	–	–	–	Cuellar-Bermudez et al. (2015)
Haptophyta									
Isochrysis galbana	13.7	2.9	0.6	2.1	–	0.8	24.6	5.8	Batista et al. (2013)
Diacronema vlkianum	10.1	1.8	0.3	0.1	0.8	1.3	23.0	5.9	Batista et al. (2013)
Heterokontophyta									
Chaetoceros muelleri									
Nannochloropsis oculata	20.3	2.9	3.0	0.1	1.5	–	31.2	0.4	Matos et al. (2016)

Nannochloropsis gaditana	53.0	4.6	3.1	0.3	1.2	–	–	–	Matos et al. (2016)
Phaeodactylum tricornutum	16.8	1.6	3.0		0.4	1.0	34.1	0.9	Matos et al. (2016)
Schizochytrium sp.	28.6	–	–	–	–	0.2	0.3	47.5	Wang et al. (2018)
Myzozoa									
Crypthecodinium cohnii	26.2	15.4	–	–	–	–	–	29.0	Sun et al. (2017)
Rhodophyta									
Porphyridium cruentum	30.5	1.3	2.5	0.9	0.3	38.3	7.0	–	Matos et al. (2016)

C16:0, palmitic acid; C18:1 ω9, oleic acid; C18:2 ω6, linoleic acid (LA); C18:3 ω3, α-linolenic acid (ALA); C18:3 ω6, γ-linolenic acid (GLA); C20:4 ω6, arachidonic acid (AA); C20:5 ω3, eicosapentaenoic acid (EPA); C22:6 ω3, docosahexaenoic acid (DHA)

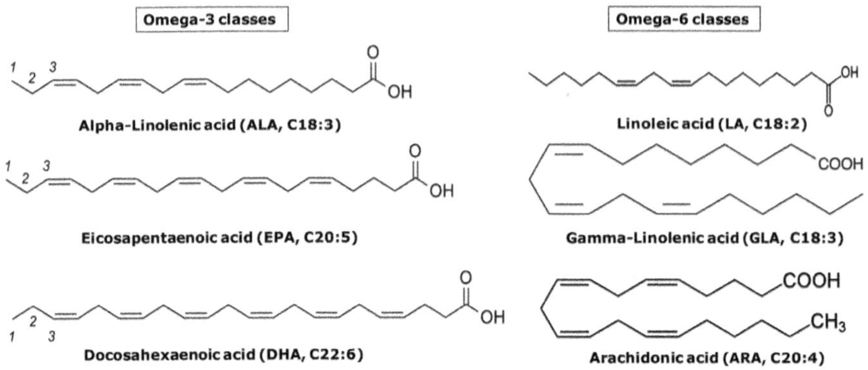

Fig. 11.2 Chemical structures of the most commonly PUFAs found in microalgae. (Source: Matos 2017)

Table 11.3 Main fatty acid proportion and omega-3/omega-6 ratio of selected microalgae

Microalgae	ΣSFA	ΣMUFA	ΣPUFA-ω3	ΣPUFA-ω6	ω-3/ω-6
Freshwater species					
Spirulina platensis	52.8	14.8	0.3	28.8	0.12
Spirulina maxima	44.9	15.7	2.2	37.0	0.06
Chlorella vulgaris[a]	30.7	9.2	40.2	18.0	2.10
Chlorella vulgaris[b]	35.9	24.0	27.8	12.2	2.27
Chlorella vulgaris[c]	14.1	52.9	7.8	24.5	0.32
Haematococcus pluvialis	21.6	37.5	16.1	24.6	0.70
Micractinium reisseri	17.1	40.1	6.4	30.8	0.21
Nannochloris bacillaris	19.5	50.9	11.6	18.0	0.65
Tetracystis sp.	22.8	49.0	11.8	17.8	0.67
Marine species					
Nannochloropsis oculata	29.8	33.6	31.8	4.6	6.88
Phaeodactylum tricornutum	30.0	27.5	36.5	5.9	6.10
Diacronema vlkianum	25.6	25.9	38.7	9.6	4.10
Isochrysis galbana	33.8	21.3	32.7	12.1	3.60
Porphyridium cruentum	35.5	5.3	7.9	51.1	0.15

Sources: [a]Matos et al. (2016), [b]Batista et al. (2013) and [c]Tibbetts et al. (2015b)

popularity of DHA + AA allows including it in infant milk formulations, which is recommended by various health and nutrition organizations (Burlingame et al. 2009).

Microalgae are the primarily source of long-chain polyunsaturated fatty acids (LC-PUFAs) that display important functions in the human metabolism. For human health, PUFA ω-3 and ω-6 are entirely derived from the diet. Since Western diet contains massive quantities of ω-6, ω-3/ω-6 fatty acid ratio of up 1:25 has been reported in the literature (Martin et al. 2006; Kleiner et al. 2015), and nutrition experts have recommended that an ω-3/ω-6 fatty acid of ≤1:5 is desirable (Burlingame et al. 2009). Therefore, nutritionists emphasize the need to consume seafood (notably fish), oleaginous seeds (linseed, Brazil nut, walnut, hazelnut, among others) and algal oil in order to supply PUFAs (Lemahieu et al. 2013; Matos 2017; da Silva Vaz et al. 2016).

11.4 General Growth Conditions Affecting Lipid Metabolism

11.4.1 Temperature

Light and temperature are probably the most important and well-studied factors influencing the lipid and fatty acid composition of algae. Changes to the lipids of photosynthetic tissues (and other organisms) as a response to different temperatures and/or light conditions have been recently studied (Simionato et al. 2011). It is believed that many of the lipid changes alter the physical properties of membranes which allows their unimpaired functioning in important physiological processes including photosynthesis, respiration, and membrane transport (Guschina and Harwood 2013).

When algae are cultured in elevated temperature, algal cells tend to synthesize higher SFAs than PUFAs. This is the case, for example, of the changes in fatty acid profile of *Nostoc flagelliforme*; when exposed to elevated temperatures (i.e., 15, 20, 25, and 30 °C), the degree of unsaturation decreased from 1.28 (15 °C) to 1.21 (30 °C) (Liu et al. 2005).

The optimum temperature for the growth of *Spirulina* is between the range of 30 and 35 °C. In this regard, outdoor pilot-scale cultivation of *Spirulina* sp. LEB-18 in two different geographic locations in Brazil (northeast and southeast) with distinct temperature (27.1–37.3 °C northeast and 19.0–25.1 °C southeast) was performed, and higher SFAs (59.2%) than (43.5%) contents in *Spirulina* sp. LEB-18 was achieved when cultivated in northeast of Brazil (de Jesus et al. 2018).

The effect of cold-water temperature on the lipid and fatty acid proportion in diatom *Porosira glacialis* has been also studied (Svenning et al. 2019). Results showed that lower temperatures (−2 °C) lead to increased levels of lipid (28.4 °C) and PUFAs (82% of total fatty acids), while at 12 °C algae produced a lipid content of 19.5% linked with higher SFAs.

11.4.2 Light

Light intensity also influences algal lipid metabolism and, therefore, lipid composition. In general, continuous light intensity (i.e., 24 h) usually lead to oxidative damage of PUFAs (Matos et al. 2017; Nwoba et al. 2019). In *Nannochloropsis* sp., the degree of unsaturation of fatty acids decreased with increasing irradiance, especially the percentage of total ω-3 fatty acids (from 29 to 8% of total fatty acids) mainly due to a decrease of eicosapentaenoic acid (Simionato et al. 2011). In general, moderate light, i.e., 12:12 light/dark is the optimum light cycle that will stimulate the algal biomass production and polar lipids, is the major lipid class. However, the stimulation of lipid and SFA synthesis are normally expected as a result of an increase in light intensity (Bonnefond et al. 2016).

11.4.3 Salt Concentrations

Salinity is usually growth limiting at the extremes of salt tolerance in some microalgae species, and every microalga has an optimum salinity range. The effect of salinity on microalgal growth relates to osmoregulation, which in microalgae is achieved through diverse strategies (Moheimani et al. 2015; Borowitzka et al. 1990). Osmoregulatory metabolites are organic substances produced by microalgae that, when the latter are exposed to water stress conditions, respond appropriately to the changes in extracellular water activity (Kumar et al. 2015). Microalgae main osmoregulators (function as intracellular osmotic regulators) are as follows: (a) polyhydric alcohols (i.e., glycerol, mannitol, or sorbitol), (b) variety of glycosides (i.e., galactosyl glycerides, floridoside, and isofloridoside), and (c) amino acids (i.e., glutamic acid and proline) (Lawton et al. 2015).

Some algae exhibit an excellent ability to tolerate high salt concentrations linked with elevated lipid content, including halotolerant *Amphora* sp. and *Navicula* sp. (salinity range from 35 to 129 ppt) and lipid content of about 40.0% and halophilic *Dunaliella salina* (salinity range from 35 to 233 pp) that produced a lipid content of 56.2% (Ishika et al. 2017b) (Table 11.4). The genus *Dunaliella*, growing in the wide range of salinities, is a good example as well as useful model in studying mechanisms of such resistance (Borowitzka 2018; Borowitzka and Siva 2007). In these algal cells, a considerably higher ratio of C_{18} (mostly unsaturated) to C_{16} (mostly saturated) fatty acids has been synthesized (Davis et al. 2015).

In another case, an addition of salt concentration from 0.5 to 1.0 M NaCl in *Dunaliella tertiolecta* medium resulted in an enhanced lipid content (from 60% to 67%). The further increase in lipid content up to 70% has been achieved when 0.5 or 1.0 M NaCl was added at mid-log phase or the end of log phase during cultivation with the initial NaCl concentration of 1.0 M (Takagi and Yoshida 2006).

11.4.4 pH

Lipids are also affected during growth of algae at extreme pH. Thus, it has been shown in *Chlorella* sp. that alkaline pH lead to increase the triacylglycerol accumulation and decreased the relative level of membrane lipids, for example, PUFAs (Lourenço 2006). Fatty acids in the polar lipids were more saturated in the unidentified *Chlamydomonas* sp. than those in *Chlamydomonas reinhardtii* when grown

Table 11.4 Average biomass and lipid productivity of salinity tolerant microalgae

Microalgae species		Salinity range (ppt)	Biomass productivity (mg L^{-1} day^{-1})	Lipid content (% of ash free dry weight)	Lipid productivity (mg L^{-1} day^{-1})
Marine	*Chrysotila carterae*	35–43	21.1	30.8	9.9
	Chaetoceros muelleri	35–55	20.5	35.4	6.9
	Nannochloropsis sp.	35–55	25.1	29.1	10.2
	Phaeodactylum tricornutum	35–65	17.7	34.6	9.1
	Tisochrysis lutea	35–75	14.4	36.0	7.8
	Tetraselmis suecica	35–109	32.0	32.9	13.6
Halotolerant	*Amphora* sp.	35–129	25.0	37.2	10.4
	Navicula sp.	35–129	25.8	43.7	11.9
Halophilic	*Dunaliella salina*	35–233	15.4	56.2	9.4

Source: Ishika et al. (2018)

under the same conditions (Tatsuzawa et al. 1996). The triacylglycerol content was also higher in *Chlamydomonas* sp. grown at pH 1.0 than that in the cells cultivated at higher pH 8.0. The increase in SFAs in membrane lipids of *Chlamydomonas* was proposed to be an adaptive reaction to low pH in order to decrease membrane lipid fluidity (Tatsuzawa et al. 1996; Guschina and Harwood 2013).

11.4.5 Nutrients

Nutrient availability affects significantly the lipid composition of algae, and a number of broad effects of nutrient limitation have been reported as important modulators of algal lipid biosynthesis (Chen et al. 2015). It is accepted that when algal growth slows down as a result of nutrient deficiency, and there is no requirement for the synthesis of new membrane compounds, the cells can transfer fatty acids into their storage lipids before conditions improve. There are many examples showing that algal species deficient in nutrients can more than double their lipid content and fatty acid composition (Isleten-Hosoglu et al. 2012).

Nutrient deficiency has been shown to affect markedly the lipid composition of microalgae under silicon, nitrogen, sulfur, or phosphorus limitation (Cheirsilp and Torpee 2012; Khademi et al. 2018). Largely, an increase in SFA accumulation and a decrease of PUFAs have been noticed in most of the nutrient-limited microalgae cultures. An increase in triacylglycerol percentages (from 69% to 75% of total lipids) together with PLs (from 6% to 8%) was reported for the microalga *Phaeodactylum tricornutum* as a result of reduced nitrogen concentration (Guschina and Harwood 2013).

11.5 Extraction of Lipids from Microalgal Biomass

Extraction has been defined as a material transportation process of one phase to another with the purpose of separating one or more substances. In the case of microalgae biomass, crude lipids are separated by a solvent, or a co-solvent, mixture from the cell biomass. During dissolution, two separate substances (the oil and the solvent) form a molecular mixture. Dissolution involves two endothermic processes and one exothermic one. First, lipid molecules separate into isolated molecules. This is an endothermic process, and its energy is small for nonpolar TAGs and slightly higher for polar lipids (i.e., PLs and GLs). Second, the separated lipid molecules are next dispersed into solvent, and energy is required to dissociate the solvent molecules. In the third process, which is exothermic, the dispersed lipid molecules interact with neighboring solvent molecules (Halim et al. 2012; Grima et al. 2013).

For complete extraction, all the linkages between the lipids and other nonpolar lipid cell components must be disrupted, and, at the same time, the disruption agent used must not cause any degradation of the lipid extracted. There are three main types of associations in which lipids participate: (*i*) hydrophobic or van de Waals interactions, in which neutral or nonpolar lipids, such as TAGs, are bound by relatively weak forces through their hydrocarbon chains to other lipids and to hydrophobic sites of proteins; (*ii*) hydrogen bonding and electrostatic association where polar lipids are bound to proteins; and (*iii*) covalent association (Grima et al. 2013).

The extraction of lipids from microalgae requires attention to their polarity. Polarity is related to the distribution of lipids within the algal cell and the association of lipids to nonlipid molecules. The microalgae species and the time of harvesting will determine the lipid composition of algae. Algae harvested in the exponential growth phase will normally contain more polar lipids (GLs and PLs) than those harvested in the late stationary phase of growth, which generally contain more neutral lipids (TAGs). In this regard, the lipid fractionation procedure and the selection of solvent will depend on the particular classes of lipids present. Overall, neutral lipids are recovered by nonpolar solvents, whereas polar lipids are recovered by polar ones (Halim et al. 2011; Medina et al. 1998; Grima et al. 2013).

11.5.1 Influence of Microalgae Cell Wall on Lipid Extraction

One of the most important factors before lipid extraction from algal biomass is the properties of the cell wall and plasma membrane of microalgae. As they play important role in the solvent extraction process, the presence of cell wall may prevent direct contact between the solvent and cell membrane and impede extraction. Knowledge of the cell wall chemistry of specific microalgae may therefore be useful because it can determine the correct implementation of pretreatment procedures (Grima et al. 2013; Kim et al. 2016).

In microalgae, the cell wall gives cells rigidity and strength, offering protection against mechanical stress. It is mainly composed of an extracellular polymeric structure based of polysaccharides, proteoglycans, peptides, proteins, and associated inorganic elements. The cell walls of marine algae, for instance, are negatively charged, which are often sulfated (Cronshaw et al. 1958; Kloareg and Quatrano 1988). In other cases, cell wall may be silicious (in the case of *Bacillariophyceae*) or calcified (as in the coccolithophores algae) (Balch 2004). Some microalgae species are naked, lacking of cell wall, for example, *Dunaliella salina*, *Isochrysis galbana*, or *Ochromonas danica*. On the contrary, the presence of insoluble nonhydrolyzable biopolymers in the outer cell walls of *Nannochloropsis* species is unusually resistant to drastic nonoxidative chemical treatment (Scholz et al. 2014). Additionally, the spherical shape and size (1–2 μm) of *Nannochloropsis* make them especially difficult to lyse (Beacham et al. 2014). Similar resistant cell wall is found in *Haematococcus pluvialis*, due to the presence of thick sporopollenin wall structure (Mendes-Pinto et al. 2001; Hagen et al. 2002). Given the large variety of cell wall structures that can be found in microalgae, it is necessary to carry out a complete characterization of them before setting up an eventual pretreatment unit for biomass (Kim et al. 2016).

11.5.2 Biomass Pretreatment for Lipid Extraction

Lipids from microalgae may often be extracted in the wet state directly after harvesting. The cells do not need to be homogenized since they are readily broken on suspension in the extracting solvent (the organic solvent seems to create channels through the cell wall). In some cases, however, breakage of the cell wall and plasma membrane may be necessary to reduce the extraction time and solvent consumption, allowing the solvent to easily penetrate into the cell and release cell contents into the bulk medium for increasing the lipid yield (Grima et al. 2013).

This process can be accomplished by using one of the well-established cell-disruption techniques (Günerken et al. 2015). A variety of disruption methods (mechanical or nonmechanical) have been currently applied for algal cell wall disruption. For instance, mechanical forces such as solid-shear forces (e.g., high speed homogenization, bead mill), liquid-shear forces (e.g., microfluidization, high pressure homogenization), energy transfer through waves (e.g., microwave, ultrasonication), currents (e.g., pulsed electric field, nonthermal plasma) or heat (e.g., autoclaving, thermolysis), and non-mechanical methods such as enzymatic cell lysis and chemical cell disruption (McMillan et al. 2013; Prabakaran et al. 2011; Günerken et al. 2015).

For large-scale cell disruption for the disintegration of algae, the mechanical method of shearing the cells with a bead mill appears more suitable. Bead mills have been used to disintegrate the algae *Chlorella vulgaris* (Postma et al. 2015), *Nannochloropsis oculata* (Montalescot et al. 2015), and *Haematococcus pluvialis* (Razon and Tan 2011). Ultrasound or sonication microalgal disruption pretreatment is another liquid-shear method that has also been applied for microalgae with less resistant cell wall and normally at laboratory scale. This disruption technique works at high acoustic power breaking down the cell wall, after the algal suspension induced at sound frequencies above 20 kHz (Kurokawa et al. 2016; Zheng et al. 2011; Gerde et al. 2012).

11.5.3 Ideal Solvent to Extract Lipids from Microalgae

The ideal solvent characteristics must follow several important factors: they should be volatile (for easy removal later), free from toxic or reactive impurities (to avoid reaction with the lipids), able to form a two-phase system with water (to remove nonlipids), and be ineffective extractors of undesirable components (e.g., proteolipid protein, small molecules). A solvent with high solubility at elevated temperature and low solubility at ambient temperature may be desirable because phase separation of oil from solvent would occur without necessity of evaporation. The lower alcohols exhibit this characteristic. In addition, and most important, the solvent must be nontoxic to workers. When the concern is only about the oil for biofuels, it is desirable to use solvents, which extract only TAGs and leave polar lipids (PLs and GLs), fatty acids, and pigments in the spent biomass. To date, the cost is a favorable factor for hexane use; however, concerns about its availability, tighter emission restrictions, safety, and its testing as a hazardous air pollutant have stimulated interest in alternative solvents (Grima et al. 2013; Medina et al. 1998).

Lipids may also be retained in living matter by mechanical confinement. For instance, in the case of algal cell wall with poor permeability to solvents, the lipid recovery yield may be increased by adding a small amount of water to the extraction phase. Water causes swelling of the cellular structures rich in polysaccharides, thereby increasing the extent of dispersion of the living systems and facilitating access of extraction solvents to the lipids (Grima et al. 2013).

11.5.4 Extraction Using Organic Solvent

The use of organic solvents to extract lipids from microalgal biomass is widely utilized and includes hexane, chloroform, and methanol as the main solvents. One of the first and most efficient extraction systems used for microorganisms was the mixture of chloroform-methanol 2:1 v/v (Folch et al. 1957). This procedure was simplified and improved by Bligh and Dyer method that has been used for extracting lipids over the last 60 years. This method uses a single-phase extraction, chloroform-methanol-water 1:2:0.8 (v/v/v/), and then diluted it with a chloroform and water mixture to form a biphasic system in a proportion 1:1:0.9 (v/v/v/), chosen in such a manner that the lower layer is practically 100% chloroform and contains the purified lipids and the upper layer nearly all methanol-water and contains the nonlipids (Bligh and Dyer 1959). The method is applicable for many biological materials including animal, vegetal, or microorganism origin and generally takes only 1 or 2 h extraction time at room temperature. In general, this method has been utilized for microalgae as a routine procedure of lipid extraction at analytical scale (Grima et al. 2013).

When a microalgal cell is exposed to a nonpolar organic solvent, such as hexane or chloroform, the organic solvent penetrates through the cell membrane into the cytoplasm (step 1) and interacts with the neutral lipids (step 2) to form an organic solvent-lipid complex (step 3). This organic solvent-lipid complex, driven by a concentration gradient, diffuses across the cell membrane (step 4) and the static organic solvent film surrounding the cell (step 5) into the bulk organic solvent (Fig. 11.3).

A suitable organic solvent should be insoluble in water, preferentially solubilize the lipid compound, have a low boiling point to facilitate its removal after extraction, and have a considerably different density than water (Mercer and Armenta 2011). Furthermore, for process cost-effectiveness, it should be easily sourced and reusable as well as with low price. Due to these qualities, hexane is typically the solvent of choice for large-scale lipid extractions (Cheng et al. 2011), although it has been also used at analytical laboratory scale. As examples, a wet lipid extraction procedure with hexane was developed that was capable of extracting 59.3% of lipids from wet mixed *Chlorella* and *Scenedesmus* sp. biomass (84% moisture) via acid and base hydrolysis (90 °C and ambient pressure), and 76% of those extracted lipids were isolated, by further processing, and converted to FAMEs (Sathish and Sims

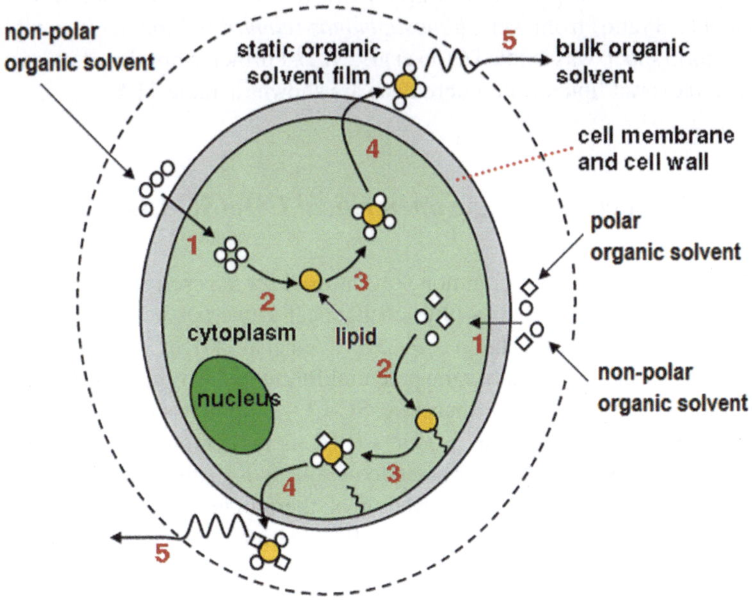

Fig. 11.3 Schematic diagram of the proposed organic solvent mechanism on algal lipid extraction. (Source: Based on Feller (2017) and further improved redrawn from Halim et al. (2012))

Table 11.5 Examples of lipid content in microalgae extracted with organic solvents

References	Technique	Solvent	Lipid content (%)	Microalgae
Sathish and Sims (2012)	Wet lipid extraction with acid/base hydrolysis	Hexane	59.3	*Chlorella*/*Scenedesmus* sp.
Yoo et al. (2012)	Wet lipid extraction with osmotic shock assisted	Hexane/methanol (7:3 v/v)	9.0	*Chlamydomonas reinhardtii*
Taher et al. (2014)	Soxhlet apparatus	Hexane	22.0	*Scenedesmus* sp.
Šoštarič et al. (2012)	Lyophilized biomass in combination with grinding and microwaves	Chloroform/ methanol (1:1 v/v)	9.6	*Chlorella vulgaris*
Halim et al. (2011)	Static hexane lipid extraction	Hexane/ isopropanol (3:2 v/v)	7.1	*Chlorococcum* sp.
Dejoye Tanzi et al. (2013)	Simultaneous distillation and extraction process	Terpene solvent (p-cymene, d-limonene)	21.4	*Nannochloropsis oculata*
Morioka et al. (2014)	Soxhlet apparatus	Hexane	23.3	*Chlorella* sp.
Carvalho Júnior et al. (2011)	Soxhlet apparatus	Methanol/HCl/ Chloroform (10:1:1)	23.0	*Nannochloropsis oculata*

2012). In another study, direct wet lipid extraction (9.3%) with optimized hexane/methanol (7:3 ratio) from wet *Chlamydomonas reinhardtii* biomass in combination with osmotic shock was studied (Yoo et al. 2012). Further examples of using organic solvents to extract lipids from microalgae are shown in Table 11.5.

11.5.5 Extraction Using Supercritical CO_2 (SC-CO_2)

Carbon dioxide is the most common solvent used for supercritical fluid extraction, mainly due to its advantageous physicochemical properties near the critical point and its chemical inertness (Halim et al. 2011; Yen et al. 2015). SC-CO_2 extraction is an emerging technology that has the potential for replacing organic solvents largely utilized for lipid extraction. Furthermore, SC-CO_2 method has received wide acceptance for extracting high-value products (PUFAs and carotenoid pigments) from microalgae biomass, which is attributed to its selectivity, solvent-free residue occurrence, rapid separation, and preservation of thermolabile products (Mercer and Armenta 2011; Reverchon and De Marco 2006; Herrero and Ibáñez 2015). The fluid

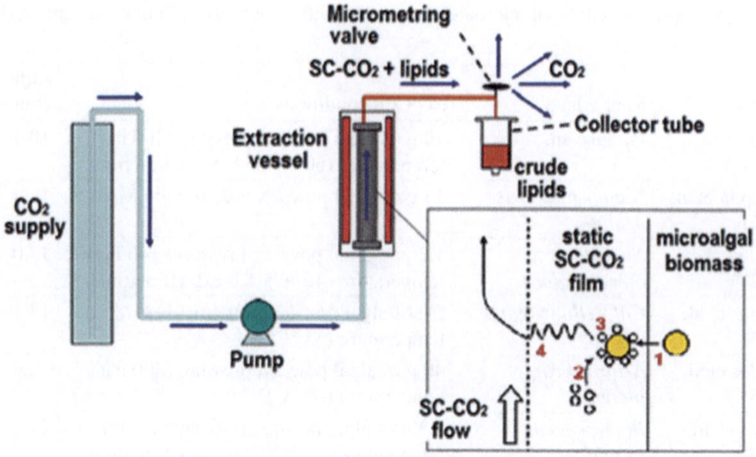

Fig. 11.4 Schematic diagram of a laboratory-scale supercritical ($SC\text{-}CO_2$) extraction system combined with the proposed $SC\text{-}CO_2$ microalgal lipid extraction mechanism. (Adapted from Feller 2017)

is called supercritical because it is above of its critical temperature (Tc) and critical pressure (Pc). When the temperature and pressure of a fluid are increased above their critical values (Tc and Pc), the fluid goes to the supercritical region (Mendes et al. 2003).

Typically, the extraction system using supercritical fluids consists of four components: gas supply, high-pressure pump, extraction vessel, and collector. Figure 11.4 shows a supercritical extraction system using CO_2 as solvent.

During the lipid extraction process, the microalgal biomass are packed inside the extraction vessel. The fluid travels on the surface of the biomass, and the lipid fraction is desorbed from the microalgal biomass. Immediately upon dissolution, the $SC\text{-}CO_2$ encloses the lipids to form a $SC\text{-}CO_2$ lipid complex. The complex, driven by concentration gradient, diffuses across the static $SC\text{-}CO_2$ film and enters the bulk $SC\text{-}CO_2$ flow. The $SC\text{-}CO_2$-lipid mixture then leaves the extraction vessel to enter the collection vessel, where a micrometering valve is used to rapidly depressurize the incoming fluid. Upon complete depressurization, the $SC\text{-}CO_2$ returns to gaseous state, and the extracted crude lipids precipitate in the collector vessel. As crude lipids derived from supercritical fluid extraction are free from any extraction solvent, it does not need to undergo an extraction solvent removal step (Mendes et al. 2003; Halim et al. 2012). Examples of $SC\text{-}CO_2$ applied to lipid extraction from microalgae are given in Table 11.6.

Table 11.6 Lipid content from microalgae extracted with supercritical fluid using carbon dioxide ($SC\text{-}CO_2$)

References	Microalgae	$SC\text{-}CO_2$ conditions	Lipid content (%)
Cheng et al. (2011)	*Pavlova* sp.	10 g of algal powder, pressure (360 bar), temperature (60 °C), 6 h extraction time.	10.4
Crampon et al. (2013)	*Nannochloropsis oculata*	13 g of algal powder, pressure (45 Mpa)	19.8
Hegel et al. (2017)	*Neochloris oleoabundans*	1.2 g of algal powder, pressure (140 bar), temperature (40 °C), 1 h extraction time.	17.0
Mendes et al. (1995)	*Chlorella vulgaris*	5 g of algal powder, pressure (35.0 Mpa), temperature (55 °C).	13.3
Mendes et al. (2005)	*Arthrospira platensis*	40 g of algal powder, pressure (300 bar), temperature (60 °C).	0.6
Paudel et al. (2015)	*Botryococcus braunii*	0.3 g of algal powder, pressure (7.2 MPa), temperature (35 °C), 2 h extraction time.	24.0
Santana et al. (2012)	*Botryococcus braunii*	5 g of algal powder, pressure (250 bar), temperature (80 °C).	14.1
Tang et al. (2011)	*Schizochytrium limacinum*	5 g of algal powder, pressure (40 MPa), temperature (40 °C), 2 h extraction time.	33.0
Viguera et al. (2016)	*Chlorella protothecoides*	2.5 g of algal powder, pressure (300 bar), temperature (70 °C).	21.0
Patil et al. (2018)	*Nannochloropsis salina*	30 g of algal powder, pressure (340 bar), temperature (60 °C), 1 h extraction time.	31.3
Morioka et al. (2014)	*Chlorella* sp.	10 g of algal powder, pressure (306 bar), temperature (60 °C), 4 h extraction time.	4.5

11.5.6 Extraction Using Subcritical Fluids

Apart from carbon dioxide, other compressed solvents including ethane, propane, *n*-butane, dimethyl ether, etc. can be used to extract lipophilic compounds, specially PUFAs and carotenoids from microalgae (dos Santos Freitas et al. 2008; Goto et al. 2015; Feller et al. 2018). The use of these gases as compressed fluids to replace carbon dioxide usage differs from the work in the subcritical region as they are compressed liquids at least one order of magnitude below the critical point. In addition, compressed solvents such as liquefied *n*-butane have gentle vapor pressure, close chemical structure, low price, and classified as authorized solvent for foodstuff production without limitation (Novello et al. 2015; Rapinel et al. 2017).

A brief review of the state-of-the-art extraction and fractionation of selected lipids using near critical fluids reported that propane and dimethyl ether can be used without co-solvents to extract polar lipids and other complex lipids from a range of raw materials, including dairy powders, eggs, biomass from fermentation, and wet marine biomass (Catchpole et al. 2009). The use of $SC\text{-}CO_2$ together with compressed solvent dimethyl ether (DME) as environmental friendly extraction processes for bioactive compounds (lipids and carotenoids) from wet algae has been studied (Goto et al. 2015). Using *n*-butane as subcritical solvent, higher content of

PUFAs ω3 and ω6 in three marine microalgae (*Nannochloropsis oculata*, *Phaeodactylum tricornutum*, and *Porphyridium cruentum*) was obtained in comparison to supercritical CO_2 (Feller et al. 2018), exemplifying the potential application of compressed solvents for the extraction of lipophilic compounds from microalgae.

Some reports have documented the efficiency of subcritical water extraction (SWE) for the extraction of intracellular lipids from microalgae (Reddy et al. 2014; Rodríguez-Meizoso et al. 2010; Ju et al. 2013). Recent developments in subcritical extraction solvents utilize subcritical alcohol solvents as a single-step process for extraction of algal lipids converting them into biodiesel (Aghilinategh et al. 2019; Thiruvenkadam et al. 2019). A single-step extraction and conversion of wet *Nannochloropsis gaditana* to fatty acid methyl esters under subcritical methanol extraction (SME) has been demonstrated (Sitthithanaboon et al. 2015). In this case, SME works at elevated temperature (225 °C) under high pressure (600–900 psi), and the methanol hydrolysis and free fatty acids esterification are simultaneously processed in a subcritical state allowing the production of in situ algal biodiesel. This process does not require a catalyst and any drying pretreatment steps, being a cost-saving and environmentally friendly process.

11.5.7 Extraction Using Ionic Liquids

Ionic liquids (ILs) are salts comprised of relatively large asymmetric organic cation coupled with smaller inorganic or organic anions that remain liquid at moderate to room temperatures (0–140 °C) (Young et al. 2010). The cation is generally composed of a nitrogen containing a ring structure (e.g., imidazolium or pyridine) to which a broad range of functional side groups can be attached. Depending upon the structure of the side group added, the polarity of the resulting IL can vary. The anions can vary from single ions such as chloride to larger complex ions such as $[N(SO_2CF_3)_2]^-$ (Fauzi and Amin 2012; Young et al. 2010). Table 11.7 represents cation and anions commonly used to prepare ILs.

The anion and cation can be modified to alter the relative hydrophobic/hydrophilic nature of the resulting IL (Freire et al. 2007). Since ILs are charged species that strongly self-associate, they also possess extremely low vapor pressure. Combined, these features contribute to their growing reputation as "green solvents" that possess relatively no vapor pressure, low toxicity, and capacity to be tailored for a specific solubility, polarity, electrical conductivity, and relative hydrophobicity (Ventura et al. 2017).

ILs have excellent properties for use in cellulosic biomass treatment and thus have been shown to extract lipids from *Chlorella vulgaris* as these algae contain lignocellulosic materials in their cell wall (Pan et al. 2016). In addition, hydrogen bonds of microalgae cell walls are affected by the ions of ILs, which modify the cell wall structure allowing to extract lipids more easily. In addition, there is a possibility to blend two ILs in order to change the algal cell wall facilitating the

Table 11.7 Commonly used cation and anions in ionic liquids (ILs)

Cation	Molecular structure	Anions	Molecular structure
Pyridinium	$[C_5H_5NH]^+$	Trifluoromethanesulfonate	$[CF_3SO_3^-]$
Imidazole	$[C_3N_2H_4]$	Methyl sulfate	$[CH_4O_4S]$
Pyrrolidinium	$[C_4H_{10}N]^+$	Methane sulfonate	$[CH_3O_3S^-]$
Pyrazolium	$[C_3H_5N_2]$	Tetrachloroaluminate	$[AlCl_4]^-$
Ammonium	$[NH_4^+]$	Hexafluoroantimonate	$[F_6Sb]^-$
Cholinium ethanoate	$[C_9H_{21}NO_5]$	Hexafluorophosphate	$[PF_6]^-$
Phosphonium	$[PH_4^+]$	Tetrafluoroborate	$[BF_4]^-$
		Chloride	$[Cl]^-$
		Bromide	$[Br]^-$
		Iodide	$[I]^-$
		Thiocyanate	$[SCN]^-$
		Nitrate	$[NO_3]^-$
		Hydrogen sulfate	$[HSO_4]^-$
		Dihydrogen phosphate	$[H_2PO_4]^-$

Source: Fauzi and Amin (2012)

permeability of IL through the algal membrane for maximum lipid extraction (Choi et al. 2014b; Vanda et al. 2018). In one study (Choi et al. 2014b), authors investigated the yield-enhancing effect of ILs on the extraction of lipids and fatty acids from *Chlorella vulgaris*. The fatty acid content of the *Chlorella vulgaris* was above 200 mg/g cell using 1-ethyl-3-methylimidazolium acetate ([EMIM] [Ac]), 1-ethyl-3-methylimidazolium diethyl phosphate ([EMIM] [DEP]), 1-ethyl-3-methylimidazolium tetrafluoroborate ([EMIM] [BF4]), and 1-ethyl-3-methylimidazolium chlorides ([EMIM] [Cl]), while the yield by hexane-methanol solvent was 185.4 mg/g cell. In another study (Pan et al. 2016), the lipid content of three algal species (*Chlorella sorokiniana*, *Nannochloropsis salina*, and *Galdieria sulphuraria*) was extracted with microwave-assisted extraction in combination with 1-butyl-3-methylimidazolium hydrogen sulfate ([BMIM][HSO4]), a cationic based IL.

In some cases, IL can also act as catalyst in the conversion of algal rich-lipid biomass to biodiesel, and this technology is promising (Zhao and Baker 2013). In this case, ILs act as homogeneous catalyst because they are in the same phase with reactants. In addition, the ability to recycle ILs at the end of reaction and easing the separation of products are benefits obtained if heterogeneous catalyst is used (Fauzi and Amin 2012).

Another class of ILs are switchable solvents (SS) that have the ability to change their neutral state or ionic state after an introduction of an external stimulus. They are also known as reversible ionic liquids (RILs) as the ionic solvent can be restored to its original nontoxic state due to the reversibility of the reaction (Fauzi and Amin 2012). The DBU-alkyl carbonate salt is the most known switchable solvent and is usually obtained from a 1:1 mixture of two neutral liquid components, 1,8-diazabic yclo-[5.4.0]-undec-7-ene (DBU) and an alcohol whose chain length can be chosen to finely tune the polarity of the nonionic form and the meting point of the ionic

Table 11.8 Lipid content of microalgae extracted with some emerging technologies

References	Microalgae	Lipid yield (mg/g)	Emerging technology and conditions
Feller et al. (2018)	*Phaeodactylum tricornutum* *Nannochloropsis oculata* *Porphyridium cruentum*	13 17 3	Subcritical *n*-butane solvent, 20 g biomass, 300 bar, 40 °C, 2 h extraction time
Thiruvenkadam et al. (2019)	*Chlorella pyrenoidosa*	14	Subcritical methanol extraction, solvent to algae ratio (3:1), 160 °C, 3 min reaction time
Sitthithanaboon et al. (2015)	*Nannochloropsis gaditana*	140	Subcritical methanol extraction, solvent to algae ratio (6:1), 225 °C, 90 min reaction time
Choi et al. (2014b)	*Chlorella vulgaris*	17–255	Ionic liquids (cations based on imidazolium), 120 °C, 2 h extraction time
Kim et al. (2012)	*Chlorella vulgaris*	119–125	Ionic liquid and methanol system (1:1) volume ratio, 65 °C, 18 h reaction time
Pan et al. (2016)	*Chlorella sorokiniana* *Nannochloropsis salina* *Galdieria sulphuraria*	230 100 190	Ionic liquid solvent ([BMIM, HSO_4]), 120 °C, 60 min reaction time under microwave irradiation (800 W)
Samorì et al. (2010)	*Botryococcus braunii*	160	Extraction with switchable-polarity solvents (1,8-diazabicyclo-[5.4.0]-undec-7-ene (DBU)/methanol)
Choi et al. (2014a)	*Chlorella vulgaris*	223	Molten-salt/ionic liquid mixtures blended as (FeC13·H2O:[Emin]OAc = 5:1)

form (Samorì et al. 2010). Despite the increasing interest toward SS, there are few studies that have shown the application of SS for algal lipid extraction. In particular, BDU/octanol and DBU/ethanol SS were tested for the extraction of lipids from freeze-dried *Botryococcus braunii* algae and compared with *n*-hexane and chloroform/methanol, and results have shown that DBU/octanol exhibited higher lipid yields (16.0%) in comparison to *n*-hexane (7.8%) (Samorì et al. 2010). Further examples of some emerging technologies used to extract lipids from microalgae are shown in Table 11.8.

11.6 Biodiesel from Microalgae

The growing need for energy and the adverse effects of the indiscriminate uses of fossil fuel are well known. This concern has been the engine for developing alternative research on renewable energy, including biodiesel from microalgae. Biodiesel production has been gradually increased and considered as an alternative to fossil fuels linked with reduction of greenhouse gas emissions. The use of biofuel has been a fundamental tool for the reduction of CO_2 emissions in accomplishment with commitments such as the Kyoto Protocol (Grubb et al. 1997).

Biodiesel synthesis is conducted through the transesterification reaction or hydro-esterification (hydrolysis followed by the esterification of the fatty acids) and the triglycerides being the fundamental raw material. The main sources of triglycerides for biodiesel production are vegetable oils such as soy, rapeseed, palm, and sunflower (Huber and Corma 2007; Vieitez et al. 2008a, b). However, concerns about potentially uses of these edible materials have been raising up, mainly due to more than 95% of worldwide biodiesel is produced from edible oils (Brennan and Owende 2010).

Advantages of biodiesel include renewability, local or domestic origin, low or no sulfur content, no aromatic content, biodegradability, reduction of most regulated exhaust emissions, inherent lubricity, high flash point (leading to safer handling), and compatibility with the existing fuel distribution infrastructure. Technical issues that beset biodiesel include poor cold flow properties and low oxidative stability as well as a slight increase in regulated nitrogen oxide (NO_x) exhaust emissions (Knothe 2011). Besides these issues, other aspects such as feedstock supply and availability, food vs. fuel, and economics have affected biodiesel.

Liquid biofuels can be categorized into different generations according to the type of the raw materials. First-generation biofuels come from vegetable oils (edible raw materials), while the second-generation ones come from nonedible raw materials, such as sugarcane waste. Third-generation biofuels, for instance, are derived from microorganisms including micro- and macro-algae, which proves to have adequate properties as biofuel (Daroch et al. 2013; Enamala et al. 2018).

In this way, cultivation of microalgae for biodiesel production possesses some advantages over plant crops, such as the following: (*i*) microalgae do not compete for land plant crops, being able to grow in nonarable such as arid and semi-arid regions, (*ii*) microalgae have rapid growth with short life cycle, generally 7 days, (*iii*) cultivation of microalgae does not require herbicides or pesticides, and (*iv*) microalgae can be grown in a variety of environmental conditions, including wastewaters often not suitable for other plant crops (Patil et al. 2008; Rodolfi et al. 2009; Huang et al. 2010).

A comparison of oil productivity in oleaginous microalgae and different oily plant materials are shown in Table 11.9.

11.6.1 Transesterification of Algal Oil for Biodiesel

Biodiesel is a biofuel consisting of methyl esters that are derived from organic oils, vegetable oils, or animal fats. The biodiesel from algal oil in itself is not significantly different from biodiesel produced from vegetable oils (Carvalho Júnior et al. 2011; Martínez et al. 2017). The vegetable oils and fats as alternative engine fuels

Table 11.9 Potential of oil production from microalgae and crop plants

Crop	Lipid content (%)	Gallons of oil/acre/year
Corn	44	68
Soy	18	252
Jatropha	28	296
Canola	41	388
Sunflower	40	428
Palm	36	2144
Microalgae	30–50	Up to 6000 (with future technology)

Source: Adapted from Waltz (2009) and Mata et al. (2010)

are all generally viscous (ASTM D6751 = 1.9–6.0 mm^2/s at 40 °C), with viscosities higher than petroleum diesel fuel (ADTM D975 = 1.3–2.4 mm^2/s at 40 °C) (Demirbas 2008a). Biodiesel is produced by transesterifying the parent oil or fat to achieve a viscosity close to that of petrodiesel. The chemical conversion of the oil into its corresponding fatty ester (biodiesel) is called transesterification (Demirbas 2011; Haas and Wagner 2011). The purpose of the transesterification process is to lower the viscosity of the oil (Demirbas and Demirbas 2011). The transesterification reaction proceeds with or without a catalyst using primary or secondary monohydric aliphatic alcohols having one to four carbon atoms. The biodiesel transesterification reaction is very simple according to the following demonstration (Demirbas 2008a):

$$\text{Triglyceride} + 3\ \text{Methanol} \xrightarrow{\text{Catalyst}} \text{Glycerine} + 3\ \text{Methyl esters}\left(\text{Biodiesel}\right)$$

An excess of methanol is normally used to force the reaction to favor the right side of the equation. The process of making biodiesel occurs as follows: (*i*) triglyceride, methanol, and catalyst are placed in a controlled reaction chamber to undergo transesterification, (*ii*) the initial products is placed in a separator apparatus to remove the glycerin byproduct, (*iii*) the excess methanol is recovered from the methyl esters through evaporation, and (*iv*) the final biodiesel is rinsed with water, neutralized and dried (Xu et al. 2006; Sati et al. 2019).

Xu et al. (2006) studied the biodiesel production from heterotrophic cultivation of *Chlorella protothecoides*. Biomass was powdering in a mortar prior of lipid extraction conducted in a Soxhlet apparatus with *n*-hexane, giving a lipid content of 55% of dry weight. Biodiesel was obtained via acidic transesterification. The optimum process combination was 100% catalyst quantity (based on oil weight) with 56:1 molar ratio of methanol to oil at a temperature of 30 °C, which reduced product-specific gravity from an initial value of 0.91 to a final value of 0.84 in about 4 h of reaction time. Diesel has specific gravity between 0.82 and 0.95, while vegetable oils typically have a specific gravity from 0.90 to about 0.92, depending on its fatty acid composition and measured temperature (Demirbas 2008b). Figure 11.5 exemplifies the process flow for biodiesel production from microalgae.

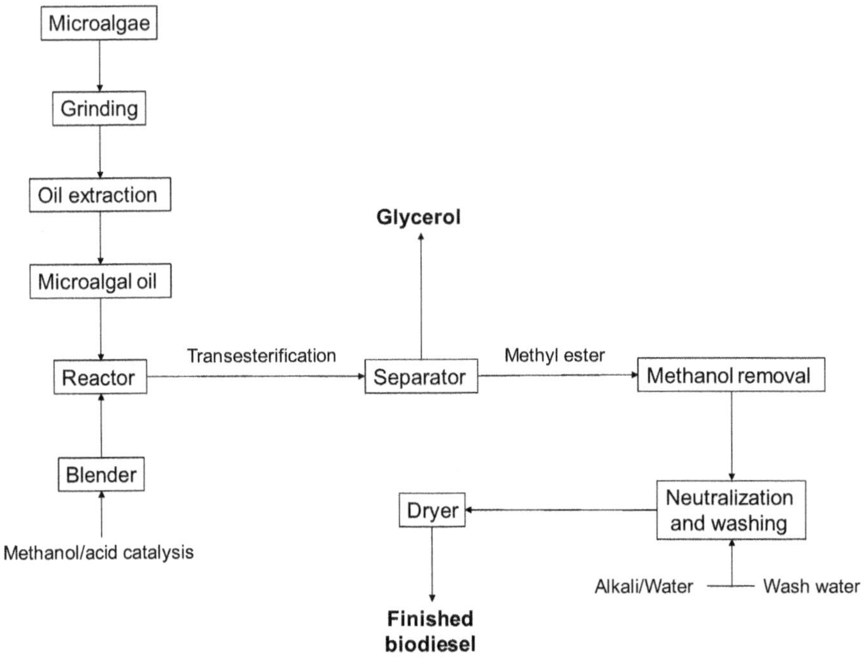

Fig. 11.5 Process flow for algal biodiesel production. (Source: Redrawn and adapted from Demirbas and Demirbas (2010))

Biodiesel standards and specifications such as American Society for Testing and Materials (ASTM D6751) and European Committee for Standardization (EN 14214) have been developed in the United States and Europe, respectively, as well as elsewhere around the world to ensure biofuel quality (Knothe 2011).

The biofuel properties (cetane number, kinematic viscosity, oxidative stability, cold flow) are largely determined in biodiesel ensuring biofuel quality (Knothe 2006; Knothe et al. 2015). Cetane number (CN) is related with chemical structural chains, and longer hydrocarbon chains are "ideal" for diesel fuels (Knothe et al. 2003). Viscosity is the major reason why vegetable oils are transesterified to biodiesel, as the viscosity of biodiesel is closer to that of petrodiesel (Knothe and Steidley 2005). Cold flow is another important biodiesel parameter, generally characterized by the temperature in which biodiesel starts to change from fluid to solid state. Oxidative stability, together with cold flow parameter, has been one of the major issues affecting its more widespread uses and commercialization. While saturated fatty acid chains are problematic for cold flow, unsaturated fatty acid chains, especially PUFA chains, cause the low oxidative stability of many biodiesel fuels (Knothe 2005, 2011).

The microelements Na, K, S, P, Ca, and Mg can carry over from the feedstock to the fuel (or, in some cases such as Na or K, be traces of catalyst). Comprehensive evaluation of some chemical microelements in marine microalgae for biodiesel production has been conducted by Silva et al. (2015). Such information is essential as these microelements can negatively affect combustion giving by ash exhaust emission (Knothe et al. 2006).

11.6.2 Economic Aspect of Algal Biodiesel

Biodiesel production and use increased exponentially in the late 2000s with additional legislative and regulatory incentives. There are few numbers of economic feasibility studies on microalgae oil for biodiesel production (Demirbas 2017; Gallagher 2011; Li et al. 2012; Stephens et al. 2010; Mascarelli 2009). Currently, microalgae biofuel has not been deemed economically feasible compared to the conventional agricultural biomass (Knothe 2011; Waltz 2009).

The concept of using oily algae to make biofuel was first discussed more than 50 years ago, but a concerted effort began with the oil crisis in the 1970s (Hu et al. 2008). The US Department of Energy (DOE) from 1978 to 1996 devoted \$25 million to algal fuels research in its aquatic species program at the National Renewable Energy Lab (NREL) in Golden, Colorado. The program, however, yielded important advances that set the stage for algal biofuel research today (Waltz 2009).

According to Waltz (2009), an informal calculation conducted by scientists at Sandia National Laboratories (SNL, United States) estimate that biodiesel from microalgae grown on open system may cost \$10–20 per gallon (4.4 L), while closed systems the values reach up to \$20–40 per gallon, which is much higher than conventional diesel fuel (\$2–3 per gallon). Comparatively, biodiesel from palm oil costs around \$4–5 per gallon, whereas biodiesel produced from rapeseed and soybean oils may cost up to \$5–6 per gallon (https://www.globalccsinstitute.com/).

A comprehensive techno-economic analysis and life cycle assessment for algal fuels and feed production in a 100-ha facility has been conducted by Beal et al. (2015) with ten distinct scenarios. Authors demonstrated that many progress has been made in the field of algal biofuels and bioproducts. Nevertheless, the large capital cost and high risk associated with a large-scale production facility still deter realized algae production at commodity scale. Moreover, to make algal biofuel realistic, further improvements and potential advances should be made on (*i*) producing additional high-value co-products, such as omega-3 supplements, pharmaceuticals, or cosmetics applications, (*ii*) improving biomass productivity over 25 g/m^2-day, (*iii*) developing automated technologies to reduce labor costs, (*iv*) creating advanced pond litter materials and pond designs to reduce capital costs below \$2/m^2, (*v*) integrating algae production with low-cost renewable electricity, (*vi*) using atmospheric carbon and/or waste forms of carbon, nitrogen, and phosphorous, (*vii*) reducing pipe costs using canals or low-cost materials, and (*viii*) downstream algal processing improvements.

11.7 Concluding Remarks and Future Prospects

Knowledge of how various cultivation conditions as well as modulation of temperature, light, pH, salinity, and nutrients highly affects algal lipid content and fatty acid composition. The fatty acid profile will determine the application of algal lipid in terms of optimal biofuel production or bioproduct destination.

Several "green solvents" have been applied to extract lipids and fatty acids, especially polyunsaturated, from microalgae, including supercritical CO_2, subcritical fluids, and ionic liquids. Despite their greener appeal, organic solvent extraction, especially *n*-hexane, has proved to be the most workable approach so far.

Direct transesterification of wet algal biomass seems to be the most suitable method for processing lipids from microalgae for biofuel purpose. Lipid extraction followed by conversion into fatty acid methyl esters may be very useful when considering a prior fractionation of lipid classes in such a way that the enriched SFA and MUFA fractions would be exploited for biofuels and the high-grade PUFA fractions can be sold as nutraceutical at much higher prices than biofuels.

Many algal species such as *Isochrysis galbana*, *Nannochloropsis*, and *Schizochytrium*, which have received some attention in the literature, have high amount of PUFAs (C18-C22 carbons chain), and they are more suitable for food, feed, and fine chemical industry applications. On the other hand, *Chlorella* and *Scenedesmus* species have relatively content of C16-C18 fatty acids, which are the most common fatty acid chain length in vegetable oils, and thus may be used toward biodiesel production. Nevertheless, biofuel properties in terms of cetane number, viscosity, cold flow, oxidative stability, lubricity, and exhaust emissions must be rigorously conducted in order to validate algal biodiesel.

References

Aghilinategh M, Barati M, Hamadanian M (2019) Supercritical methanol for one put biodiesel production from chlorella vulgaris microalgae in the presence of CaO/TiO_2 nano-photocatalyst and subcritical water. Biomass Bioenergy 123:34–40

Balch WM (2004) Re-evaluation of the physiological ecology of coccolithophores. In: Coccolithophores. Springer

Batista AP, Gouveia L, Bandarra NM, Franco JM, Raymundo A (2013) Comparison of microalgal biomass profiles as novel functional ingredient for food products. Algal Res 2:164–173

Beacham TA, Bradley C, White DA, Bond P, Ali ST (2014) Lipid productivity and cell wall ultrastructure of six strains of Nannochloropsis: implications for biofuel production and downstream processing. Algal Res 6:64–69

Beal CM, Gerber LN, Sills DL, Huntley ME, Machesky SC, Walsh MJ, Tester JW, Archibald I, Granados J, Greene CH (2015) Algal biofuel production for fuels and feed in a 100-ha facility: a comprehensive techno-economic analysis and life cycle assessment. Algal Res 10:266–279

Bligh EG, Dyer WJ (1959) A rapid method of total lipid extraction and purification. Can J Biochem Physiol 37:911–917

Bondioli P, Bella LD, Rivolta G, Zittelli GC, Bassi N, Rodolfi L, Casini D, Prussi M, Chiaramonti D, Tredici MR (2012) Oil production by the marine microalgae Nannochloropsis sp. F&M-M24 and Tetraselmis suecica F&M-M33. Bioresour Technol 114:567–572

Bonnefond H, Moelants N, Talec A, Bernard O, Sciandra A (2016) Concomitant effects of light and temperature diel variations on the growth rate and lipid production of Dunaliella salina. Algal Res 14:72–78
Bonnefond H, Moelants N, Talec A, Mayzaud P, Bernard O, Sciandra A (2017) Coupling and uncoupling of triglyceride and beta-carotene production by Dunaliella salina under nitrogen limitation and starvation. Biotechnol Biofuels 10:25
Borowitzka MA (2018) The 'stress' concept in microalgal biology—homeostasis, acclimation and adaptation. J Appl Phycol 30:2815–2825
Borowitzka MA, Siva CJ (2007) The taxonomy of the genus Dunaliella (Chlorophyta, Dunaliellales) with emphasis on the marine and halophilic species. J Appl Phycol 19:567–590
Borowitzka MA, Borowitzka LJ, Kessly D (1990) Effects of salinity increase on carotenoid accumulation in the green alga Dunaliella salina. J Appl Phycol 2:111–119
Bredda EH, Da Rós PCM, Pedro GA, de Castro HF, Silva MB (2018) Nannochloropsis gaditana and Dunaliella salina as feedstock for biodiesel production: lipid production and biofuel quality. J Adv Biol Biotechnol 20(3):1–10
Brennan L, Owende P (2010) Biofuels from microalgae—a review of technologies for production, processing, and extractions of biofuels and co-products. Renew Sust Energ Rev 14:557–577
Burlingame B, Nishida C, Uauy R, Weisell R (2009) Fats and fatty acids in human nutrition: introduction. Ann Nutr Metab 55:5–7
Byrd SM, Burkholder JAM, Zimba PV (2017) Environmental stressors and lipid production by Dunaliella spp. I. Salinity. J Exp Mar Biol Ecol 487:18–32
Catchpole OJ, Tallon SJ, Eltringham WE, Grey JB, Fenton KA, Vagi EM, Vyssotski MV, MacKenzie AN, Ryan J, Zhu Y (2009) The extraction and fractionation of specialty lipids using near critical fluids. J Supercrit Fluids 47:591–597
Cheirsilp B, Torpee S (2012) Enhanced growth and lipid production of microalgae under mixotrophic culture condition: effect of light intensity, glucose concentration and fed-batch cultivation. Bioresour Technol 110:510–516
Chen Y, Tang X, Kapoore RV, Changan X, Vaidyanathan S (2015) Influence of nutrient status on the accumulation of biomass and lipid in Nannochloropsis salina and Dunaliella salina. Energy Convers Manag 106:61–72
Cheng C-H, Tz-Bang D, Pi H-C, Jang S-M, Lin Y-H, Lee H-T (2011) Comparative study of lipid extraction from microalgae by organic solvent and supercritical CO2. Bioresour Technol 102:10151–10153
Chisti Y (2007) Biodiesel from microalgae. Biotechnol Adv 25:294–306
Chisti Y (2008) Biodiesel from microalgae beats bioethanol. Trends Biotechnol 26:126–131
Choi S-A, Lee J-S, You-Kwan O, Jeong M-J, Kim SW, Park J-Y (2014a) Lipid extraction from Chlorella vulgaris by molten-salt/ionic-liquid mixtures. Algal Res 3:44–48
Choi S-A, You-Kwan O, Jeong M-J, Kim SW, Lee J-S, Park J-Y (2014b) Effects of ionic liquid mixtures on lipid extraction from Chlorella vulgaris. Renew Energy 65:169–174
Colla LM, Reinehr CO, Reichert C, Costa JAV (2007) Production of biomass and nutraceutical compounds by Spirulina platensis under different temperature and nitrogen regimes. Bioresour Technol 98:1489–1493
Costa JAV, De Morais MG (2011) The role of biochemical engineering in the production of biofuels from microalgae. Bioresour Technol 102:2–9
Crampon C, Mouahid A, Toudji S-AA, Lépine O, Badens E (2013) Influence of pretreatment on supercritical CO_2 extraction from Nannochloropsis oculata. J Supercrit Fluids 79:337–344
Cronshaw J, Myers A, Preston RD (1958) A chemical and physical investigation of the cell walls of some marine algae. Biochim Biophys Acta 27:89–103
Cuellar-Bermudez SP, Romero-Ogawa MA, Vannela R, Lai YJS, Rittmann BE, Parra-Saldivar R (2015) Effects of light intensity and carbon dioxide on lipids and fatty acids produced by Synechocystis sp. PCC6803 during continuous flow. Algal Res 12:10–16
da Silva Vaz B, Moreira JB, de Morais MG, Costa JAV (2016) Microalgae as a new source of bioactive compounds in food supplements. Curr Opin Food Sci 7:73–77

Daroch M, Geng S, Wang G (2013) Recent advances in liquid biofuel production from algal feedstocks. Appl Energy 102:1371–1381
Davis RW, Carvalho BJ, Jones HDT, Singh S (2015) The role of photo-osmotic adaptation in semi-continuous culture and lipid particle release from Dunaliella viridis. J Appl Phycol 27:109–123
de Jesus CS, da Silva Uebel L, Costa SS, Miranda AL, de Morais EG, de Morais MG, Costa JAV, Nunes IL, de Souza Ferreira E, Druzian JI (2018) Outdoor pilot-scale cultivation of Spirulina sp. LEB-18 in different geographic locations for evaluating its growth and chemical composition. Bioresour Technol 256:86–94
Dejoye Tanzi C, Vian MA, Chemat F (2013) New procedure for extraction of algal lipids from wet biomass: a green clean and scalable process. Bioresour Technol 134:271–275
Demirbas A (2008a) Production of biodiesel from algae oils. Energy Sources A: Recovery Util Environ Eff 31:163–168
Demirbas A (2008b) Relationships derived from physical properties of vegetable oil and biodiesel fuels. Fuel 87:1743–1748
Demirbas MF (2011) Biofuels from algae for sustainable development. Appl Energy 88:3473–3480
Demirbas A (2017) Production economics of high-quality microalgae. Energy Sources B: Econ Plan Policy 12:395–401
Demirbas A, Demirbas MF (2010) Algae energy: algae as a new source of biodiesel. Springer
Demirbas A, Demirbas MF (2011) Importance of algae oil as a source of biodiesel. Energy Convers Manag 52:163–170
dos Santos Freitas L, de Oliveira JV, Dariva C, Jacques RA, Caramão EB (2008) Extraction of grape seed oil using compressed carbon dioxide and propane: extraction yields and characterization of free glycerol compounds. J Agric Food Chem 56:2558–2564
Draaisma RB, Wijffels RH, Ellen Slegers PM, Brentner LB, Roy A, Barbosa MJ (2013) Food commodities from microalgae. Curr Opin Biotechnol 24:169–177
Enamala MK, Enamala S, Chavali M, Donepudi J, Yadavalli R, Kolapalli B, Aradhyula TV, Velpuri J, Kuppam C (2018) Production of biofuels from microalgae—a review on cultivation, harvesting, lipid extraction, and numerous applications of microalgae. Renew Sust Energ Rev 94:49–68
Fauzi AHM, Amin NAS (2012) An overview of ionic liquids as solvents in biodiesel synthesis. Renew Sust Energ Rev 16:5770–5786
Feller R (2017) Microalgae biomass as a source of natural compounds: chemical characterization and new approaches for lipid extraction and culture harvesting. Universidade Federal de Santa Catarina
Feller R, Matos ÂP, Mazzutti S, Moecke EHS, Tres MV, Derner RB, Oliveira JV, Junior AF (2018) Polyunsaturated ω-3 and ω-6 fatty acids, total carotenoids and antioxidant activity of three marine microalgae extracts obtained by supercritical CO_2 and subcritical n-butane. J Supercrit Fluids 133:437–443
Folch J, Lees M, Stanley GHS (1957) A simple method for the isolation and purification of total lipides from animal tissues. J Biol Chem 226:497–509
Freire MG, Santos LMNBF, Fernandes AM, Coutinho JAP, Marrucho IM (2007) An overview of the mutual solubilities of water–imidazolium-based ionic liquids systems. Fluid Phase Equilib 261:449–454
Gallagher BJ (2011) The economics of producing biodiesel from algae. Renew Energy 36:158–162
Gerde JA, Montalbo-Lomboy M, Yao L, Grewell D, Wang T (2012) Evaluation of microalgae cell disruption by ultrasonic treatment. Bioresour Technol 125:175–181
Goto M, Kanda H, Diono W, Machmudah S (2015) Extraction of carotenoids and lipids from algae by supercritical CO_2 and subcritical dimethyl ether. J Supercrit Fluids 96:245–251
Grima EM, González MJI, Giménez AG (2013) Solvent extraction for microalgae lipids. In: Algae for biofuels and energy. Springer
Grubb M, Vrolijk C, Brack D (1997) The Kyoto Protocol: a guide and assessment. Royal Institute of International Affairs Energy and Environmental Programme
Günerken E, D'Hondt E, Eppink MHM, Garcia-Gonzalez L, Elst K, Wijffels RH (2015) Cell disruption for microalgae biorefineries. Biotechnol Adv 33:243–260

Guschina IA, Harwood JL (2013) Algal lipids and their metabolism. In: Algae for biofuels and energy. Springer
Haas MJ, Wagner K (2011) Simplifying biodiesel production: the direct or in situ transesterification of algal biomass. Eur J Lipid Sci Technol 113:1219–1229
Hagen C, Siegmund S, Braune W (2002) Ultrastructural and chemical changes in the cell wall of Haematococcus pluvialis (Volvocales, Chlorophyta) during aplanospore formation. Eur J Phycol 37:217–226
Halim R, Gladman B, Danquah MK, Webley PA (2011) Oil extraction from microalgae for biodiesel production. Bioresour Technol 102:178–185
Halim R, Danquah MK, Webley PA (2012) Extraction of oil from microalgae for biodiesel production: a review. Biotechnol Adv 30:709–732
Harwood JL (1998) Membrane lipids in algae. In: Lipids in photosynthesis: structure, function and genetics. Springer
Hegedűs A, Mocan A, Barbu-Tudoran L, Coman C, Drugă B, Sicora C, Dragoş N (2015) Morphological, biochemical, and phylogenetic assessments of eight Botryococcus terribilis strains collected from freshwaters of Transylvania. J Appl Phycol 27:865–878
Hegel P, Martín L, Popovich C, Damiani C, Pancaldi S, Pereda S, Leonardi P (2017) Biodiesel production from Neochloris oleoabundans by supercritical technology. Chem Eng Process Process Intensif 121:232–239
Herrero M, Ibáñez E (2015) Green processes and sustainability: an overview on the extraction of high added-value products from seaweeds and microalgae. J Supercrit Fluids 96:211–216
Hu Q, Sommerfeld M, Jarvis E, Ghirardi M, Posewitz M, Seibert M, Darzins A (2008) Microalgal triacylglycerols as feedstocks for biofuel production: perspectives and advances. Plant J 54:621–639
Huang GH, Chen F, Wei D, Zhang XW, Chen G (2010) Biodiesel production by microalgal biotechnology. Appl Energy 87:38–46
Huber GW, Corma A (2007) Synergies between bio-and oil refineries for the production of fuels from biomass. Angew Chem Int Ed 46:7184–7201
Ishika T, Moheimani NR, Bahri PA (2017a) Sustainable saline microalgae co-cultivation for biofuel production: a critical review. Renew Sust Energ Rev 78:356–368
Ishika T, Moheimani NR, Bahri PA, Laird DW, Blair S, Parlevliet D (2017b) Halo-adapted microalgae for fucoxanthin production: effect of incremental increase in salinity. Algal Res 28:66–73
Ishika T, Bahri PA, Laird DW, Moheimani NR (2018) The effect of gradual increase in salinity on the biomass productivity and biochemical composition of several marine, halotolerant, and halophilic microalgae. J Appl Phycol 30:1453–1464
Isleten-Hosoglu M, Gultepe I, Elibol M (2012) Optimization of carbon and nitrogen sources for biomass and lipid production by Chlorella saccharophila under heterotrophic conditions and development of Nile red fluorescence based method for quantification of its neutral lipid content. Biochem Eng J 61:11–19
James GO, Hocart CH, Hillier W, Chen H, Farzaneh Kordbacheh G, Price D, Djordjevic MA (2011) Fatty acid profiling of Chlamydomonas reinhardtii under nitrogen deprivation. Bioresour Technol 102:3343–3351
Ju Y-H, Huynh LH, Tsigie YA, Ho Q-P (2013) Synthesis of biodiesel in subcritical water and methanol. Fuel 105:266–271
Júnior C, Rui M, Vargas JVC, Ramos LP, Marino CEB, Torres JCL (2011) Microalgae biodiesel via in situ methanolysis. J Chem Technol Biotechnol 86:1418–1427
Khademi H, Morowvat MH, Ghasemi Y (2018) Effects of nitrogen and sulfur deprivation on β-carotene and fatty acid content of Dunaliella salina. Trends Pharmacol Sci 4:187–196
Kim Y-H, Choi Y-K, Park J, Lee S, Yang Y-H, Kim HJ, Park T-J, Kim YH, Lee SH (2012) Ionic liquid-mediated extraction of lipids from algal biomass. Bioresour Technol 109:312–315
Kim D-Y, Vijayan D, Praveenkumar R, Han J-I, Lee K, Park J-Y, Chang W-S, Lee J-S, You-Kwan O (2016) Cell-wall disruption and lipid/astaxanthin extraction from microalgae: Chlorella and Haematococcus. Bioresour Technol 199:300–310

Kleiner AC, Cladis DP, Santerre CR (2015) A comparison of actual versus stated label amounts of EPA and DHA in commercial omega-3 dietary supplements in the United States. J Sci Food Agric 95:1260–1267
Kloareg B, Quatrano RS (1988) Structure of the cell walls of marine algae and ecophysiological functions of the matrix polysaccharides. Oceanogr Mar Biol Annu Rev 26:259–315
Knothe G (2005) Dependence of biodiesel fuel properties on the structure of fatty acid alkyl esters. Fuel Process Technol 86:1059–1070
Knothe G (2006) Analyzing biodiesel: standards and other methods. J Am Oil Chem Soc 83:823–833
Knothe G (2011) A technical evaluation of biodiesel from vegetable oils vs. algae. Will algae-derived biodiesel perform? Green Chem 13:3048–3065
Knothe G, Steidley KR (2005) Kinematic viscosity of biodiesel fuel components and related compounds. Influence of compound structure and comparison to petrodiesel fuel components. Fuel 84:1059–1065
Knothe G, Matheaus AC, Ryan III TW (2003) Cetane numbers of branched and straight-chain fatty esters determined in an ignition quality tester. Fuel 82:971–975
Knothe G, Sharp CA, Ryan TW (2006) Exhaust emissions of biodiesel, petrodiesel, neat methyl esters, and alkanes in a new technology engine. Energy Fuel 20:403–408
Knothe G, Krahl J, Van Gerpen J (2015) The biodiesel handbook. Elsevier
Koller M, Muhr A, Braunegg G (2014) Microalgae as versatile cellular factories for valued products. Algal Res 6:52–63
Kumar K, Mishra SK, Shrivastav A, Park MS, Yang J-W (2015) Recent trends in the mass cultivation of algae in raceway ponds. Renew Sust Energ Rev 51:875–885
Kurokawa M, King PM, Xiaoge W, Joyce EM, Mason TJ, Yamamoto K (2016) Effect of sonication frequency on the disruption of algae. Ultrason Sonochem 31:157–162
Lari Z, Moradi-kheibari N, Ahmadzadeh H, Abrishamchi P, Moheimani NR, Murry MA (2016) Bioprocess engineering of microalgae to optimize lipid production through nutrient management. J Appl Phycol 28:3235–3250
Lawton RJ, de Nys R, Magnusson ME, Paul NA (2015) The effect of salinity on the biomass productivity, protein and lipid composition of a freshwater macroalga. Algal Res 12:213–220
Lemahieu C, Bruneel C, Termote-Verhalle R, Muylaert K, Buyse J, Foubert I (2013) Impact of feed supplementation with different omega-3 rich microalgae species on enrichment of eggs of laying hens. Food Chem 141:4051–4059
Li R, Watanabe MM (2001) Fatty acid profiles and their chemotaxonomy in planktonic species of Anabaena (Cyanobacteria) with straight trichomes. Phytochemistry 57:727–731
Li Y, Moheimani NR, Schenk PM (2012) Current research and perspectives of microalgal biofuels in Australia. Biofuels 3:427–439
Liu X-J, Jiang Y, Chen F (2005) Fatty acid profile of the edible filamentous cyanobacterium Nostoc flagelliforme at different temperatures and developmental stages in liquid suspension culture. Process Biochem 40:371–377
Lourenço SO (2006) Cultivo de microalgas marinhas: princípios e aplicações. RiMa São Carlos
Martin CA, Vivian V, de Almeida M, Ruiz R, Visentainer JEL, Matshushita M, de Souza NE, Visentainer JV (2006) Ácidos graxos poliinsaturados ômega-3 e ômega-6: importância e ocorrência em alimentos. Rev Nutr 19(6). https://doi.org/10.1590/S1415-52732006000600011
Martínez N, Callejas N, Morais EG, Vieira JA, Costa IJ, Vieitez I (2017) Obtaining biodiesel from microalgae oil using ultrasound-assisted in-situ alkaline transesterification. Fuel 202:512–519
Mascarelli AL (2009) Gold rush for algae. Nature 461:460–461. Nature Publishing Group
Mata TM, Martins AA, Caetano NS (2010) Microalgae for biodiesel production and other applications: a review. Renew Sust Energ Rev 14:217–232
Matos ÂP (2017) The impact of microalgae in food science and technology. J Am Oil Chem Soc 94:1333–1350
Matos ÂP, Feller R, Moecke EHS, Sant'Anna ES (2015) Biomass, lipid productivities and fatty acids composition of marine Nannochloropsis gaditana cultured in desalination concentrate. Bioresour Technol 197:48–55

Matos ÂP, Feller R, Moecke EHS, Vladimir J, de Oliveira A, Junior F, Derner RB, Sant'Anna ES (2016) Chemical characterization of six microalgae with potential utility for food application. J Am Oil Chem Soc 93:963–972

Matos ÂP, Cavanholi MG, Moecke EHS, Sant'Anna ES (2017) Effects of different photoperiod and trophic conditions on biomass, protein and lipid production by the marine alga Nannochloropsis gaditana at optimal concentration of desalination concentrate. Bioresour Technol 224:490–497

McMillan JR, Watson IA, Ali M, Jaafar W (2013) Evaluation and comparison of algal cell disruption methods: microwave, waterbath, blender, ultrasonic and laser treatment. Appl Energy 103:128–134

Medina AR, Grima EM, Giménez AG, González MJI (1998) Downstream processing of algal polyunsaturated fatty acids. Biotechnol Adv 16:517–580

Mendes RL, Fernandes HL, Coelho JP, Reis EC, Cabral JMS, Novais JM, Palavra AF (1995) Supercritical CO_2 extraction of carotenoids and other lipids from Chlorella vulgaris. Food Chem 53:99–103

Mendes RL, Nobre BP, Cardoso MT, Pereira AP, Palavra AF (2003) Supercritical carbon dioxide extraction of compounds with pharmaceutical importance from microalgae. Inorg Chim Acta 356:328–334

Mendes RL, Reis AD, Pereira AP, Cardoso MT, Palavra AF, Coelho JP (2005) Supercritical CO_2 extraction of γ-linolenic acid (GLA) from the cyanobacterium Arthrospira (Spirulina)maxima: experiments and modeling. Chem Eng J 105:147–151

Mendes-Pinto MM, Raposo MFJ, Bowen J, Young AJ, Morais R (2001) Evaluation of different cell disruption processes on encysted cells of Haematococcus pluvialis: effects on astaxanthin recovery and implications for bio-availability. J Appl Phycol 13:19–24

Menegazzo ML, Fonseca GG (2019) Biomass recovery and lipid extraction processes for microalgae biofuels production: a review. Renew Sust Energ Rev 107:87–107

Mercer P, Armenta RE (2011) Developments in oil extraction from microalgae. Eur J Lipid Sci Technol 113:539–547

Moheimani NR, McHenry MP, De Boer K, Bahri PA (2015) Biomass and biofuels from microalgae. In: Biofuel and biorefinery technologies. Springer-Verlag GmbH

Montalescot V, Rinaldi T, Touchard R, Jubeau S, Frappart M, Jaouen P, Bourseau P, Marchal L (2015) Optimization of bead milling parameters for the cell disruption of microalgae: process modeling and application to Porphyridium cruentum and Nannochloropsis oculata. Bioresour Technol 196:339–346

Morioka LR, Ito ÂP, Matos GO, Sant'Anna ES (2014) Floculação de Chlorella sp. produzida em concentrado de dessalinização e estudo de método de extração de lipídeos intracelulares. Química Nova 37:44–49

Novello Z, Scapinello J, Dal Magro J, Zin G, Di Luccio M, Tres MV, Vladimir Oliveira J (2015) Extraction, chemical characterization and antioxidant activity of andiroba seeds oil obtained from pressurized n-butane. Ind Crop Prod 76:697–701

Nwoba EG, Parlevliet DA, Laird DW, Alameh K, Moheimani NR (2019) Light management technologies for increasing algal photobioreactor efficiency. Algal Res 39:101433

Pan J, Muppaneni T, Sun Y, Reddy HK, Jie F, Xiuyang L, Deng S (2016) Microwave-assisted extraction of lipids from microalgae using an ionic liquid solvent [BMIM][HSO4]. Fuel 178:49–55

Patil V, Tran K-Q, Giselrød HR (2008) Towards sustainable production of biofuels from microalgae. Int J Mol Sci 9:1188–1195

Patil PD, Dandamudi KPR, Wang J, Deng Q, Deng S (2018) Extraction of bio-oils from algae with supercritical carbon dioxide and co-solvents. J Supercrit Fluids 135:60–68

Paudel A, Jessop MJ, Stubbins SH, Champagne P, Jessop PG (2015) Extraction of lipids from microalgae using CO_2-expanded methanol and liquid CO_2. Bioresour Technol 184:286–290

Pohl P, Zurheide F (1982) Fat production in freshwater and marine algae. Mar Algae Pharm Sci 2:65–80

Postma PR, Miron TL, Olivieri G, Barbosa MJ, Wijffels RH, Eppink MHM (2015) Mild disintegration of the green microalgae Chlorella vulgaris using bead milling. Bioresour Technol 184:297–304

Prabakaran P, Ravindran AD (2011) A comparative study on effective cell disruption methods for lipid extraction from microalgae. Lett Appl Microbiol 53:150–154

Rapinel V, Rombaut N, Rakotomanomana N, Vallageas A, Cravotto G, Chemat F (2017) An original approach for lipophilic natural products extraction: use of liquefied n-butane as alternative solvent to n-hexane. LWT-Food Sci Technol 85:524–533

Ratledge C, Cohen Z (2008) Microbial and algal oils: do they have a future for biodiesel or as commodity oils? Lipid Technol 20:155–160

Razon LF, Tan RR (2011) Net energy analysis of the production of biodiesel and biogas from the microalgae: Haematococcus pluvialis and Nannochloropsis. Appl Energy 88:3507–3514

Reddy HK, Muppaneni T, Sun Y, Li Y, Ponnusamy S, Patil PD, Dailey P, Tanner Schaub F, Holguin O, Dungan B, Cooke P, Lammers P, Voorhies W, Xiuyang L, Deng S (2014) Subcritical water extraction of lipids from wet algae for biodiesel production. Fuel 133:73–81

Reverchon E, De Marco I (2006) Supercritical fluid extraction and fractionation of natural matter. J Supercrit Fluids 38:146–166

Rizwan M, Mujtaba G, Lee K (2017) Effects of iron sources on the growth and lipid/carbohydrate production of marine microalga Dunaliella tertiolecta. Biotechnol Bioprocess Eng 22:68–75

Rodolfi L, Zittelli GC, Bassi N, Padovani G, Biondi N, Bonini G, Tredici MR (2009) Microalgae for oil: strain selection, induction of lipid synthesis and outdoor mass cultivation in a low-cost photobioreactor. Biotechnol Bioeng 102:100–112

Rodríguez-Meizoso I, Jaime L, Santoyo S, Señoráns FJ, Cifuentes A, Ibáñez E (2010) Subcritical water extraction and characterization of bioactive compounds from Haematococcus pluvialis microalga. J Pharm Biomed Anal 51:456–463

Ruangsomboon S (2015) Effects of different media and nitrogen sources and levels on growth and lipid of green microalga Botryococcus braunii KMITL and its biodiesel properties based on fatty acid composition. Bioresour Technol 191:377–384

Sajjadi B, Chen W-Y, Raman AAA, Ibrahim S (2018) Microalgae lipid and biomass for biofuel production: a comprehensive review on lipid enhancement strategies and their effects on fatty acid composition. Renew Sust Energ Rev 97:200–232

Samorì C, Torri C, Samorì G, Fabbri D, Galletti P, Guerrini F, Pistocchi R, Tagliavini E (2010) Extraction of hydrocarbons from microalga Botryococcus braunii with switchable solvents. Bioresour Technol 101:3274–3279

Santana A, Jesus S, Larrayoz MA, Filho RM (2012) Supercritical carbon dioxide extraction of algal lipids for the biodiesel production. Procedia Eng 42:1755–1761

Sathish A, Sims RC (2012) Biodiesel from mixed culture algae via a wet lipid extraction procedure. Bioresour Technol 118:643–647

Sati H, Mitra M, Mishra S, Baredar P (2019) Microalgal lipid extraction strategies for biodiesel production: a review. Algal Res 38:101413

Scholz MJ, Weiss TL, Jinkerson RE, Jing J, Roth R, Goodenough U, Posewitz MC, Gerken HG (2014) Ultrastructure and composition of the Nannochloropsis gaditana cell wall. Eukaryot Cell 13:1450–1464

Silva BF, Wendt EV, Castro JC, Elcana A, de Oliveira A, Carrim JI, Vieira JDG, Sassi R, da Costa Sassi CF, da Silva ALV, de Oliveira Barboza GF, Filho NRA (2015) Analysis of some chemical elements in marine microalgae for biodiesel production and other uses. Algal Res 9:312–321

Simionato D, Sforza E, Carpinelli EC, Bertucco A, Giacometti GM, Morosinotto T (2011) Acclimation of Nannochloropsis gaditana to different illumination regimes: effects on lipids accumulation. Bioresour Technol 102:6026–6032

Sitthithanaboon W, Reddy HK, Muppaneni T, Ponnusamy S, Punsuvon V, Holguim F, Dungan B, Deng S (2015) Single-step conversion of wet Nannochloropsis gaditana to biodiesel under subcritical methanol conditions. Fuel 147:253–259

Šoštarič M, Klinar D, Bricelj M, Golob J, Berovič M, Likozar B (2012) Growth, lipid extraction and thermal degradation of the microalga Chlorella vulgaris. New Biotechnol 29:325–331

Stephens E, Ross IL, King Z, Mussgnug JH, Kruse O, Posten C, Borowitzka MA, Hankamer B (2010) An economic and technical evaluation of microalgal biofuels. Nat Biotechnol 28:126

Sun D, Zhang Z, Mao X, Tao W, Jiang Y, Liu J, Chen F (2017) Light enhanced the accumulation of total fatty acids (TFA) and docosahexaenoic acid (DHA) in a newly isolated heterotrophic microalga Crypthecodinium sp. SUN. Bioresour Technol 228:227–234

Svenning JB, Dalheim L, Eilertsen HC, Vasskog T (2019) Temperature dependent growth rate, lipid content and fatty acid composition of the marine cold-water diatom Porosira glacialis. Algal Res 37:11–16

Syrpas M, Bukauskaitė J, Paškauskas R, Bašinskienė L, Venskutonis PR (2018) Recovery of lipophilic products from wild cyanobacteria (Aphanizomenon flos-aquae) isolated from the Curonian Lagoon by means of supercritical carbon dioxide extraction. Algal Res 35:10–21

Taher H, Al-Zuhair S, Al-Marzouqi AH, Haik Y, Farid M (2014) Effective extraction of microalgae lipids from wet biomass for biodiesel production. Biomass Bioenergy 66:159–167

Takagi M, Yoshida T (2006) Effect of salt concentration on intracellular accumulation of lipids and triacylglyceride in marine microalgae Dunaliella cells. J Biosci Bioeng 101:223–226

Tang S, Qin C, Wang H, Li S, Tian S (2011) Study on supercritical extraction of lipids and enrichment of DHA from oil-rich microalgae. J Supercrit Fluids 57:44–49

Tatsuzawa H, Takizawa E, Wada M, Yamamoto Y (1996) Fatty acid and lipid composition of the acidophilic green alga Chlamydomonas sp. 1. J Phycol 32:598–601

Thiruvenkadam S, Izhar S, Hiroyuki Y, Harun R (2019) One-step microalgal biodiesel production from Chlorella pyrenoidosa using subcritical methanol extraction (SCM) technology. Biomass Bioenergy 120:265–272

Tibbetts SM, Melanson RJ, Park KC, Banskota AH, Stefanova R, McGinn PJ (2015a) Nutritional evaluation of whole and lipid-extracted biomass of the microalga Scenedesmus sp. AMDD isolated in Saskatchewan, Canada for animal feeds: proximate, amino acid, fatty acid, carotenoid and elemental composition. Curr Biotechnol 4:530–546

Tibbetts SM, Whitney CG, MacPherson MJ, Bhatti S, Banskota AH, Stefanova R, McGinn PJ (2015b) Biochemical characterization of microalgal biomass from freshwater species isolated in Alberta, Canada for animal feed applications. Algal Res 11:435–447

Vanda H, Dai Y, Wilson EG, Verpoorte R, Choi YH (2018) Green solvents from ionic liquids and deep eutectic solvents to natural deep eutectic solvents. C R Chim 21:628–638

Ventura SPM, Silva FAE, Quental MV, Mondal D, Freire MG, Coutinho JAP (2017) Ionic-liquid-mediated extraction and separation processes for bioactive compounds: past, present, and future trends. Chem Rev 117:6984–7052

Vieitez I, da Silva C, Alckmin I, Borges GR, Corazza FC, Vladimir Oliveira J, Grompone MA, Jachmanián I (2008a) Effect of temperature on the continuous synthesis of soybean esters under supercritical ethanol. Energy Fuel 23:558–563

Vieitez I, da Silva C, Borges GR, Corazza FC, Vladimir Oliveira J, Grompone MA, Jachmanián I (2008b) Continuous production of soybean biodiesel in supercritical ethanol-water mixtures. Energy Fuel 22:2805–2809

Vigani M, Parisi C, Rodríguez-Cerezo E, Barbosa MJ, Sijtsma L, Ploeg M, Enzing C (2015) Food and feed products from micro-algae: market opportunities and challenges for the EU. Trends Food Sci Technol 42:81–92

Viguera M, Marti A, Masca F, Prieto C, Calvo L (2016) The process parameters and solid conditions that affect the supercritical CO_2 extraction of the lipids produced by microalgae. J Supercrit Fluids 113:16–22

Waltz E (2009) Biotech's green gold? Nat Biotechnol 27:15–18. Nature Publishing Group

Wang K, Sun T, Cui J, Liu L, Bi Y, Pei G, Chen L, Zhang W (2018) Screening of chemical modulators for lipid accumulation in Schizochytrium sp. S31. Bioresour Technol 260:124–129

Xu H, Miao X, Qingyu W (2006) High quality biodiesel production from a microalga Chlorella protothecoides by heterotrophic growth in fermenters. J Biotechnol 126:499–507

Yao S, Jingquan L, Sárossy Z, Baggesen C, Brandt A, An Y (2016) Neutral lipid production in Dunaliella salina during osmotic stress and adaptation. J Appl Phycol 28:2167–2175

Yen H-W, Yang S-C, Chen C-H, Jesisca, Chang J-S (2015) Supercritical fluid extraction of valuable compounds from microalgal biomass. Bioresour Technol 184:291–296

Yoo G, Park W-K, Kim CW, Choi Y-E, Yang J-W (2012) Direct lipid extraction from wet Chlamydomonas reinhardtii biomass using osmotic shock. Bioresour Technol 123:717–722
Young G, Nippgen F, Titterbrandt S, Cooney MJ (2010) Lipid extraction from biomass using co-solvent mixtures of ionic liquids and polar covalent molecules. Sep Purif Technol 72:118–121
Zhao H, Baker GA (2013) Ionic liquids and deep eutectic solvents for biodiesel synthesis: a review. J Chem Technol Biotechnol 88:3–12
Zheng H, Yin J, Gao Z, Huang H, Ji X, Dou C (2011) Disruption of Chlorella vulgaris cells for the release of biodiesel-producing lipids: a comparison of grinding, ultrasonication, bead milling, enzymatic lysis, and microwaves. Appl Biochem Biotechnol 164:1215–1224

Chapter 12
Microalgae Superfoods

Michele Greque de Morais, Luciane Maria Colla, and Jorg Alberto Vieira Costa

Abstract The demand for healthy foods and a balanced diet has increased the consumption of functional foods. Superfoods are foods that contain one or more ingredients with higher nutritional and energy value than that of traditional foods. These products have been associated with health promotion, well-being, and quality of life. In this scenario, microalgae are gaining prominence since they contain several compounds with biological activity. These microorganisms are a natural source of bioactive compounds, such as vitamins, essential amino acids, polyunsaturated fatty acids, minerals, carotenoids, enzymes, and fibers. These compounds contribute nutraceutical, anti-inflammatory, anti-aging, hypocholesterolemic, antioxidant, immunosuppressive, photoprotective, and neurotransmitter activities to the microalgae biomass. Given this, the objective of this chapter is to present the benefits and use of microalgae as superfoods, especially *Spirulina* and *Chlorella*, also addressing the relevant nutrients related to the benefit of the addition of microalgae in the diet and the scenario of microalgal biomass production for human consumption.

Keywords Bioactive compounds · Biomass microalgal · *Chlorella* · Health promotion · *Spirulina*

M. G. Morais
Laboratory of Microbiology and Biochemistry, College of Chemistry and Food Engineering, Federal University of Rio Grande, Rio Grande, RS, Brazil

L. M. Colla (✉)
Laboratory of Biochemistry and Bioprocess, Faculty of Engineering and Architecture, University of Passo Fundo, Passo Fundo, RS, Brazil
e-mail: lmcolla@upf.br migreque@yahoo.com.br

J. A. V. Costa (✉)
Laboratory of Biochemical Engineering, College of Chemistry and Food Engineering, Federal University of Rio Grande, Rio Grande, RS, Brazil

J. A. V. Costa et al. (eds.), *Microalgal Bioengineering*,
https://doi.org/10.1007/978-3-031-61253-4_12

12.1 Introduction

Knowledge about the impact of food on human health has resulted in the search for new foods with functional characteristics. These foods are generally defined as those containing one or more ingredients with higher nutritional and energy value than traditional foods (Matos et al. 2017; Van Den Driessche et al. 2018). In the context of functional foods, there are some known as superfoods.

The concept of superfoods includes food, safety, and nutrition ideas linked to foods containing high levels of essential nutrients and other health-enhancing compounds, such as vitamins, minerals, phytonutrients, and antioxidants (Varzakas et al. 2018). These foods are considered an increasingly significant category within the scope of healthy foods, functional nutrition, and critical consumption. Superfoods are recognized for their nutritional properties, their impact on improving quality of life and reducing disease risks, and their traditional use histories by ancient communities (Garthe and Maughan 2018; Matos et al. 2017; Van Den Driessche et al. 2018). There is still no standard definition of the "superfood" term; it is not a legal or regulated category, and it is still considered by many as a marketing tool (Lunn 2006; Van Den Driessche et al. 2018).

The growing demand for superfoods has driven research into the physiological effects of high biological-value components of natural origin. Within this scenario, microalgae are increasingly studied (Freitas et al. 2019; Matos et al. 2017; Santos et al. 2016). Microalgae, including *Nostoc*, *Spirulina*, and *Chlorella,* have been consumed as food for thousands of years in different countries, such as China, Africa, and Mexico (Pulz and Gross 2004; Spolaore et al. 2006).

Microalgae are a natural source of bioactive compounds, such as vitamins, essential amino acids, polyunsaturated fatty acids, minerals, carotenoids, enzymes, and fiber (Matos et al. 2017). These microorganisms are considered superfoods because the nutritional profile of their biomass is superior to that of foods such as vegetables and grains (Morsy et al. 2014).

Among the factors that highlight the use of microalgae as superfoods are the presence of vitamins A, B1, B12, C, D, and E, riboflavin, niacin, pantothenic acid, and folic acid in the microalgal biomass (Gupta and Abu-Ghannam 2011). Another important feature is that superfoods have antioxidant, antibacterial, antiviral, and antifungal properties (Kaur et al. 2018; Volk and Furkert 2006). Microalgal biomass has prebiotic, neuroprotective, anti-inflammatory, immunomodulatory, antidiabetic, anticoagulant, and anticancer effects (Andrade et al. 2018; Charoensiddhi et al. 2018). The consumption of microalgal biomass contributes to maintaining calcium, sodium, magnesium, phosphorus, potassium, iron, zinc, and iodine levels in the body (Cofrades et al. 2013). In addition to the factors mentioned above, the addition of microalgae to the diet contributes to the reduction of the risks of colorectal cancer (since the algae dietary fiber favors the growth and maintenance of intestinal flora) (Ścieszka and Klewicka 2018) and breast cancer due to biomass iodine availability (Gupta and Abu-Ghannam 2011).

In this context, the purpose of this chapter is to present the benefits and use of microalgae as superfoods. The main aspects of how *Spirulina* and *Chlorella* are used in human food, the relevant nutrients related to the benefit of the addition of microalgae in the diet, and the scenario of microalgal biomass production for human consumption will be approached and discussed. In addition, examples of the development of functional foods added with microalgae and their effect on technological, sensory, and nutritional characteristics in the products will be presented.

12.2 Functional Foods and Superfoods

Several studies have contributed to increasing awareness of the population correlating diet and health, particularly the functionality of food-based biocompounds (Varzakas et al. 2018). Health has been a growing concern, which has motivated the demand to consume functional foods and superfoods. These foods have been associated with health promotion, improved well-being, and quality of life, reducing the risk of disease development (Koyande et al. 2019).

For Kozonova and Pavarova (2015), functional food is any modified food or ingredient that promotes health benefits beyond the traditional nutrients it contains. The definition "natural or processed food that has known or unknown biologically active compounds, and which in definite, effective and nontoxic amounts provide a clinically proven and documented health benefit for the prevention, control or treatment of chronic disease" was launched in 2012 by the Functional Food Center at United States (EUA) (Martirosyan and Singh 2015) to define functional foods. According to Santini et al. (2017), nutraceuticals consist of food or part of it of either vegetable or animal origin and have pharmaceutical benefits beyond nutritional value.

Superfoods may be fresh, processed foods with bioactive compounds, or the active ingredients may be in their isolated or concentrated form as food supplements. Enrichment of the food product with bioactive compounds may be achieved by the addition of proteins, pigments, amino acids, fatty acids, vitamins, or carbohydrates. Superfoods have spurred new research evaluating different sources for the production of these foods (Brockman and Beeren 2011).

In this context, superfoods may include milk, cheese, and eggs enriched with omega-3, probiotic yogurt, fruit juices, beverages with increased antioxidant concentrations, dietary fiber-enriched cereals, and grains and vegetable oils with added essential fatty acids (McClements et al. 2015).

The growing importance of consuming functional foods has stimulated studies on the physiological effects of bioactive components of natural origin. These compounds may act together to alter one or more physiological processes in the body to benefit health. Pigments (phycocyanin and carotenoids) and polyphenols are bioactive compounds with antioxidant properties that act by eliminating free radicals, preventing premature aging (Brewer 2011). The synergy of polyphenols, monounsaturated fatty acids, and fibers may aid in cardiovascular protection (Casas et al.

2018) and digestive system health (Alqurashi et al. 2017). In addition, essential amino acids in conjunction with minerals are vital for muscle contraction and regeneration (Alqurashi et al. 2017). *Lactobacillus* and *Bifidobacterium* probiotics and prebiotics, such as oligosaccharides and dietary fibers, contribute to the improvement of gastrointestinal tract health (Tripathi and Giri 2014; Varzakas et al. 2018). For the control of triglycerides, eicosapentaenoic acid and docosahexaenoic acid are the most studied compounds (Adarme-Vega et al. 2012; Chauton et al. 2015).

Epidemiological investigations have focused mainly on the ingestion of vegetables and fruits for disease prevention (Aune et al. 2017). Microalgae are considered superfoods because they contain several compounds with biological activity. These compounds have nutraceutical, antimicrobial, anti-inflammatory, anti-aging, hypocholesterolemic, antioxidant, immunosuppressive, protective, photoprotective, and neurotransmitter activities (Sathasivam et al. 2019) and therefore should be further explored when present in the biomass of these microorganisms.

Therefore, superfoods can aid in quality of life through diet and are effective alternatives similar to any other traditional treatment for the prevention of various diseases. For this, the benefits and advantages of superfoods should be disseminated broadly, so that there is awareness of the importance of the consumption of these foods.

12.3 Nutrients and Functional Ingredients in Microalgae

Functional foods or superfoods have nutrients with additional functions in basic nutrition. Bernal et al. (2011) classify functional ingredients as lipids (fatty acids, sterols, terpenes, glycerolipids, sphingolipids, and carotenoids), vitamins, proteins, peptides and amino acids, carbohydrates, glycosides, and phenolic compounds.

Research has reported the benefits of the chemical composition of microalgae and their compounds (Ambrosi et al. 2008; Kaur et al. 2018; Matos et al. 2017; Sathasivam et al. 2019), seeking to demonstrate the potential of microalgae products in the context of functional and nutraceutical foods. According to Andrade et al. (2018) and Sathasivam et al. (2019), the chemical composition of microalgae includes long chain polyunsaturated fatty acids, phenolic compounds, proteins, amino acids, peptides, vitamins, polysaccharides, and pigments. These compounds are responsible for the functional activity of microalgae. The microalgae *Spirulina* and *Chlorella* are the most cited and used to supply these biocompounds.

According to Chew et al. (2017), proteins are one of the most important constituents of microalgae due to their various applications in food and feed. Kaur et al. (2018) mentioned that microalgae could contribute up to 18% of protein sources by the middle of this century, and other compounds extracted from them could gain importance in this market. Comparing *Chlorella* and *Spirulina* as protein sources, Muys et al. (2019) concluded that the microalgae had similar protein contents (47%, obtained from different commercial crops). However, the amino acid quality and *Spirulina* digestibility were higher than those of *Chlorella*. Kaur et al. (2018) state that both microalgae have balanced amino acid profiles according to WHO/FAO/UNU recommendations regarding human requirements, comparable to conventional sources, such as eggs and soy. This protein content and the amino acid profile vary depending on the culture conditions of the microorganism.

Among the carbohydrates present in microalgae are β-1,3-glucans, gelling agents, starch, and cellulose (Matos et al. 2017). When presented in fiber form (cellulose or other grades), these compounds can be considered functional ingredients due to their prebiotic activity.

Three classes of pigments are normally present: carotenoids, chlorophylls, and phycobiliproteins (Chew et al. 2017). Di Lena et al. (2018) analyzed the profile of carotenoids in microalgae produced on an industrial scale (*Porphyridium cruentum*, *Isochrysis galbana*, *Phaeodactylum tricornutum*, *Tetraselmis suecica,* and *Nannochloropsis gaditana*). It has been shown that the microalgae presented mostly free-form carotenoids, with a prevalence of xanthophylls and carotenes, with beta-carotene being the most common among the species.

In addition to carotenoids, phycocyanin and chlorophyll are mentioned as photosynthetic microalgal pigments. Phycocyanin is a blue soluble protein pigment formed by two protein subunits in the phycobiliprotein class (Kamble et al. 2013). This pigment can be used as a colorant in food and cosmetics (Chen et al. 2018), and it is known for its high antioxidant capacity and anticarcinogenic and anti-inflammatory properties (Raposo et al. 2013).

Microalgal pigments are studied due to their antioxidant activities, with phycocyanin and carotenoids being the most explored. In addition, phenolic compounds also exhibit antioxidant action (Colla et al. 2007). Pigment concentrations found in microalgae biomass depend on culture conditions and the type of microalgae and are present in both eukaryotic and cyanobacterial species (Colla et al. 2007; Morowvat and Ghasemi 2016).

Abd El-Baky et al. (2008) found that concentrations of sodium nitrate added to the culture medium (0–2.5 g/L) altered the amounts of chlorophyll, carotenoids, phenolic compounds, and phycocyanin in the microalgae *Spirulina maxima*. High concentrations of chlorophyll, phenolic compounds, and phycocyanin were obtained with the highest nitrate concentrations, while the inverse effect was observed for carotenoids and tocopherols.

12.4 Microalgae Applied in Foods

Microalgae are unicellular or multicellular organisms capable of developing in diverse habitats, under nutrient limitation conditions, and in the presence of carbon dioxide (CO_2) as a carbon source using photosynthesis. By manipulating cultivation conditions, such as variations in temperature, light, pH, salinity, humidity, and/or nutritional limitation, microalgae can accumulate different macromolecules (Costa et al. 2017).

Microalgae are traditionally classified according to their structural characteristics (eukaryotic and prokaryotic), type of reserve metabolites, cell wall components, and pigments (Ścieszka and Klewicka 2018). These microorganisms are divided into the classes *Chlorophyceae* (green algae), *Bacillariophyceae* (diatoms), *Chrysophyceae* (yellow algae), *Cyanophyceae* (cyanobacteria), *Rhodophyceae* (red algae), and *Phaeophyceae* (brown algae) (Bharathiraja et al. 2015).

Microalgae are commonly marketed as capsules, tablets, powders, and liquids or even added to develop functional products. Microalgae biomass is also added to candy, gum, pasta, breakfast cereals, wine (Koyande et al. 2019), cookies, cakes, instant soup, cereal bars, pudding, and snacks (Fig. 12.1).

Microalgae widely used in food products include *Spirulina platensis*, *Chlorella* sp., *Dunaliella terticola*, *Dunaliella salina*, and *Aphanizomenon flos-aquae* due to their high protein content and nutritional value (Soletto et al. 2005). In recent years, *Chlorella* and *Spirulina* have become increasingly frequent in the global microalgae market, as they are gaining popularity in supermarkets and health food stores (García et al. 2017; Sathasivam et al. 2019).

Spirulina is a filamentous microalga belonging to the *Cyanophyceae* class, which presents prokaryotic and multicellular structures (Fox 1996). This microalga develops in tropical climates, tolerating a pH of 10–11. The filaments have a variable length (usually 50–500 µm) and a diameter close to 3–12 µm. Cell dimensions, degree of curl, and filament length vary by species (Ciferri and Tiboni 1985). *Spirulina*'s chemical composition is as follows: 55–70% proteins, 15–25% carbohydrates, 5–6% lipids, and 2.2–4.8% minerals (Hosseini et al. 2013).

Fig. 12.1 Foods developed with the addition of biomass of *Spirulina* sp. LEB 18 (unpublished data): cookies (**a**), cakes (**b**), instant soup (**c**), cereal bar (**d**), pudding (**e**), and snacks (**f**)

Spirulina is considered a superfood mainly because of its high protein content. In addition to its nutritional contribution, this microalga has antioxidant potential from its phycobiliproteins, which act by sequestering free radicals (Raja et al. 2018). *Spirulina* protein has up to 90% digestibility using casein as a reference. The high digestibility has been attributed to the absence of a cell wall (Reboleira et al. 2019). The biomass of this microalga is also an excellent source of vitamin B1 and γ-linolenic acid, which has antitumor and anticancer effects, contributing to the nutraceutical potential of its biomass (Raja et al. 2018).

Chlorella is a unicellular eukaryotic microalgae of the class *Chlorophyceae* that has a spherical structure with a diameter of 2–10 μm and plant-like structural elements (Safi et al. 2014). The cell wall of this microalga is rigid, protecting against invaders and factors that promote a hostile environment. *Chlorella* biomass contains photosynthetic pigments, such as chlorophyll a and chlorophyll b, located in the chloroplasts. This microalga is representative of the phytoplankton of lakes and ponds but can also be found in soils (Qiao et al. 2009).

Chlorella is often used as a dietary supplement. This microalga is capable of accumulating a high content of monounsaturated/fatty acids (30–60%) (Yusof et al. 2011) and is considered a source of carotenoids (up to 0.2%). These compounds have beneficial effects on human health, such as anti-inflammatory, antioxidant, anticoagulant, antimicrobial, antitumor, and anticancer effects (Vaz et al. 2016). Therefore, the use of *Spirulina* and *Chlorella* as superfoods has been growing continuously. The potential of these microalgae is beyond their nutritional value because they have different biological activities that can prevent the emergence of chronic diseases, promoting health and well-being.

12.5 Foods with Added Microalgae

Many functional foods with added microalgae have been reported in the literature. Morais et al. (2006) developed cookies with added *Spirulina*. Chocolate cookies were prepared with the addition of 1.0%, 3.0%, and 5.0% *Spirulina platensis* biomass. The biscuit added with *S. platensis* showed higher digestibility (86.9%) and a 7.7% higher protein content than those without the added biomass. The attributes of color, crunchiness, chewability, and softness showed no significant difference between the cookies with and without the addition of microalgae.

Peron et al. (2018) developed gluten-free pasta with 1–5% added *Spirulina platensis* biomass. The protein content in the pasta with 5% *Spirulina* was higher (10.2%) than in the control (8.3%). The microalgae-based food may be considered a high-protein product compared to the gluten-free pasta on the market, which provide an average of 7 g protein/100 g pasta.

Maccari et al. (2008) developed probiotic fermented milk with added microalgae *Spirulina platensis* at concentrations of 0%, 0.125%, and 0.25%. The yogurts differed only in the pink and green color attributes, the aroma and taste of *Spirulina*, and sandiness ($p < 0.05$), indicating that consumers can perceive sensory differences without affecting acceptability.

Varga et al. (2002) evaluated the addition of *Spirulina* at a concentration of 3 g/L in fermented milk containing *Streptococcus thermophilus*, *Lactobacillus acidophilus,* and *Bifidobacterium* spp. during storage at 4 °C and 15 °C. The authors determined the viability of the microorganisms and concluded that the addition of *Spirulina* to fermented milk allowed for the viability of the microorganisms to be maintained.

Santos et al. (2016) developed chocolate-flavored shakes with added *Spirulina* for the elderly. The shake supplemented with *Spirulina* presented higher ash and protein contents than the shake developed without the microalgal biomass. Both products had comparable levels of these compounds, which are important for the elderly to consume. In addition, the developed products showed high acceptability when compared to the commercial product. Lucas et al. (2018) developed snacks containing 2.6% *Spirulina* sp. LEB 18. *Spirulina* addition increased protein, lipid, and mineral contents by 22.6%, 28.1%, and 46.4%, respectively, without affecting the physical parameters ($p > 0.05$) of expansion and hardness. The biomass pigments influenced the color of the snacks without affecting the product acceptability index (82%).

Carvalho et al. (2016) produced electrolyte-replenishing, muscle-recovering, and muscle-enhancing products. The authors found that the 0.5% *Spirulina*-added electrolyte deposition showed a 0.35% (w/w) increase in mineral content compared to that of the non-added *Spirulina* product. Compared to commercial products, Carvalho et al. (2017) found that the acceptability of the replenishers was equal to or higher for the developed products with added *Spirulina* biomass.

Freitas et al. (2019) developed meal-replacement shakes. The addition of *Spirulina* sp. LEB 18 caused an 8.6% reduction in carbohydrate content and a 29.3% increase in lipid content in the shake with added *Spirulina* compared to the product without the microalgae. The products with added *Spirulina* had higher acceptability averages than those of the commercial products.

Therefore, numerous food products can be supplemented with microalgae. Although some sensory properties are affected by the taste of microalgae, the current consumer understands the need for consuming functional foods for their health benefits. Moreover, consumption habits help consumers adapt to different flavors, which is a sociocultural issue.

12.6 World Production of Microalgae for Human Consumption

Algae have been exploited for centuries as food and feed and the proposed use as alternative protein sources to meet global food demand emerged in the 1950s, thus generating a new market to be explored (Burlew 1953; Vigani et al. 2015). From then on, Japan began the first industrial-scale production of *Chlorella* microalgae species for human consumption (Vigani et al. 2015), and the first large-scale production of *Spirulina* appeared in the 1970s on Lake Texcoco, near Mexico City (Mexico) (Borowitzka 2013). During the 1970s and 1980s, industrial production of microalgae such as *Spirulina* and *Chlorella* for human consumption expanded to the USA, China, Taiwan, Australia, India, Israel, Switzerland, France, Portugal, Mongolia, Brazil, Canada, New Zealand, Myanmar, Thailand, Sweden, Mongolia, the Netherlands, and Germany, with production increasing in recent years (Vigani et al. 2015; Andrade et al. 2018).

Among the companies that sell microalgae biomass for human consumption are Cyanotech Corporation (EUA), Parry Nutraceuticals (India), Fuqing King Dnarmsa *Spirulina* Co., Ltd. (China), *Spirulina* Mater © (Chile), Dongtai City *Spirulina* Bioengineering Co. Ltd. (China), Boonsom *Spirulina* Farm (Thailand), Olson Nutrition Ltd. (Brazil), Fazenda Tamanduá (Brazil), Vital Brazil (Brazil), Earthrise Nutritionals, LLC (USA), Hainan-DIC Microalgae (China), Taiwan *Chlorella* Manufacturing (Japan), Bluetec Naturals (China), Sun *Chlorella* (Japan), and Allmicroalgae Natural Products S.A. (Portugal), among others. The largest producers of Spirulina are Earthrise Nutritionals, LLC (USA) and Hainan-DIC Microalgae (China), which are both owned by the DIC Corporation (DIC) and produce approximately 550 and 350 tons of biomass per year, respectively (L. DIC LIFETEC Co), while Taiwan *Chlorella* Manufacturing produces 70% of the *Chlorella* biomass in the global market (Taiwan Chlorella Manufacturing).

The microalgae consumer market as a superfood is among the four promising potential scenarios for using microalgal biomass (energy, feed and food, chemicals and materials, and pharmaceuticals and personal care). Within these markets, supply and demand for these products, particularly those of superfoods, are expected to grow substantially in the future (Barsanti and Gualtieri 2018).

The commercial production of microalgae products still faces challenges to be overcome (Caporgno and Mathys 2018; Kaur et al. 2018). These challenges include consumer acceptance of other more traditional food sources, as well as time and cost barriers to obtaining and approving new products by regulatory agencies. In addition, it is still difficult to guarantee proven market demand and competitive market value for products such as microalgae-based superfoods (Matos 2017; Caporgno and Mathys 2018).

Among the most important considerations for microalgae superfood production is the reduction of the costs of biomass production, which is directly influenced by the appropriate choice of cultivation, harvesting, drying, and processing systems. It is noteworthy that the extraction of higher purity products, such as carotenoids, phycobilins, fatty acids, sterols, polysaccharides, and other compounds, still presents high production costs (Costa et al. 2019; Kaur et al. 2018), which makes market prices for microalgae products overvalued (Matos 2017). One way to substantially reduce production costs and thus lower the final cost to the consumer of microalgae superfoods is to incorporate biorefinery processes (Suganya et al. 2016).

The strategy of a photobiorefinery for economically viable production of microalgae biomass is applied in the integrated production of different products and coproducts, with the full use of the generated biomass. Strategies that can be proposed in microalgal biorefinery models include cultivation using local waters and nutrient recycling for biomass production and extraction of high value products and the transformation of residual biomass into biofuels (Zhu 2015; Suganya et al. 2016).

12.7 Conclusions

Microalgae contain several bioactive molecules with biological activities related mainly to the antioxidant, anti-inflammatory, and antimicrobial potential of the biomass. Many studies have been conducted to develop novel foods with the addition of microalgae biomass, such as *Spirulina* and *Chlorella*. Studies show that the biomass of these microalgae contributes to improving the nutritional properties of foods and digestibility without interfering with the acceptability of products. In addition, these microalgae are currently widely known and marketed worldwide. To increase the viability of producing microalgae superfoods, it is important to constantly seek to reduce biomass production costs, as the extraction of higher purity products still has high costs. For this, biorefinery techniques can be applied in the integrated production of different products and coproducts, with the full use of the generated biomass. Thus, *Spirulina* and *Chlorella* are candidates with high potential for developing superfoods given their nutritional profiles, the possibility of reducing production costs, and consumer demand for a food that promotes health and well-being.

References

Abd El-Baky HH, El Baz FK, & El-Baroty GS (2008) Characterization of nutraceutical compounds in blue green alga Spirulina maxima. J Med Plant Res, v2(10):292–300

Adarme-Vega TC, Lim DK, Timmins M et al (2012) Microalgal biofactories: a promising approach towards sustainable omega-3 fatty acid production. Microb Cell Factories 11:1–10. https://doi.org/10.1186/1475-2859-11-96

Alqurashi RM, Alarifi SN, Walton GE et al (2017) In vitro approaches to assess the effects of açai (Euterpe oleracea) digestion on polyphenol availability and the subsequent impact on the faecal microbiota. Food Chem 234:190–198. https://doi.org/10.1016/j.foodchem.2017.04.164

Ambrosi MA, Reinehr CO, Bertolin TE et al (2008) Propriedades de saúde de *Spirulina* spp. Rev Ciênc Farm Básica Apl 29(2):109–117

Andrade LM, Andrade CJ, Dias M et al (2018) *Chlorella* and *Spirulina* microalgae as sources of functional foods, nutraceuticals, and food supplements: an overview. MOJ Food Process Technol. https://doi.org/10.15406/mojfpt.2018.06.00144

Aune D, Giovannucci E, Boffetta P et al (2017) Fruit and vegetable intake and the risk of cardiovascular disease, total cancer and all-cause mortality-a systematic review and dose-response meta-analysis of prospective studies. Int J Epidemiol 46(3):1029–1056. https://doi.org/10.1093/ije/dyw319

Barsanti L, Gualtieri P (2018) Is exploitation of microalgae economically and energetically sustainable? Algal Res 31:107–115. https://doi.org/10.1016/j.algal.2018.02.001

Bernal J, Mendiola JA, Ibáñez E et al (2011) Advanced analysis of nutraceuticals. J Pharm Biomed Anal 55:758–774. https://doi.org/10.1016/j.jpba.2010.11.033

Bharathiraja B, Chakravarthy M, Ranjith Kumar R et al (2015) Aquatic biomass (algae) as a future feed stock for bio-refineries: a review on cultivation, processing and products. Renew Sust Energ Rev 47:634–653. https://doi.org/10.1016/j.rser.2015.03.047

Borowitzka MA (2013) High-value products from microalgae-their development and commercialisation. J Appl Phycol 25:743–756. https://doi.org/10.1007/s10811-013-9983-9

Brewer MS (2011) Natural antioxidants: sources, compounds, mechanisms of action, and potential applications. Compr Rev Food Sci Food Saf 10:221–247. https://doi.org/10.1111/j.1541-4337.2011.00156.x

Brockman C, Beeren CJM (2011) Additives in dairy foods; consumer perceptions of additives in dairy products. In: Fuquay JW, Fox PF, McSweeney PLH (eds) Encyclopedia of dairy sciences, 2nd edn. Academic, Washington, pp 41–48

Burlew JS (1953) Algal culture from laboratory to pilot plant. AIBS Bull. https://doi.org/10.1093/aibsbulletin/3.5.11

Caporgno MP, Mathys A (2018) Trends in microalgae incorporation into innovative food products with potential health benefits. Front Nutr 5:1–10. https://doi.org/10.3389/fnut.2018.00058

Carvalho LF, Moreira JB, Oliveira MZ et al (2017) Novel food supplements formulated with *Spirulina* to meet athletes' needs. Braz Arch Biol Technol. https://doi.org/10.1590/1678-4324-2017160656

Carvalho LF, Moreira JB, Oliveira MZ et al (2016) Acceptance of novel foods for physical activity practitioners formulated with *Spirulina*. IOSR J Biotechnol Biochem 2(4):7–11

Casas R, Castro-Barquero S, Estruch R et al (2018) Nutrition and cardiovascular health. Int J Mol Sci. https://doi.org/10.3390/ijms19123988

Charoensiddhi S, Lorbeer AJ, Franco CMM et al (2018) Process and economic feasibility for the production of functional food from the brown alga *Ecklonia radiata*. Algal Res 29:80–91. https://doi.org/10.1016/j.algal.2017.11.022

Chauton MS, Reitan KI, Norsker NH et al (2015) A technoeconomic analysis of industrial production of marine microalgae as a source of EPA and DHA-rich raw material for aquafeed: research challenges and possibilities. Aquaculture 436(20):95–103. https://doi.org/10.1016/j.aquaculture.2014.10.038

Chen KH, Wang SSS, Show PL et al (2018) A rapid and efficient technique for direct extraction of C-phycocyanin from highly turbid *Spirulina platensis* algae using hydrophobic interaction chromatography in stirred fluidized bed. Biochem Eng J 140:47–56. https://doi.org/10.1016/j.bej.2018.09.005

Chew KW, Yap JY, Show PL et al (2017) Microalgae biorefinery: high value products perspectives. Bioresour Technol 229:53–62. https://doi.org/10.1016/j.biortech.2017.01.006

Ciferri O, Tiboni O (1985) The biochemistry and industrial potential of *Spirulina*. Ann Rev Microbiol 39:503–526. https://doi.org/10.1146/annurev.mi.39.100185.002443

Cofrades S, Serdaroğlu M, Jimenez-Colmenero F (2013) Design of healthier foods and beverages containing whole algae. In: Dominguez H (ed) Functional ingredients from algae for foods and nutraceuticals. Woodhead Publishing, Oxford, pp 609–633. https://doi.org/10.1533/9780857098689.4.609

Colla LM, Reinehr CO, Reichert C et al (2007) Production of biomass and nutraceutical compounds by *Spirulina platensis* under different temperature and nitrogen regimes. Bioresour Technol 98:1489–1493

Costa JAV, Freitas BCB, Santos TD et al (2019) Open pond systems for microalgal culture. In: Pandey A, Chang J-S, Soccol CR, Lee DJ, Chisti Y (eds) Biofuels from algae, 2nd edn. Elsevier, New York, pp 199–223. https://doi.org/10.1016/b978-0-444-64192-2.00009-3

Costa JAV, Moraes L, Moreira JB et al (2017) Microalgae-based biorefineries as a promising approach to biofuel production. In: Tripathi BN, Kumar D (eds) Prospects and challenges in algal biotechnology, 2nd edn. Springer Singapore, pp 113–140

Di Lena G, Casini I, Lucarini M et al (2018) Carotenoid profiling of five microalgae species from large-scale production. Food Res Int 120:810–818. https://doi.org/10.1016/j.foodres.2018.11.043
Fox DR (1996) Spirulina: production & potential. Edisud, France, p 232
Freitas BCB, Santos TD, Moreira JB et al (2019) Novel foods: a meal replacement shake and a high-calorie food supplemented with *Spirulina* biomass. Int Food Res J 2(1):59–65
García JL, Vicente M, Galán B (2017) Microalgae, old sustainable food and fashion nutraceuticals. Microb Biotechnol 10(5):1017–1024. https://doi.org/10.1111/1751-7915.12800
Garthe I, Maughan RJ (2018) Athletes and supplements: prevalence and perspectives. Int J Sport Nutr Exerc Metab 28:126–138. https://doi.org/10.1123/ijsnem.2017-0429
Gupta S, Abu-Ghannam N (2011) Bioactive potential and possible health effects of edible brown seaweeds. Trends Food Sci Technol 22:315–326. https://doi.org/10.1016/j.tifs.2011.03.011
Hosseini SM, Khosravi-Darani K, Mozafari MR (2013) Nutritional and medical applications of *Spirulina* microalgae. Mini-Rev Med Chem 13(8):1231–1237. https://doi.org/10.2174/1389557511313080009
Kamble SP, Gaikar RB, Padalia RB et al (2013) Extraction and purification of C-phycocyanin from dry *Spirulina* powder and evaluating its antioxidant, anticoagulation and prevention of DNA damage activity. J Appl Pharm Sci 3(8):14959–14153. https://doi.org/10.7324/JAPS.2013.3826
Kaur KK, Allahbadia G, Singh M (2018) Utilizing microalgae addition in novel food products which might improve health—a review. CPQ Nutr 3(1):1–15
Koyande AK, Chew KW, Rambabu K et al (2019) Microalgae: a potential alternative to health supplementation for humans. Food Sci Human Wellness 8:16–24. https://doi.org/10.1016/j.fshw.2019.03.001
Kozonova J, Povarova N (2015) Leading trends in the European functional food market. Refrig Technol 51:92–96
Lucas BF, Morais MG, Santos TD et al (2018) *Spirulina* for snack enrichment: nutritional, physical and sensory evaluations. LWT-Food Sci Technol 90:270–276. https://doi.org/10.1016/j.lwt.2017.12.032
Lunn J (2006) Superfoods. Nutr Bull 31:171–172. https://doi.org/10.1111/j.1467-3010.2006.00578.x
Maccari E, Armigliatto M, França T (2008) Produção de iogurte probiótico adicionado de microalga *Spirulina platensis*. Trabalho de conclusão de curso em engenharia de Alimentos, Universidade de Passo Fundo
Martirosyan DM, Singh JA (2015) A new definition of functional food by FFC: what makes a new definition unique? Funct Food Health Dis 5(6):209–223
Matos AP (2017) The impact of microalgae in food science and technology. J Am Oil Chem Soc 94(11):1333–1350. https://doi.org/10.1007/s11746-017-3050-7
Matos J, Cardoso C, Bandarra NM et al (2017) Microalgae as healthy ingredients for functional food: a review. Food Funct 8:2672–2685. https://doi.org/10.1039/c7fo00409e
McClements DJ, Li F, Xiao H (2015) The nutraceutical bioavailability classification scheme: classifying nutraceuticals according to factors limiting their oral bioavailability. Annu Rev Food Sci Technol 6:299–327. https://doi.org/10.1146/annurev-food-032814-014043
Morais M, Miranda M, Costa JAV (2006) Biscoitos de chocolate enriquecidos com *Spirulina platensis*: características físico-químicas, sensoriais e digestibilidade. Aliment Nutr 17(3):323–328
Morowvat MH, Ghasemi Y (2016) Evaluation of antioxidant properties of some naturally isolated microalgae: identification and characterization of the most efficient strain. Biocatal Agric Biotechnol 8:263–269. https://doi.org/10.1016/j.bcab.2016.09.010
Morsy OM, Sharoba AM, EL-Desouky AI et al (2014) Production and evaluation of some extruded food products using *Spirulina* algae. Ann Agric Sci 52:329–342
Muys M, Sui Y, Schwaiger B et al (2019) High variability in nutritional value and safety of commercially available *Chlorella* and *Spirulina* biomass indicates the need for smart production strategies. Bioresour Technol 275:247–257. https://doi.org/10.1016/j.biortech.2018.12.059
Peron BC, Colla LM, Colla E et al (2018) Massa alimentícia isenta de glúten com adição de spirulina platensis: avaliação tecnológica e caracterização físico-química. In: Vieira V (ed) Desafios

da Ciência e Tecnologia de Alimentos. Atena Editora, Ponta Grossa. https://doi.org/10.22533/at.ed.178182208

Pulz O, Gross W (2004) Valuable products from biotechnology of microalgae. Appl Microbiol Biotechnol 65:635–648. https://doi.org/10.1007/s00253-004-1647-x

Qiao H, Wang G, Zhang X (2009) Isolation and characterization of *Chlorella sorokiniana* GXNN01 (Chlorophyta) with the properties of heterotrophic and microaerobic growth. J Phycol 45:1153–1162. https://doi.org/10.1111/j.1529-8817.2009.00736.x

Raja R, Coelho A, Hemaiswarya S et al (2018) Applications of microalgal paste and powder as food and feed: an update using text mining tool. Beni-Suef Univ J Basic Appl Sci 7:740–747. https://doi.org/10.1016/j.bjbas.2018.10.004

Raposo MFDJ, Morais RMD, Morais AMD (2013) Health applications of bioactive compounds from marine microalgae. Life Sci 93(15):479–486. https://doi.org/10.1016/j.lfs.2013.08.002

Reboleira J, Freitas R, Pinteus S et al (2019) Spirulina. In: Nabavi SM, Sanches A (eds) Nonvitamin and nonmineral nutritional supplements. Academic, Cambridge, MA, pp 409–413. https://doi.org/10.1016/B978-0-12-812491-8.00055-2

Safi C, Zebib B, Merah O et al (2014) Morphology, composition, production, processing and applications of Chlorella vulgaris: a review. Renew Sust Energ Rev 35:265–278. https://doi.org/10.1016/j.rser.2014.04.007

Santini A, Tenore GC, Novellino E (2017) Nutraceuticals: a paradigm of proactive medicine. Eur J Pharm Sci 96:53–61. https://doi.org/10.1016/j.ejps.2016.09.003

Santos TD, Freitas BCB, Moreira JB et al (2016) Development of powdered food with the addition of *Spirulina* for food supplementation of the elderly population. Innov Food Sci Emerg Technol 37:216–220. https://doi.org/10.1016/j.ifset.2016.07.016

Sathasivam R, Radhakrishnan R, Hashem A et al (2019) Microalgae metabolites: a rich source for food and medicine. Saudi J Biol Sci 26(4):709–722. https://doi.org/10.1016/j.sjbs.2017.11.003

Ścieszka S, Klewicka E (2018) Algae in food—a general review. Crit Rev Food Sci Nutr 18:1–10. https://doi.org/10.1080/10408398.2018

Soletto D, Binaghi L, Lodi A et al (2005) Batch and fed-batch cultivations of *Spirulina platensis* using ammonium sulphate and urea as nitrogen sources. Aquaculture 243:217–224. https://doi.org/10.1016/j.aquaculture.2004.10.005

Spolaore P, Joannis-Cassan C, Duran E et al (2006) Commercial applications of microalgae. J Biosci Bioeng 101:87–96. https://doi.org/10.1263/jbb.101.87

Suganya T, Varman M, Masjuki HH et al (2016) Macroalgae and microalgae as a potential source for commercial applications along with biofuels production: a biorefinery approach. Renew Sust Energ Rev 55:909–941. https://doi.org/10.1016/j.rser.2015.11.026

Tripathi MK, Giri SK (2014) Probiotic functional foods: survival of probiotics during processing and storage. J Funct Foods 9:225–241. https://doi.org/10.1016/j.jff.2014.04.030

Van Den Driessche JJ, Plat J, Mensink RP (2018) Effects of superfoods on risk factors of metabolic syndrome: a systematic review of human intervention trials. Food Funct 9:1944–1966. https://doi.org/10.1039/c7fo01792h

Varga L, Szigeti J, Kovács R, Földes T et al (2002) Influence of a *Spirulina* platensis biomass on the microflora of fermented ABT milks during storage (R1). J Dairy Sci 85:1031–1038. https://doi.org/10.3168/jds.S0022-0302(02)74163-5

Varzakas T, Kandylis P, Dimitrellou D et al (2018) Innovative and fortified food: probiotics, prebiotics, GMOs, and superfood. In: Ali ME, Nizar NNA (eds) Preparation and processing of religious and cultural foods. Woodhead Publishing, Oxford, pp 67–129. https://doi.org/10.1016/B978-0-08-101892-7.00006-7

Vaz BS, Moreira JB, Morais MG et al (2016) Microalgae as a new source of bioactive compounds in food supplements. Curr Opin Food Sci 7:73–77. https://doi.org/10.1016/j.cofs.2015.12.006

Vigani M, Parisi C, Rodríguez-Cerezo E et al (2015) Food and feed products from micro-algae: market opportunities and challenges for the EU. Trends Food Sci Technol 42:81–92. https://doi.org/10.1016/j.tifs.2014.12.004

Volk RB, Furkert FH (2006) Antialgal, antibacterial and antifungal activity of two metabolites produced and excreted by cyanobacteria during growth. Microbiol Res 161:180–186. https://doi.org/10.1016/j.micres.2005.08.005

Yusof YAM, Basari JMH, Mukti NA et al (2011) Fatty acids composition of microalgae Chlorella vulgaris can be modulated by varying carbon dioxide concentration in outdoor culture. Afr J Biotechnol 10:13536–13542. https://doi.org/10.5897/AJB11.1602

Zhu L (2015) Biorefinery as a promising approach to promote microalgae industry: an innovative framework. Renew Sust Energ Rev 41:1376–1384. https://doi.org/10.1016/j.rser.2014.09.040

Index

J. A. V. Costa et al. (eds.), *Microalgal Bioengineering*,
https://doi.org/10.1007/978-3-031-61253-4

GPSR Compliance

The European Union's (EU) General Product Safety Regulation (GPSR) is a set of rules that requires consumer products to be safe and our obligations to ensure this.

If you have any concerns about our products, you can contact us on ProductSafety@springernature.com

In case Publisher is established outside the EU, the EU authorized representative is:

Springer Nature Customer Service Center GmbH
Europaplatz 3
69115 Heidelberg, Germany

Batch number: 10761599

Printed by Printforce, the Netherlands